P9-DUP-384

FRONTIERS IN CLINICAL NEUROSCIENCE

A Symposium in Abel Lajtha's Honour

ADVANCES IN EXPERIMENTAL MEDICINE AND BIOLOGY

Recent Volumes in this Series

Volume 533
RETINAL DEGENERATIONS: Mechanisms and Experimental Theory
Edited by Matthew M. LaVail, Joe G. Hollyfield, and Robert E. Anderson

Volume 534
TISSUE ENGINEERING, STEM CELLS, AND GENE THERAPIES
Edited by Y. Murat Elçin

Volume 535
GLYCOBIOLOGY AND MEDICINE
Edited by John S. Axford

Volume 536
CHEMORECEPTION: From Cellular Signaling to Functional Plasticity
Edited by Jean-Marc Pequignot, Constancio Gonzalez, Colin A. Nurse, Nanduri R. Prabhakar, and Yvette Dalmaz

Volume 537
MATHEMATICAL MODELING IN NUTRITION AND THE HEALTH SCIENCES
Edited by Janet A. Novotny, Michael H. Green, and Ray C. Boston

Volume 538
MOLECULAR AND CELLULAR ASPECTS OF MUSCLE CONTRACTION
Edited by Haruo Sugi

Volume 539
BLADDER DISEASE: Research Concepts and Clinical Applications
Edited by Anthony A. Atala and Debra Slade

Volume 540
OXYGEN TRANSPORT TO TISSUE, VOLUME XXV
Edited by Maureen S. Thorniley, David K. Harrison, and Philip E. James

Volume 541
FRONTIERS IN CLINICAL NEUROSCIENCE: Neurodegeneration and Neuroprotection
Edited by László Vécsei

Volume 542
QUALITY OF FRESH AND PROCESSED FOODS
Edited by Fereidoon Shahidi, Arthur M. Spanier, Chi-Tang Ho, and Terry Braggins

---

FRONTIERS IN CLINICAL NEUROSCIENCE

Neurodegeneration and Neuroprotection

A Symposium in Abel Lajtha's Honour

Edited by

László Vécsei
University of Szeged
Szeged, Hungary

Kluwer Academic / Plenum Publishers
New York, Boston, Dordrecht, London, Moscow

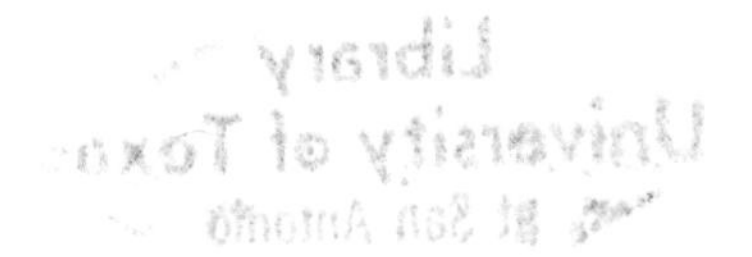

ISBN 0-306-48038-7

233 Spring Street, New York, New York 10013

http://www.wkap.nl/

10 9 8 7 6 5 4 3 2

A C.I.P. record for this book is available from the Library of Congress

Printed in the United States of America

AN INTRODUCTION TO DR. ABEL LAJTHA

Abel Lajtha enjoyed the benefits of being raised in a family in which, besides his world renowned father, László Lajtha (a musician, composer, and professor), his beloved mother created a spirit of culture, peace, and deep love. Abel Lajtha, after receiving his Ph.D. at the Péter Pázmány University of Budapest, began working with Albert Szent-Györgyi in the Department of Biochemistry. After World War II Hungary was occupied by the Russian army, and when the Communists came into power his father was removed from his job and Abel and his brother László Lajtha, Jr., a well respected cancer researcher, left the country. He went to Italy (Naples) on a postdoctoral fellowship and in 1949 settled down in the United States. When Abel married his fiancée Marie his parents could not get passports to travel to Vienna, 250 km from Budapest, where the couple was staying. This very happy event, his marriage to Marie, enriched his entire life as Marie recognized, understood, and actively supported his dedication to science and made every effort to ensure that despite this commitment their lives together would be very happy.

Dr. Lajtha is one of the founders of modern neurochemistry and a founding member of several journals and societies. Over the past fifty-five years, he has earned his reputation as a first-class scientist, editor of several books and journals, and his *Handbook of Neurochemistry*.

Dr. Lajtha has been an editor of about 20 journals, has served on several advisory boards, and has organized numerous conferences and symposia. He has received several honours, including an honorary M.D. degree from the University of Padua, has been President of the International Society for Neurochemistry and of the American Society for Neurochemistry, and has been the Editor-in-Chief of the journal *Neurochemical Research* since its foundation in 1975. He was elected as foreign or corresponding member of the Slovenian Academy of Sciences, the Hungarian Academy of Sciences, the Indian Academy of Neuroscience, and the Armenian National Academy of Sciences, and is an honorary member of institutes and societies in many countries. He has published about 660 journal articles and 90 reviews or chapters. He was fortunate to practice science during the last half of the previous century, when many basic neurochemical concepts could be experimentally addressed for the first time. He has personally pioneered several discoveries. However, when we asked him what pleases him most about his scientific accomplishments, he answered (very characteristically of him), "most rewarding was the feeling that I was helping young scientists from all over the world experience the pleasure of performing creative research." In his life there has been no change whatsoever in his commitment to quality science—that is what he has always stood for.

In 1963, Dr. Lajtha was appointed Director of the New York State Research Institute for Neurochemistry and Drug Addiction at Ward's Island, New York City, with a staff of about 60 people. Scientists from many countries spent profitable time with Dr. Lajtha in his lab. He was appointed Research Professor of Psychiatry at New York University School of Medicine in 1971. His Institute later merged with the N.S. Kline Institute for Psychiatric Research, where he is now Director of the Center for Neurochemistry. At Ward's Island and at the N.S. Kline Institute Dr. Lajtha contributed to neurochemical research in many important ways.

Neurochemical research at the bench level and collaboration with other colleagues has always been important to him. It was an additional pleasure for him to have informal contacts with foreign scientists of different cultural backgrounds. Such contacts and collaborations, carried out in several laboratories, have fostered many long-lasting friendships. The Hungarians have been extremely proud of him, his role in supporting Hungarian scientists, and his role in developing the study of neurochemistry. Not only the Hungarians, but also colleagues from all over the world, are very grateful to him for his help.

His personality is unique. He has many friends, *real* friends, and almost every day he plays tennis—his forehand is still very good, he very rarely makes unforced error, and is in very good physical shape. At the age of 80, he continues to derive a great deal of pleasure and satisfaction from working in the laboratory writing grant applications and setting up collaborations and making friends. There are a number of young scientists being trained at this time in his laboratory, and he is now working on the third edition of his *Handbook of Neurochemistry*—now planned at about 25 volumes with 500 or more chapters.

On behalf of the scientific community we hope that Abel can keep this spirit alive for another twenty years. Clearly, the respect and admiration that we have for him and his scientific accomplishments and leadership are shared across the ocean. On this, the year of his 80th birthday, his colleagues worldwide hail a remarkable career and offer our hope and expectation that there is much more to come.

His many pupils and friends are honoured to have the opportunity to contribute to this Festschrift in recognition of the work of Dr. Lajtha as a scientist, a teacher, and a good friend.

E. Sylvester Vizi (Budapest) and L. Battistin (Padua)

PREFACE

Over the last decade, the considerable progress made in biochemistry, molecular biology, genetics and neuropharmacology has revealed some of the intimate mechanisms of the neurodegenerative disorders. There is increasing evidence linking genetic defects affecting mitochondria to Alzheimer's disease, Parkinson's disease, amyotrophic lateral sclerosis, Huntington's disease and some other neurological disorders. Advances in knowledge are fueled through improved animal models that use mitochondrial toxins, excitotoxins, and transgenic animals. Therapeutic studies in these models have strengthened the possibility for effective treatments in man. By defining the pathomechanisms, we hope to be in the position to prevent cell death by protecting neurons. Indeed serious preclinical and clinical research is going on in the field of neuroprotection in stroke, Parkinson's disease, epilepsy, demyelinating disorders and dementia. Based on these scientific ideas, the Symposium in honour of Professor Abel Lajtha was organized by the European Society for Clinical Neuropharmacology (ESCNP) and the Danube Symposium for Neurological Sciences in Budapest, Hungary, October 24–25, 2002. Professor Lajtha was born in Budapest in 1922 and his home town is an ideal venue for strengthening the bonds between Western and Eastern European Neuroscientists. Professor Peter Riederer (Wurzburg) held the 2002 special "Dezso Miskolczy Memorial Lecture" in Abel Lajtha's Honour. Thanks are due to the invited speakers of this Symposium for their excellent contribution.

László Vécsei

CONTENTS

ADVANCES IN NEUROPROTECTION RESEARCH FOR NEURODEGENERATIVE DISEASES

Mario E. Götz and Peter Riederer[1]

1. INTRODUCTION

Recent advances in science enable new and closer insights into brain structure and function. The extent of CNS damage due to ischemia, or neurodegeneration can be followed by the use of modern brain imaging technology such as positron emission tomography (PET) or single-photon emission computed tomography (SPECT) and radiolabeled tracers. *Ex vivo* the systematic use of gene expression arrays is becoming more and more important to select sensitive genes as targets for neuroprotection. And gene therapy is considered as an alternative approach to trigger neuroprotection in experimental models of neurodegeneration. At the same time as these modern technologies pave their way, new promising pharmacological intervention concepts to halt disease progression of Parkinson's and Alzheimer's disease, and to diminish ischemia reperfusion injury, have emerged.

This review summarizes some of the recent results of experimental preclinical research in the field of neuroprotection against excitotoxic and oxidative stress mediated neuronal damage. These pathogenetic mechanisms are considered important for the progression of chronic neurodegenerative diseases and stroke induced brain damage. Despite considerable scientific progress many more efforts have to be undertaken however, to understand the complex time dependent and thus fluctuating interactions of excitotoxicity, mitochondrial damage and oxidative stress as causative factors for neuronal apoptotic and necrotic death.

Although, it has to be always kept in mind that the complexity of the human neuropathology cannot easily be mimicked by animal models, cell culture experiments

---

1 Mario E. Götz, Institute of Pharmacology and Toxicology, 97078 Würzburg, Germany. Peter Riederer, Department of Psychiatry, Head Division of Clinical Neurochemistry, 97080 Würzburg, Germany.

and *in vivo* data give a hint for the assessment of new therapeutic concepts in clinical trials. With preclinical research a significant number of promising drugs has been developed that in animal studies confer neuroprotection but await clinical trials. A failure to induce neuroprotection in man with currently available drugs would suggest that either the animal models employed are not truly representative of the diseases, or that a single drug would not be effective enough for a successful neuroprotection.

This implies that combinations of drugs with each of it having a well defined pharmacologic dose response on specific, but different molecular targets, ought to be investigated in future *in vitro* and *in vivo* studies. This concept has long been accepted for the therapy of cardiovascular diseases, AIDS, cancer and neuropsychiatric disorders. It is important to mention that a variety of pathogenetic mechanisms exists, but that these mechanisms are linked and not necessarily independent of each other. Some of them might be activated only for certain periods within the disease progression. So it is crucial to offer the correct remedy at the correct time period when the pathogenetic mechanisms are active. To find out these time frames suitable for therapeutic intervention will be the task for future prospective trials of neuroprotective drugs utilising new brain imaging technology.

Here we want to highlight some of the pharmacological concepts that might be worth to proceed in their preclinical developments. Ideally, relevant drugs in focus should first be able to alleviate disease syndromes and second to decrease or even halt disease progression.

In animal models for neurodegeneration iron chelators, nitric oxide synthase inhibitors, monoamine oxidase inhibitors, antioxidants, dopamine agonists and glutamate receptor blockers have demonstrated mild to moderate protection[1, 2] and in clinical studies selegiline, a selective MAO-B inhibitor as well as several dopamine agonists delay the necessity for L-DOPA therapy in parkinsonian patients[3]. The most promising classes of drugs for the treatment of age-related chronic neurodegenerative diseases such as Alzheimer's and Parkinson's disease that will be disussed in the following might currently be propargylamines, dopamine agonists, glutamate receptor antagonists, and antioxidant compounds such as estradiol derivatives, coenzyme Q, vitamin A, vitamin C, vitamin E, and 6,8-dithiooctanoic acid, known as lipoic acid.

2. PROPARGYLAMINES

N-Methyl-N-propargyl-1-phenyl-2(R)-propylamine might be the most prominent propargylamine, better known as selegiline. Selegiline, a selective MAO-B inhibitor, has proved to be as well an effective radical scavenger,[4-6] increases neurotrophic factor synthesis[7] in cultured astrocytes and inhibits apoptosis.[8]

Interestingly, several clinical prospective studies have shown that selegiline delays the need for L-DOPAin newly diagnosed parkinsonian patients.[9-14] In contrast, treatment of Alzheimer's disease sufferers with selegiline does not result in a clinically measurable benefit as reported by a meta-analysis of several clinical studies.[15]

To date however, beside direct symptomatic effects of selegiline in PD which may be related to decreased monoamine turnover, the nature of the potentially neuroprotective effect could not yet be elucidated.[16, 17]

Further developments in medicinal chemistry generated a class of propargylamines that are not metabolized to methamphetamines and amphetamines, being thus devoid of indirect sympatomimetic effects in contrast to selegiline metabolites.

CGP 3466B is a propargylamine derivative devoid of MAO-inhibitory and indirect sympatomimetic effects that shows remarkable neuroprotection *in vitro*[18] and following MPTP administration to monkeys.[19] Biochemical investigations have shown that CGP 3466B binds to glyceraldehyde-3-phosphate-dehydrogenase[20] and upregulates the antiapoptotic protein L-isoaspartyl-methyl-transferase.[21] Clinical studies may follow.

N-Propargyl-1(R)-aminoindan, rasagiline is a ten times more potent irreversible inhibitor of MAO-B as compared with selegiline.[22, 23] Interestingly, rasagiline appears to be as well a more potent antiapoptotic agent in PC12 cells following growth factor withdrawal, N-methyl(R)-salsolinol, or peroxynitrite.[24-27] This is linked to altered gene expression involving growth factors and antiapoptotic proteins such as bcl-2, and is connected to the preservation of mitochondrial membrane potential and thus inhibition of caspases.[26-28] Reserpine induced ptosis is reversed by rasagiline at doses above 2 mg / kg i.p., which inhibit MAO-A as well as MAO-B. Combination of rasagiline (10 mg / kg i.p.) with L-DOPA or L-tryptophan (50 mg / kg i.p.), or with fluoxetine (10 mg/kg p.o.) does not induce the behavioural hyperactivity syndrome which is seen following inhibition of both MAO-A and MAO-B by tranylcypromine together with the monoamine precursors. Rasagiline does not modify CNS monoamine tissue levels or monoamine-induced behavioural syndromes at doses which selectively inhibit MAO-B but not MAO-A.[29] A multi center placebo-controlled randomized double blind study, conducted for 12 weeks in 70 patients with Parkinson's disease demonstrated a good safety (up to 2 mg / d) and a beneficial effect in fluctuating patients with Parkinson's disease when given as an add-on to chronic L-DOPA therapy.[30] Another, multicenter, 26 week, parallel-group, randomized, double-blind, placebo-controlled clinical trial concludes that already rasagiline 1 mg per day is effective as monotherapy for patients with early PD not requiring L-DOPA.[31]

N-propargyl-3(R)-aminoindan-5-yl-ethyl-methyl-carbamate (TV 3326) is an equally interesting future candidate for the therapy of Alzheimer's diesease since it offers symptomatic benefit by ameliorating cognitive functions in the elderly, perhaps via its potential to inhibit acetylcholine esterase,[32] and on the other hand is still a potent irreversible MAO-B inhibitor.[33-36]

3. DOPAMINE RECEPTOR AGONISTS

Nowadays, L-DOPA treatment of parkinsonian patients is complemented by dopamine receptor agonists. The rationale for this is to prolong the time period patients are free of L-DOPA-induced dyskinesias. Beyond the symptomatic benefits that have been unequivocally proven by many long term clinical trials[37-44] several potentially neuroprotective effects of dopamine agonists could be discovered.[45, 46] The L-DOPA sparing effects by using dopamine receptor agonists as an adjuct therapy to L-DOPA may result in a decrease of hydrogen peroxide formation because less dopamine has to be metabolized by monoamine oxidases.

Moreover, dopamine receptor agonists including ropinirole, pramipexole, pergolide, lisuride, bromocriptine, cabergoline, and α-dihydroergocryptine in clinical use show direct superoxide and hydroxyl radical scavenging, i.e. antioxidant and neuroprotective properties *in vitro* and *in vivo*.[47-53] By slowing presynaptic dopamine synthesis and release, an effect ascribed to presynaptic dopamine D_2-receptor stimulation, dopamine turnover is minimized providing long term protection from possibly deleterious effects of L-DOPA metabolites.

In fact, the use of modern imaging techniques such as ^{18}F-DOPA-PET or ^{123}J-β-CIT-SPECT, demonstrating presynaptic DOPA binding and dopamine transporter labeling *in vivo*, respectively, may prove or disprove neuroprotective effects of several dopamine receptor agonists in the near future.[54, 55] Recent preliminary interim analysis of ongoing prospective studies indeed report a tendency towards slower progression of dopaminergic cell degeneration in parkinsonian patients administered pergolide, or pramipexole, or ropinirole.[40, 56, 57]

4. DERIVATIVES OF ESTRADIOL

Oxidative stress, a cellular imbalance between production and elimination of reactive oxygen species (ROS), such as superoxide, hydrogen peroxide, hydroxyl radical, and peroxynitrite, is considered to be of major pathophysiological relevance for a variety of pathological processes, including ischemia-reperfusion injury and chronic progressive neurodegenerative diseases.[58] This hypothesis has prompted research efforts to identify compounds which might act as antioxidants, i.e. compounds that antagonize the deleterious actions of ROS on biomolecules. The modes of action of these compounds could be either to directly scavenge ROS or to trigger protective mechanisms inside the cell, resulting in an improved defense against ROS.

Based on the phenolic A-ring, estrogens are potent free radical scavengers.[59-62] Their lipophilic character is a prerequisite for membrane interaction and penetration. In order to exert protective properties at least a phenolic moiety is necessary. Furthermore, bulky alkyl substituents in 2- and 4-positions have been introduced to estrogens to increase the antioxidant efficacy.[63]

It is hypothesized that these estrogen derivatives interfere with the redox state of membrane proteins and lipids thereby protecting the cell from more severe damage leading to death. In line with this hypothesis are data reporting that 17β-estradiol protects rat cortical synaptosomes from amyloid peptide A β_{25-35} and $FeSO_4$ - induced membrane lipid peroxidation and prevents the oxidative stress-related impairment of Na^+/K^+-ATPase activity, glutamate transport and glucose transport.[64]

In our studies we tested estra-1,3,5(10),8-tetraene-3,17α-diol, J 811, a $\Delta^{8,9}$-dehydro derivative of 17 β-estradiol, known to exhibit antioxidative activity by altering iron redox state and inhibiting the formation of superoxide anion radicals *in vitro*. The advantage of this compound is based on a more potent radical scavenging activity than that of the naturally occurring 17 β-estradiol.[65] The stronger antioxidant properties of J811 as compared to 17 β-estradiol might be linked to the more extended delocalized π-electron system capable of stabilizing radical adducts.[66, 67] The future will show us whether estradiol derivatives devoid of hormonal function can replace vitamin E, a naturally occurring lipophilic antioxidant as potentially neuroprotective agents. At least *in vitro*, vitamin E is in some experimental conditions more potent as antioxidant than 17β-estradiol.[68]

5. VITAMIN E

Vitamin E is the term used for eight naturally occurring fat-soluble nutrients.[69] Four compounds bear a saturated phytyl side chain and differ only with respect to number and position of methyl groups at the chromanol ring (α-, β-, γ-, and δ-tocopherols). However, α- tocopherol predominates in many species. The phytyl side chain in the 2-position facilitates incorporation and retention of tocopherol in biomembranes, while the active site of radical scavenging is the 6-hydroxyl group of the chromanol ring.[70, 71] The stereoisomer R,R,R-tocopherol is the only stereoisomer out of eigh possible ones to occur in nature.

The most widely accepted physiological function of tocopherol is its role as a scavenger of free radicals. Thus it prevents oxidant injury to polyunsaturated fatty acids and thiol rich proteins in cellular membranes and cytoskeleton. It is thought to preserve the structure and functional integrity of subcellular organelles.[72]

The absorption, transport and metabolism of tocopherol in animals has been reviewed on several occasions.[73, 74] Tocopherol is transferred from circulating lipoproteins to the brain, spinal cord and peripheral nerves and muscle by unknown mechanisms.[75] There is no uniform distribution of tocopherol in the central and peripheral nervous system.[76] In contrast to other brain regions, the cerebellum is particularly active in the utilization of tocopherol.[77] During experimental tocopherol deficiency, nerve tissue retains a greater percentage of tocopherol than do serum, liver and adipose tissue.[78] Morphological and functional studies performed on experimental tocopherol deficient rats have revealed axonal dystrophy and degeneration of peripheral nerve. This can be aggravated by increasing dietary polyunsaturated fatty acids providing increased quantities of peroxidisable substrate and reduced by feeding a synthetic antioxidant (ethoxyquin).[79] These experiments provide evidence in favour of an antioxidant role for tocopherol in the nervous system. In brain, tocopherol is predominantly localized in the mitochondrial, microsomal and synaptosomal fractions[80] suggesting that protection by tocopherol from peroxidative damage to subcellular membranes may be important for mitochondrial energy production or microsomal enzyme activity.

Recent *in vitro* studies on neuronal cultures support the important role of vitamin E for neuroprotection especially following amyloid beta-peptide (Aβ) mediated neuronal damage.[81, 82] Although vitamin E does not inhibit fibril formation from Aβ peptides, it protects from oxidative stress, induced by mechanisms following Aβ peptide exposure. Moreover, vitamin E may even provide neuroprotection *in vivo* through suppression of signaling events necessary for microglial activation.[83] The impressive antioxidant potentials of phenolic structures including vitamin E and estradiol derivatives in various *in vitro* models were summarized by Behl and Moosmann.[84]

A placebo-controlled, clinical trial of vitamin E in patients with moderately advanced Alzheimer's disease was conducted by the Alzheimer's Disease Cooperative Study Group. Vitamin E 2000 IU (1342 α-tocopherol equivalents / d) slows the functional deterioration leading to nursing home placement in Alzheimer subjects according to that study. These data are encouraging to design clinical trials aiming at investigating the delay of cognitive function in patients with mild cognitive disorders.[85]

In contrast, it was the Parkinson Study Group[11] that diminished hope that lipophilic antioxidants such as vitamin E (2000 IU / d) would be neuroprotective in Parkinson's disease. In the DATATOP clinical trial vitamin E did not reduce the requirement for L-DOPA. Because of this result one should however not dismiss the hypothesis of oxidative stress as a major factor for neurodegeneration in chronic diseases such as Alzheimer's or Parkinson's disease.

The potential of vitamin E to counteract oxidative stress may be restricted to certain lipophilic compartments, i. e. cellular and subcellular membranes and thus be not sufficient for radical defense of the whole cell. Vitamin E may need support for antioxidant defense by hydrophilic compounds as for example shown by studies reporting that dopamine induced PC12 cell death can be prevented by thiols,[86] that the phytoestrogen kaempferol protects PC12 and T47D cells from Aβ-toxicity,[87] and that flavonoids protect HT-4 cells from glutamate-induced cytotoxicity.[88] Interestingly, the water soluble antioxidant Trolox, a vitamin E analogue, is capable to decrease neuronal death *in vivo*.[89] *In vitro*, trolox is often used as a gold standard to measure antioxidant capacity. Many flavanols, flavonols, aflavins and anthocyanidins show two to six fold higher antioxidant capacity than trolox.[90] Their clinical effectiveness and their safety, however, remain to be determined. Vitamin E antioxidant action may be as well strongly facilitated by ascorbic acid, also called vitamin C.

6. VITAMIN C

Ascorbic acid is an extremely water-soluble and therefore cytosolic antioxidant essential for humans, primates and guinea pigs but not for rodents, which can synthesize it from glucose. Ascorbic acid serves as a cofactor in several iron-dependent hydroxylases,[91] important for collagen synthesis, (prolyl- and lysyl- hydroxylases), for carnitine biosynthesis (6-N-trimethyl-L-lysine-hydroxylase) and for catabolism of tyrosine (4-hydroxyphenyl-pyruvate-hydroxylase). Two major further functions of ascorbate are support of the synthesis of norepinephrine, and α-amidation of neurohormones, explaining in part its higher concentrations in brain and endocrine tissues (adrenal gland). The copper-containing dopamine-ß-hydroxylase catalyzes the final step in the synthesis of norepinephrine, the hydroxylation of dopamine. Ascorbate is most likely required by hydroxylases to maintain iron or copper at the active enzyme site in the reduced form, as it is necessary for hydroxylation. The semidehydroascorbate radical is not very reactive. It decays by disproportionation to ascorbate and dehydroascorbate (the latter subsequently degrades to oxalic acid and L-threonic acid), rather than acting as a reactive free radical, but ascorbic acid is readily oxidized by superoxide.[92] Many membrane proteins are sensitive to tissue redox state[93] such as the NMDA receptor, thought to be involved in neuronal degeneration in seizure and ischemia.[94, 95] This can be inhibited by ascorbate.

Furthermore, an important protective action of ascorbic acid is its ability to act synergistically with tocopherol in the inhibition of various oxidation reactions.[96-98] Packer et al.[99] have shown in pulse radiolysis studies that, in solution, the tocopheryl radical reacts rapidly with ascorbic acid to yield tocopherol. This synergism may work in liposomal membranes as well.[100] Some further studies indicate that ascorbic acid helps to maintain tissue levels of tocopherol *in vivo*.[101, 102]

Although the antioxidant ascorbic acid does not penetrate the blood brain barrier, its oxidized form, dehydroascorbic acid, enters the brain by means of facilitated transport. Unlike exogenous ascorbic acid, dehydroascorbic acid confers *in vivo*, dose-dependent neuroprotection in reperfused and non-reperfused cerebral ischemia at clinically relevant times.[103] As a naturally occurring interconvertible form of ascorbic acid with blood brain barrier permeability, dehydroscorbic acid represents a promising pharmacological therapy for stroke. Recently a dehydroascorbate reductase has been found in the rat brain that regenerates ascorbate after it is oxidized during normal aerobic metabolism. This enzyme can be found in endothelial cells, perivascular astrocytes and in neuronal cytosol and nuclei.[104] These data may speak in favour of a neuroprotective effect of dehydroascorbate following intracerebral reduction to ascorbate as long as dehydroascorbate reductase is functional.

7. VITAMIN A

Vitamin A chemically comprises all types of retinoids. However, biologically, the expression "vitamin A" is confined to retinol and its esters only. Retinol is present in animals but not in plants, which instead form vitamin A precursors, the carotenoids. Gut mucosa cells are able to synthesize retinol from carotenoids. Supplementation with ß-carotin is unlikely to cause symptoms of hypervitaminosis including dry skin, headache and symptoms resulting from increased intracranial pressure, because gut resorption and metabolism to retinol are limited. Retinol is stored as fatty acid retinylesters in chylomicrons. In this form it is transported to the liver, where it then will be bound to specific proteins following hydrolisation. Plasma levels are kept constant in a range between 50 and 60 μg / l. Following passive diffusion, retinol is released from its binding protein and is oxidized to retinal and further to retinoic acid. Retinal is important for photoreception in the retina. Retinoic acid is supposed to mainly act as a ligand for the retinoic acid receptor. Bound ot the receptor it activates nuclear gene transcription. Genes relevant for cell proliferation and cell differentiation, as well as proteins for cell adhesion are transcriptional targets. Retinoic acid is especially important for the development of the hindbrain.[105, 106] The liganded retinoic acid receptors are specific transcription factors that are essential for full expression of the dopamine D2 receptor.[107] *In vitro*, retinoic acid potentiates the protective effect of NGF against staurosporine-induced apoptosis in cultured chick neurons by increasing trkA protein expression. Thus, it was concluded that increasing the endogenous synthesis of growth factors as well as the expression of their receptors by small, blood-brain barrier-permeable drugs might constitute a promising concept for neuroprotection.[108] Given the cytotoxic effects of retinoic acid, in chronic supplementation of the adult, doses higher than 30 mg / d should not be exceeded to avoid hypervitaminosis.[109]

Apart from the promising nuclear actions, the antioxidant activity of retinoids in human lymphocytes following oral ß-carotin intake is only moderate as compared to vitamin E and vitamin C. This may be related to strict control of plasma levels of retinylesters bound to chylomicrons and thus restricted uptake by other cell types in the blood. According to a compilation of results from several epidemiological studies, vitamin C still forms the first line of defense against LPO induced by numerous oxidants *in vivo* followed by vitamin E, not vitamin A.[110]

8. COENZYME Q

Coenzyme Q_{10} is an important electron transducer in mitochondria, microsomal membranes, Golgi complex, nucleus, and cytoplasma membrane.[111] Furthermore coenzyme Q is a potent lipophilic antioxidant capable of scavenging lipophilic radicals within these membranes, as well as in cytosol and plasma when bound to lipoproteins.[112-114] The semiquinones and quinones produced upon oxidation of quinols are substrates for reductases so that the quinols are regained. This redox cycle can be interrupted if oxygen is oxidizing the hydroquinone or semiquinone yielding superoxide. Thus coenzyme Q may under certain conditions become a prooxidant, as it is the case with ascorbate. However, as long as sufficient reductive equivalents such as NADH are formed, reduction of semiquinones is assured. In the brain roughly 80-90 % of coenzyme Q are in the quinol form.[115, 116]

Recently, a highly significant correlation between the level of CoQ_{10} and the activities of both complexes I and II/III of the mitochondrial electron transfer chain has been reported.[117] By using mitochondrially transformed cells (cybrids) from PD patients a 26 % deficiency of complex I activity could be detected although cytosolic calcium and energy levels of those transformed cells were normal. These subtle alterations may reflect an increased susceptibility of cells by impaired mitochondrial electron transport under circumstances not ordinarily toxic to those cells containing mitochondria from parkinsonian patients.[118]
The redox ratio of cerebral coenzyme Q can be augmented by intraperitoneal injections of racemic mixtures of lipoic acid,[119] and in several experimental conditions producing striatal lesions in the rat by malonate, 3-nitropropionic acid, 1-methyl-4-phenyl-1,2,3,6-tetrahydropyridine (MPTP) and methamphetamine coenzyme Q confers protection.[120-124] In combination with remacide, an NMDA-receptor antagonist, coenzyme Q additively attenuates striatal atrophy in a transgenic model of Huntington's disease.[125, 126] Although coenzyme Q diminishes neuronal injury in the hippocampus following experimental ischemia induced by endothelins in the rat,[127] it cannot reverse focal and global ischemia induced cerebral damage following four vessel occlusion.[128] In Parkinson's disease alterations in the activities of complex I have been reported in substantia nigra and platelets. Deficiency of mitochondrial enzyme activities could affect electron transport which might be reflected by the platelet coenzyme Q redox state. We have determined concentrations of the reduced and oxidized forms of coenzyme Q in platelets isolated from parkinsonian patients and age- and gender-matched controls. Platelet coenzyme Q redox ratios (amount of the reduced form, the quinol, to the amount of the oxidized form, the quinone) and the ratio of the reduced form, compared with total platelet coenzyme Q, were significantly decreased in *de novo* parkinsonian patients. Platelet coenzyme Q redox ratios were further decreased by L-DOPA treatment (not significant), whilst selegiline treatment partially restored coenzyme Q redox ratios. Our results either suggest an impairment of electron transport or a higher need for reduced forms of coenzyme Q in the platelets of even *de novo* parkinsonian patients. However, the coenzyme Q redox ratio was not correlated to disease severity, as determined by the Hoehn and Yahr PD disability classification, suggesting that this parameter may not be useful as a peripheral trait marker for the severity of PD but as an early state marker of PD.[129]

In contrast to these encouraging results, similar to vitamin E, coenzyme Q monotherapy (200 mg/d) turned out not to be protective in Parkinson's disease[130] in an open label three months trial. However, to date new galenic forms of administration of coenzyme Q are considered that should increase bioavailability of coenzyme Q in the brain. Again, pharmacokinetic aspects, absorption, distribution, metabolism and excretion may have to be more thoroughly addressed and reevaluated in advance of further clinical trials[131] before final conclusions concerning the neuroprotective activity of coenzyme Q in man can be drawn.

9. LIPOIC ACID

The oxidized form of α-lipoate (thioctic acid, 1,2-dithiolane-3-pentanoic acid, 1,2-dithiolane-3-valeric acid, 6,8-dithiooctanoic acid) is a low molecular weight sub-stance that is absorbed from the diet and crosses the blood-brain barrier. α-Lipoate is taken up and reduced in cells and tissues to dihydrolipoate, which is also exported to the extracellular medium. Thus antioxidant protection is possible intra- and extracellularily.[132] Lipoate and dihydrolipoate exert antioxidant activity[133, 134] by reducing dehydroascorbate or glutathione disulfide to raise intracellular glutathione levels. The most important thiol antioxidant glutathione has to be synthesized in the brain since it will not cross the blood brain barrier following systemic administration. Dihydrolipoate is thus an interesting alternative to increase cerebral antioxidant potential in the cytosol and the extracellular space.

In addition to the antioxidant features of dihydrolipoate, it serves as a covalently bound coenzyme in α-ketoacid dehydrogenases, such as the important mitochondrial pyruvate dehydrogenase and α-ketoglutarate dehydrogenase. Because of these characteristics it is reasonable to assume a considerable neuroprotective potential. And indeed, lipoate and dihydrolipoate exert neuroprotection in experimental cerebral ischemia-reperfusion injury, excitotoxic amino acid brain injury, mitochondrial dysfunction and diabetic neuropathy.[132, 135, 136] Moreover, lipoate turned out to be effective in preventing 3,4-methylenedioxy-methamphetamine (MDMA)-induced neurotoxicity in rats[137] and improves survival in transgenic mouse models of amyotrophic lateral sclerosis and Huntington's disease.[138, 139]

Despite the broad antioxidant activity of lipoate and dihydrolipoate, it has been argued, that other modes of action might contribute to the remarkable neuroprotective effects observed in rodents and diabetic patients. There are indications that lipoic acid triggers both heat-shock and phase II responses following activation of certain signalling proteins capable of detecting oxidants and electrophiles.[140, 141] Recently, it has been reported that lipoate induces time and concentration dependently the activity of NAD(P)H:quinone oxidoreductase (NQO1) and of glutathione transferases in C6 astroglial cells. The authors conclude, that upregulation of phase II detoxication enzymes may highly contribute to lipoates neuroprotective potential.[142-143] In a first randomized study including nine patients with Alzheimer's disease lipoic acid was successfully administered for up to one year demonstrating constant scores for cognitive performance in two neuropsychological tests. Thus, lipoate might evolve as an adjunct therapeutic option in age related disorders and dementia.[144]

10. GLUTAMATE RECEPTOR ANTAGONISTS

The importance of excitatory amino acids (EAA) in physiological function of neural transmission is well known. Glutamate was the first of them to be recognized. Now a further subdivision of the glutamate receptors has taken place. Four major glutamate receptor subtypes are known: 1. the NMDA, 2. the AMPA, (α-amino-3-hydroxy-5-methyl-4-isoxazolepropanoic acid), 3. the kainate and 4. the metabotropic receptor family.[145]

The EAA are involved as neurotransmitters in the sensory input of spinal and supraspinal systems. Another important function of EAA is to contribute to the programming and execution of movements in the motor loop. A further EAA-mediated neuronal function is the process of learning and memory, where the NMDA receptor in particular is thought to be important because of its electrophysiological action in long-term potentiation, which introduces synaptic plasticity.[146]

Beside the role of EAA in physiological actions, an excitotoxic etiology of neurodegenerative diseases has been proposed. A lot of investigations show a noxious effect of EAA, which is mediated by Ca^{2+} influx into the cytoplasm.[146] Endogenous and environmental EAA receptor agonists can cause acute and chronic neurodegenerative diseases resulting in dysfunction of motion and memory. Environmental diseases such as Guam disease, neurolathyrism and mussel poisoning can be used for studying the putative role of excitotoxic substances in Parkinson's disease, amyotrophic lateral sclerosis (ALS) and dementia.

Application of kainic acid, and the more specific quinolinic acid, into the striatum is followed by Huntington-like structural and biochemical changes. Excitotoxin injection into the striatum produces locomotor hyperactivity and deficits in learning tests.[147] Quinolinic acid is an endogenous NMDA receptor agonist formed from L-tryptophan thought to be elevated in HD. Although there is still disagreement that neurodegeneration in HD is mediated by altered tryptophan/quinolinic acid ratios the excitotoxic model of HD is a solid approach to the elucidation of the role of glutamate in movement disorders.[148]

Several preclinical studies have clearly demonstrated protection from excitotoxic neuronal death.[149-151] Clinically parkinsonian patients benefit from amantadine, a synthetic tricyclic amine that exerts anticholinergic activity, releases dopamine, inhibits dopamine uptake and acts as an NMDA receptor antagonist at pharmacological doses. Especially useful is amantadine sulfate for the treatment of akinetic crisis,[152] thus is antidyskinetic. Moreover, long-term amantadine treatment improves survival suggesting neuroprotection in parkinsonian patients.[153-154] Wether this protective effect is solely attributable to NMDA receptor blockade is under debate.

In AD, impairment of memory and cognition could reflect disturbances in NMDA-mediated long term potentiation, and a decline of NMDA receptors in hippocampal and cortical regions has been found.[155] Thus, it appears that treatment of dementia with NMDA receptor antagonists, in contrast to motor disturbances, is not recommendable from a symptomatic point of view, since cognitive functions might worsen because of suppression of long-term potentiation. NMDA receptor antagonists produce adverse effects on the effectiveness of encoding and of consolidation into information storage. These adverse effects may be accentuated in the aging and demented subject.[156] Because of severe developmental neurotoxicity NMDA receptor antagonists should of course be avoided in pregnancy.[157]

Excitotoxicity also plays a role in epileptic seizures and cerebral ischemia. During ischemic injury a rapid accumulation of glutamate takes place. This induces a noxious Ca^{2+} influx into the cytoplasm. Although activation of AMPA and kainate receptors can directly mediate excitotoxic cell death following both focal and global ischemia, this process is less powerful than the NMDA receptor mediated neurotoxicity. The Ca^{2+} conductance of NMDA receptors explains the efficacy of NMDA antagonists against hypoxic damage. In addition, AMPA antagonists such as NBQX have also a protective function in hypoxia.[158] The activation of non-NMDA receptors is sufficient for epileptogenesis but the latency before the onset of convulsions, their duration and the resulting brain damage depends critically on NMDA participation.[159] The action of glutamate in epilepsia is assumed to be the consequence of NMDA receptor binding and of AMPA receptor activation as well. *In vivo* kindling studies provoke epileptic activity. *In vitro* application of glutamate on hippocampal cell cultures burst firing. These phemomena can be reduced by NMDA antagonists. Further interest is put on more subtype selective NMDA antagonists in preclinical and biological testing such as memantine, ifenprodil, and many more.[160, 161]

In contrast to NMDA receptor antagonists, whose favorable clinical actions are compromised by quite strong adverse effects, the selective AMPA antagonists of the GYKI 53784 type may gain potential clinical value in acute and chronic neurological disorders[162] if they prove to be devoid of clinically relevant adverse effects. Experiments with cultured cortical neurons suggest that kainate induced excitotoxicity not only involves kainate but also AMPA receptors.[163] Interestingly, AMPA-selective antagonists developed to date exhibit noncompetitive kinetics, still providing effective blockade even in the presence of excess extracellular glutamate.

Metabotropic glutamate receptors (mGluRs) are classified into three major groups based on sequence homology degree and mechanisms of signal transduction. Group I mGluRs (mGluR1 and –5) couple via phospholipase C to phosphoinositide turnover and Ca^{2+} release from intracellular stores, whereas group II (mGluR2 and-3) and III (mGluR4,6,7,8) receptors couple to the inhibition of adenylyl cyclases and consequently lead to reduction in cyclic AMP levels[164]. Activation of Group I receptor activation may result in excitotoxicity. In contrast, group II and III agonists downregulate transmitter release most probably by downregulation of presynaptic voltage-gated Ca^{2+}channels.[165] of possible therapeutic importance could be the findings that the selective group III mGluR agonist (+)-4-phosphonophenylglycine attenuates NMDA-induced excitotoxic neuronal death both in cortical cultures and *in vivo*.[166]

In ischemia reperfusion injury blocking Ca^{2+} entry to cells may be of superior relevance as compared to other neuroprotective strategies since downstream many enzymes are activated including proteases, lipases, endonucleases, kinases and phosphatases. They contribute to cellular damage after excitotoxic receptor activation. Calpain inhibition attenuated neuronal death triggered by exogenous excitotoxins *in vitro*[167] and following transient global ischemia in rodents.[168] Potent inhibitors of calpains decrease infarct volume after focal ischemia even when administered six hours postocclusion.[169] Cytosolic Ca^{2+} following glutamate receptor activation triggers the formation of multiple free radical species, which are deleterious to lipids, proteins and DNA. Antioxidants reduce neuronal death induced by exogenous excitotoxins in culture or by intrastriatal injections of excitotoxins *in vivo*.[170, 171] Oxidative stress follows through activation of constitutively expressed neuronal nitric oxide synthase (nNOS). In the presence of superoxide, nitric oxide forms peroxynitrite[172] that subsequently can degrade to nitrogen dioxide and hydroxyl radical. Further enhancement of oxidative stress may originate from uncoupled mitochondria following Ca^{2+} overload.[173, 174] Superoxide can be produced as well during arachidonic acid metabolism catalyzed by cyclooxygenase, COX.[175] As a matter of fact, COX-2 inhibitors decrease NMDA-induced excitotoxicity in cortical neuronal cultures.[176]

Thus, blocking excitotoxicity is linked to many different target mechanisms, including blockade of Ca^{2+} channels, receptor stimulation or inhibition, scavenging of ROS and inhibition of proteases.[177] However, inhibiting Ca^{2+} entry in an unselective way may disturb memory consolidation so that future efforts in medicinal chemistry will have to provide even more receptor subtype-selective drugs to decrease the probability for adverse effects on cognitive functions. Many radical scavengers are experimentally very promising inhibitors of oxidative stress,[178] but those few tested, such as vitamin E and coenzyme Q, were not strongly neuroprotective in clinical studies if administered as monotherapy. Protease inhibitors might generate many adverse effects, since selective proteolysis is a physiological process. Depending on the severity of stroke mediated lesions for example, excitotoxic necrosis is prominent near the ischemic core at early time points, whereas apoptosis is more prominent in penumbral brain areas at later time points. This illustrates that different pathways to cell death are activated simultaneously in susceptible neurons which might need a specific combination of drugs to be exposed to at the same time to counteract neurodegeneration. It will be the task for the future to find ideal drug candidates that can be safely combined to counterbalance excitotoxicity, oxidative stress and proteolysis. With the help of modern brain imaging technology and sophisticated neurobehavioural analysis tools, as currently utilised, scientists will be able to even more effectively prove or disprove the potential of known or novel drugs for neuroprotection in prospective clinical trials.

10. REFERENCES

1. M. Gerlach, P. Riederer, M. B. Youdim, Neuroprotective therapeutic strategies, comparison of experimental and clinical results, *Biochem. Pharmacol.* **50**(1), 1-16 (1995).
2. E. Grünblatt, R. Schlößer, M. Gerlach, and P. Riederer, Preclinical versus clinical neuroprotection, in *Parkinson's disease: Advances in Neurology* **91**, edited by A. Gordin, S. Kaakkola, and H. Teräväinen (Lippincott Williams & Wilkins, Philadelphia, 2003), pp. 309-328.
3. C. E. Clarke, and M. Guttman, Dopamine agonist monotherapy in Parkinson's disease, *Lancet* **360**(9347), 1767-1769 (2002).
4. W. Birkmayer, P. Riederer, L. Ambrozi, and M. H. Youdim, Implications of combined treatment with "Madopar" and L-deprenyl in Parkinson's disease; A long-term study, *Lancet* **1**(8009), 439-443 (1977).
5. J. Knoll, J. Dallo, and, T.T. Yen, Striatal dopamine, sexual activity and lifespan, Longevity of rats treated with (-) deprenyl, *Life Sci.* **45**(6),525-531 (1989).
6. M. C. Carrillo, K. Kitani, S. Kanai, Y. Sato, K. Miyasaka, and G. O. Ivy, (-)Deprenyl increases activities of superoxide dismutase and catalase in certain brain regions in old mice, *Life Sci.* **54**(14), 975-981 (1994).
7. I. Mizuta, M. Ohta, K. Ohta, M. Nishimura, E. Mizuta, K. Hayashi, and S. Kuno, Selegiline and desmethylselegiline stimulate NGF, BDNF, and GDNF synthesis in cultured mouse astrocytes, *Biochem. Biophys. Res. Commun.* **279**(3), 751-755 (2000).
8. W. Maruyama, and M. Naoi, Neuroprotection by (-)deprenyl and related compounds, *Mech. Aging Dev.* **111**(2-3), 189-200 (1999).
9. Parkinson Study Group, DATATOP: a multicenter controlled clinical trial in early Parkinson's disease, *Arch. Neurol.* **46**(10), 1052-1060 (1989).
10. Parkinson Study Group, Effect of deprenyl on the progression of disabilityy in early Parkinson's disease, *N. Engl. J. Med.* **321**(20), 1364-1371 (1989).
11. Parkinson Study Group, Effects of tocopherol and deprenyl on the progression of disability in early Parkinson's disease, *N. Engl. J. Med.* **328**(3), 176-183 (1993).

12. Parkinson Study Group, Mortality in DATATOP: a multicenter trial in early Parkinson's disease, *Ann. Neurol.* **43**(3), 318-325 (1998).
13. J. W. Tetrud, and J. W. Langston, The effect of deprenyl (selegiline) on the natural history of Parkinson's disease, *Science* **245**(4917), 519-522(1989).
14. H. Przuntek, B. Conrad, J. Dichgans, P. H. Kraus, P. Krauseneck, G. Pergande, U. Rinne, K. Schimrigk, J. Schnitker, and H. P. Vogel, SELEDO: a 5-year long-term trial on the effect of selegiline in early parkinsonian patients treated with levodopa. *Eur. J. Neurol.* **6**(2), 141-150 (1999).
15. J. Birks, and L. Flicker, Selegiline for Alzheimer's disease (Cochrane Review), *Cochrane Database Syst. Rev.* **1**, CD000442 (2003).
16. M. Gerlach, M. B. Youdim, and P. Riederer, Pharmacology of selegiline, *Neurology* **47**(6 Suppl 3), S137-145 (1996).
17. K. Magyar, and D. Haberle, Neuroprotective and neuronal rescue effects of selegiline: review, *Neurobiology* (BP) **7**(2), 175-190 (1999).
18. P. C. Waldmeier, A. A. Boulton, A. R. Cools, A. C. Kato, and W. G. Tatton, Neurorescuing effects of the GAPDH ligand CGP 3466B, *J. Neural Transm.* **60**(Suppl.), 197-214 (2000).
19. G. Andringa, and A. R. Cools, The neuroprotective effects of CGP 3466B in the best *in vivo* model of Parkinson's disease, the bilaterally MPTP-treated rhesus monkey, *J. Neural Transm.* **60**(Suppl.), 215-225 (2000).
20. E. Kragten, I. Lalande, K. Zimmermann, S. Roggo, P. Schindler, D. Muller, J. van Oostrum, P. Waldmeier, and P. Furst, Glyceraldehyd-3-phosphate dehydrogenase, the putative target of the antiapoptotic compounds CGP 3466 and R-(-)-deprenyl, *J. Biol. Chem.* **273**(10), 5821-5828 (1998).
21. K. J. Huebscher, J. Lee, G. Rovelli, B. Ludin, A. Matus, D. Stauffer, and P. Furst, Protein isoaspartyl methyltransferase protects from Bax-induced apoptosis, *Gene* **240**(2), 333-341 (1999).
22. M. B. Youdim, A. Gross, and J. P. Finberg, Rasagiline [N-propargyl-1R(+)-aminoindan], a selective and potent inhibitor of mitochondrial monoamine oxidase B, *Br. J. Pharmacol.* **132**(2), 500-506 (2001).
23. D. Haberle, K. Magyar, and E. Szoko, Determination of the norepinephrine level by high-performance liquid chromatography to assess the protective effect of MAO-B inhibitors against DSP-4 toxicity, *J. Chromatogr. Sci.* **40**(9), 495-499 (2002).
24. W. Maruyama, T. Yamamoto, K. Kitani, M. C. Carrillo, M. B. Youdim and M. Naoi, Mechanism underlying anti-apoptotic activity of a (-)deprenyl-related propargylamine, rasagiline, *Mech. Aging Dev.* **116**(2-3), 181-191(2000).
25. W. Maruyama, Y. Akao, M. B. Youdim, and M. Naoi, Neurotoxins induce apoptosis in dopamine neurons; protection by N-propargylamine-1-(R)- and (S)-aminoindan, rasagiline and TV1022, *J. Neural Transm.* **60**(Suppl), 171-186(2000).
26. W. Maruyama, Y. Akao, M. C. Carrillo, K. Kitani, M. B. Youdim, and M. Naoi, Neuroprotection by propargylamines in Parkinson's disease: suppression of apoptosis and induction of prosurvival genes, *Neurotoxicol. Teratol.* **24**(5), 675-682 (2002).
27. W. Maruyama, T. Takahashi, M. B. Youdim, and M. Naoi, The anti-Parkinson drug, rasagiline, prevents apoptotic DNA damage induced by peroxynitrite in human dopaminergic neuroblastoma SH-SY5Y cells, *J. Neural Transm.* **109**(4), 467-481 (2002).
28. Y. Akao, W. Maruyama, S. Shimizu, H. Yi, Y. Nakagawa, M. Shamoto-Nagai, M. B. Youdim, Y. Tsujimoto, and M. Naoi, Mitochondrial permeability transition mediates apoptosis induced by N-methyl(R)salsolinol, an endogenous neurotoxin, and is inhibited by bcl-2 and rasagiline, N-propargyl-1(R)-aminoindan, *J. Neurochem.* **82**(4), 913-923 (2002).
29. J. P. Finberg, and M. B. Youdim, Pharmacological properties of the anti-Parkinson drug rasagiline; modification of endogenous brain amines, reserpine reversal, serotonergic and dopaminergic behaviours, *Neuropharmacology* **43**(7), 1110-1118 (2002).
30. J. M. Rabey, I. Sagi, M. Huberman, E. Melamed, A. Korczyn, N. Giladi, R. Inzelberg, R. Djaldetti, C. Klein, and G. Berecz, The Rasagiline Study Group, Rasagiline mesylate, a new MAO-B inhibitor, for the treatment of Parkinson's disease: a double blind study as adjunctive therapy to levodopa, *Clin. Neuropharmacol.* **23**(6), 324-330 (2000).
31. Parkinson Study Group, A controlled trial of rasagiline in early Parkinson disease: the TEMPO Study, *Arch. Neurol.* **59**(12), 1937-1943 (2002).
32. J. Sterling, Y. Herzig, T. Goren, N. Finkelstein, D. Lerner, W. Goldenberg, I. Miskolczi, S. Molnar, F. Rantal, T. Tamas, G. Toth, A. Zagyva, A. Zekany, G. Lavian, A. Gross, R. Friedman, M. Razin, W. Huang, B. Krais, M. Chorev, M. B. Youdim, and M. Weinstock, Novel dual inhibitors of AChE and MAO derived from hydroxy aminoindan and phenethylamine as potential treatment for Alzheimer's disease. *J. Med. Chem.* **45**(24), 5260-5279 (2002).
33. M. Weinstock, N. Kirschbaum-Slager, P. Lazarovici, C. Bejar, M. B. Youdim and S. Shoham, Neuroprotective effects of novel cholinesterase inhibitors derived from rasagiline as potential anti-

Alzheimer drug, in: *Neuroprotective Agents, Ann. N. Y. Acad. Sci. U. S. A. vol. 939*, edited by W. Slikker Jr., and B. Trembly (N. Y. Acad. Sci., New York, 2001) pp. 148-161.
34. M. B. Youdim, and M. Weinstock, Molecular basis of neuroprotective activities of rasagiline and the anti-Alzheimer drug TV3326 [(N-propargyl-(3)aminoindan-5-yl)-ethyl methyl carbamate], *Cell. Mol. Neurobiol.* **21**(6), 555-573 (2001).
35. M. B. Youdim, and M. Weinstock, Novel neuroprotective anti-Alzheimer drugs with anti-depressant activity derived from the anti-Parkinson drug, rasagiline, *Mech. Ageing Dev.* **123**(8), 1081-1086 (2002).
36. M. Yogev-Falach, T. Amit, O. Bar-Am, M. Weinstock, and M. B. Youdim, Involvement of MAP kinase in the regulation of amyloid precursor protein processing by novel cholinesterase inhibitors derived from rasagiline, *FASEB J.* **16**(12), 1674-1676 (2002).
37. Parkinson Study Group, Pramipexole vs L-DOPA as initial treatment for Parkinson disease: a randomized controlled trial, *JAMA* **284**(15), 1931-1938(2000).
38. A. Lieberman, A. Ranhosky, and D. Korts, Clinical evaluation of pramipexole in advanced Parkinson's disease: results of a double-blind, placebo-controlled, parallel-group study, *Neurology* **49**(1), 162-168 (1997).
39. H. Allain, A. Destee, H. Petit, M. Patay, S. Schuck, D. Bentue-Ferrer, and P. Le Cavorzin, Five-year follow-up of early lisuride and L-DOPA combination therapy versus L-DOPA monotherapy in *de novo* Parkinson's disease, The French Lisuride Study Group, *Eur. Neurol.* **44**(1), 22-30 (2000).
40. P. Barone, D. Bravi, F. Bermejo-Pareja, R. Marconi, J. Kulisevsky, S. Malagu, R. Weiser, and N. Rost, Pergolide monotherapy in the treatment of early PD: a randomized, controlled study, *Neurology* **53**(), 573-579 (1999).
41. B. Bergamasco, L. Frattola, A. Muratorio, F. Piccoli, F. Mailland, and L. Parnetti, α-Dihydroergocryptine in the treatment of *de novo* parkinsonian patients: results of a multicentre, randomized, double-blind, placebo-controlled study, *Acta Neurol. Scand.* **101**(6), 372-380 (2000).
42. L. Battistin, P. G. Bardin, F. Ferro-Milone, C. Ravenna, V. Toso, and G. Reboldi, α-Dihydroergocryptine in Parkinson's disease: a multicentre randomized double blind parallel group study, *Acta Neurol. Scand.* **99**(1), 36-42 (1999).
43. U. K. Rinne, F. Bracco, C. Chouza, E. Dupont, O. Gershanik, J. F. Marti-Masso, J. L. Montastruc, and C. D. Marsden, Early treatment of Parkinson's disease with cabergoline delays the onset of motor complications, Results of a double-blind L-DOPA controlled trial, The PKDS009 Study Group, *Drugs* **55**(Suppl 1), 23-30 (1998).
44. A. Lledo, Dopamine agonists: the treatment for Parkinson's disease in the XXI century? *Parkinsonism Relat. Disord.* **7**(1), 51-58 (2000).
45. J. P. Bennett, and M. F. Piercey, Pramipexole – a new dopamine agonist for the treatment of Parkinson's disease, *J. Neurol. Sci.* **163**(1), 25-31 (1999).
46. P. M. Carvey, S. O. McGuire, and Z. D. Ling, Neuroprotective effects of D3 dopamine receptor agonists, *Parkinsonism & Related Disorders* **7**(3), 213-223 (2001).
47. N. Ogawa, K. Tanaka, M. Asanuma, M. Kawai, T. Masumizu, M. Kohno, and A. Mori, Bromocriptine protects mice against 6-hydroxydopamine and scavenges hydroxyl free radicals *in vitro*, *Brain Res.* **657**(1-2), 207-213 (1994).
48. T. Yoshikawa, Y. Minamiyama, Y. Naito, and M. Kondo, Antioxidant properties of bromocriptine, a dopamine agonist, *J. Neurochem.* **62**(3), 1034-1038 (1994).
49. A. Ubeda, C. Montesino, M. Paya, and M. J. Alcaraz, Iron-reducing and free-radical-scavenging properties of apomorphine and some related benzylisoquinolines. *Free Radic. Biol. Med.* **15**(2), 159-167 (1993).
50. E. E. Sam, and N. Verbeke, Free radical scavenging properties of apomorphine enantiomers and dopamine: possible implication in their mechanism of action in parkinsonism, *J. Neural Transm. Park. Dis. Dement. Sect.* **10**(2-3), 115-127 (1995).
51. M. Gassen, Y. Glinka, B. Pinchasi, and M. B. Youdim, Apomorphine is a highly potent free radical scavenger in rat brain mitochondrial fraction, *Eur. J. Pharmacol.* **308**(2), 219-225 (1996).
52. M. Iida, I. Miyazaki, K.-I. Tanaka, H. Kabuto, E. Iwata-Ichikawa, and N. Ogawa, Dopamine D2 receptor-mediated antioxidant and neuroprotective effects of ropinirole, a dopamine agonist, *Brain Res.* **838**(1-2), 51-59 (1999).
53. T. Kihara, S. Shimohama, H. Sawada, K. Honda, T. Nakamizo, R. Kanki, H. Yamashita, and A. Akaike, Protective effect of dopamine D2 agonists in cortical neurons via the phosphatidylinositol 3 kinase cascade, *J. Neurosci. Res.* **70**(3), 274-282 (2002).
54. W. D. Le, and J. Jankovic, Are dopamine receptor agonists neuroprotective in Parkinson's disease?, *Drugs Aging* **18**(6), 389-396 (2001).
55. S. Thobois, S. Guillouet, and E. Broussolle, Contributions of PET and SPECT to the understanding of the pathophysiology of Parkinson's disease, *Neuophysiol. Clin.* **31**(5), 321-340 (2001).

56. K. Marek, β-CIT/SPECT assessments of progression of Parkinson's disease in subjects participating in the CALM PD study, *Neurology* **54**(Suppl3), A90 (2000).
57. J. S. Rakshi, N. Pavese, T. Uema, K. Ito, P. K. Morrish, D. L. Bailey, and D. J. Brooks, A comparison of the progression of early Parkinson's disease in patients started on ropinirole or L-dopa: an (18)F-DOPA PET study, *J. Neural Transm.* **109**(12), 1433-1443 (2002).
58. M. E. Götz, G. Künig, P. Riederer, and M. B. Youdim, Oxidative stress: Free radical production in neural degeneration, *Pharmac. Ther.* **63**(1), 37-122 (1994).
59. A. D. Mooradian, Antioxidant properties of steroids, *J. Steroid Biochem. Mol. Biol.* **45**(6), 509-511 (1993).
60. B. Ruiz-Larrea, A. Leal, C. Martin, R. Martinez, and M. Lacort, Effects of estrogens on the redox chemistry of iron: A possible mechanism of the antioxidant action of estrogens, *Steroids* **60**(11),780-783 (1995).
61. C. Behl, M. Widmann, T. Trapp, and F. Holsboer, 17β-estradiol protects neurons from oxidative stress-induced cell death *in vitro*, *Biochem. Biophys. Res. Commun.* **216**(2), 473-482 (1995).
62. C. Behl, T. Skutella, F. Lezoualc'h, A. Post, M. Widmann, C. J. Newton, and F. Holsboer, Neuroprotection against oxidative stress by estrogens: Structure-activity relationship, *Mol. Pharm.* **51**(4), 535-541 (1997).
63. C. P. Miller, I. Jirkovsky, D. A. Hayhurst, and S. J. Adelman, *In vitro* antioxidant effects of estrogens with a hindered 3-OH function on the copper-induced oxidation of low density lipoprotein, *Steroids* **61**(5), 305-308 (1996).
64. J. N. Keller, A. Germeyer, J. G. Begley, and M. P. Mattson, 17β-estradiol attenuates oxidative impairment of synaptic Na^+/K^+-ATPase activity, glucose transport, and glutamate transport induced by amyloid β-peptide and iron, *J. Neurosci. Res.* **50**(4), 522-530 (1997).
65. W. Römer, M. Oettel, P. Droescher, and S. Schwarz, Novel "scavestrogens" and their radical scavenging effects, iron-chelating, and total antioxidative activities: Δ8,9-dehydro derivatives of 17α-estradiol and 17β-estradiol, *Steroids* **62**(3), 304-310 (1997).
66. D. Blum-Degen, M. Haas, S. Pohli, R. Harth, W. Römer, M. Oettel, P. Riederer, M. E. Götz, Scavestrogens protect IMR 32 cells from oxidative stress - induced cell death, *Toxicol. Appl. Pharmacol.* **152**(1), 49-55 (1998).
67. W. Römer, M. Oettel, B. Menzenbach, P. Droescher, S. Schwarz, Novel estrogens and their radical scavenging effects, iron-chelating, and total antioxidative activities: 17α-substituted analogs of Δ9(11)-dehydro-17β-estradiol, *Steroids* **62**(11), 688-694 (1997).
68. C. Behl, Vitamin E protects neurons against oxidative cell death *in vitro* more effectively than 17β-estradiol and induces the activity of the transcription factor NF-kappaB, *J. Neural Transm.* **107**(4), 393-407 (2000).
69. G.A. Fritsma, Vitamin E and autoxidation, *Am. J. Med. Tech.* **49**(6) 453-456 (1983).
70. J.A. Lucy, Functional and structural aspects of biological membranes: a suggested role for vitamin E in the control of membrane permeability and stability. *Ann. N. Y. Acad. Sci.* **203**, 4-11 (1972).
71. J. R. Burton, and K. U. Ingold, Autoxidation of biological molecules. The antioxidant activity of vitamin E and related chain-breaking phenolic antioxidants *in vitro*, *J. Am. Chem. Soc.* **103**, 6472-6477 (1981).
72. C. K. Chow, Vitamin E and oxidative stress, *Free Radic. Biol. Med.* **11**(2), 215-232 (1991).
73. A. Bjorneboe, G.-E. Bjorneboe, and C. A. Drevon, Absorption, transport and distribution of vitamin E, *J. Nutr.* **120**(3), 233-242 (1989).
74. C. A. Drevon, Absorption, transport and metabolism of vitamin E, *Free Radic. Res. Commun.* **14**(4), 229-246 (1991).
75. R. J. Sokol, Vitamin E and neurologic function in man, *Free Radic. Biol. Med.* **6**(2), 189-207 (1989).
76. G. T. Vatassery, C. K. Angerhofer, and C. A. Knox, Effect of age on vitamin E concentrations in various regions of the brain and a few selected peripheral tissues of the rat, and on the uptake of radioactive vitamin E by various regions of the rat brain, *J. Neurochem.* **43**(2), 409-412 (1984).
77. G. T. Vatassery, Selected aspects of the neurochemistry of vitamin E, in: *Clinical and nutritional aspects of vitamin E*, edited by, O. Hayaishi, and M. Mino, (Elsevier, Amsterdam, 1987), pp. 147-155.
78. M. A. Goss-Sampson, C. J. McEvilly, and D. P. R. Muller, Longitudinal studies of the neurobiology of vitamin E and other antioxidant systems, and neurological function in the vitamin E deficient rat, *J. Neurol. Sci.* **87**(1), 25-35 (1988).
79. E. Southam, P. K. Thomas, R. H. M. King, M. A. Goss-Sampson, and D. P. R. Muller, (1991) Experimental vitamin E deficiency in rats, morphological and functional evidence of abnormal axonal transport secondary to free radical damage, *Brain* **114**(Pt 2), 915-936 (1991).
80. G. T. Vatassery, C. K. Angerhofer, C. A. Knox, and D. S. Deshmukh, Concentrations of vitamin E in various neuroanatomical regions and subcellular fractions, and the uptake of vitamin E by specific areas, of rat brain, *Biochim. Biophys. Acta* **792**(2), 118-122 (1984).
81. D.A. Butterfield, T. Koppal, R. Subramaniam, and S. Yatin, Vitamin E as an antioxidant/free radical scavenger against amyloid β-peptide-induced oxidative stress in neocortical synaptosomal membranes and hippocampal neurons in culture: insights into Alzheimer's disease, *Rev.Neurosci.* **10**(2), 141-149 (1999).

82. S. M. Yatin, S., Varadaryjan, and D. A. Butterfield, Vitamin E prevents Alzheimer's amyloid β-peptide (1-42)-induced neuronal protein oxidation and reactive oxygen species production, *J. Alzheimers Dis.* **2**(2), 123-131 (2000).
83. Y. Li, L. Liu, S. W. Barger, R. E. Mrak, and W. S. Griffin, Vitamin E suppression of microglial activation is neuroprotective, *J. Neurosci. Res.* **66**(2), 163-170 (2001).
84. C. Behl, and B. Moosmann, Oxidative nerve cell death in Alzheimer's disease and stroke: antioxidants as neuroprotective compounds, *Biol. Chem.* **383**(3-4), 521-536 (2002).
85. M. Grundman, Vitamin E and Alzheimer disease: the basis for additional clinical trials, *Am. J. Clin. Nutr.* **71**(2), 630S-636S (2000).
86. D. Offen, I. Ziv, H. Sternin, E. Melamed, and A. Hochman, Prevention of dopamine-induced cell death by thiol antioxidant: possible implications for treatment of Parkinson's disease, *Exp. Neurol.* **141**(1), 32-39 (1996).
87. A. Roth, W. Schaffner, and C. Hertel, Phytoestrogen kaempferol (3,4',5,7-tetrahydroxyflavone) protects PC12 and T47D cells from β-amyloid-induced toxicity, *J. Neurosci. Res.* **57**(3), 399-404 (1999).
88. M. S. Kobayashi, D. Han, and L. Packer, Antioxidants and herbal extracts protect HT-4 neuronal cells against glutamate-induced cytotoxicity, *Free Radic. Res.* **32**(2), 115-124 (2000).
89. L. Iacovitti, N. D. Stull, and A. Mishizen, Neurotransmitters, KCl and antioxidants rescue striatal neurons from apoptotic cell death in culture, *Brain Res.* **816**(2), 276-285 (1999).
90. E. J. Lien, S. Ren, H.-H. Bui, and R. Wang, Quantitative structure-activity relationship analysis of phenolic antioxidants, *Free Radic. Biol. Med.* **26**(3/4), 285-294 (1999).
91. H. Padh, Vitamin C: Newer insights into its biochemical functions, *Nutr. Rev.* **49**(3), 65-70 (1991).
92. B. H. J. Bielski, and H. W. Richter, Some properties of the ascorbate free radical. *Ann. N.Y. Acad. Sci.* **258**, 231-237 (1975).
93. R. L. Levine, Oxidative modification of glutamine synthetase: characterization of the ascorbate model system, *J. Biol. Chem.* **258**(19), 11828-11833 (1983).
94. D. W. Choi, Calcium-mediated neurotoxicity: relationship to specific channel types and role in ischemic damage, *Trends Neurosci.* **11**(10), 465-469 (1988).
95. D. W. Choi, Glutamate neurotoxicity and diseases of the nervous system, *Neuron* **1**(8), 623-634(1988).
96. P. B. McCay, Vitamin E: Interactions with free radicals and ascorbate, *Ann. Rev. Nutr.* **5**, 323-340 (1985).
97. E. Niki, Antioxidants in relation to lipid peroxidation, *Chem. Phys. Lipids* **44**(2-4), 227-253 (1987).
98. E. Niki, Interaction of ascorbate and α-tocopherol, *Ann. N.Y. Acad. Sci.* **498**, 186-199 (1987).
99. J. E. Packer, T. F. Slater, and R. L. Willson, Direct observation of a free radical interaction between vitamin E and vitamin C, *Nature, Lond.* **278**(5706), 737-738 (1979).
100. M. Scarpa, A. Rigo, M. Maiorino, F. Ursini, and C. Gregolin, Formation of α-tocopherol radical and recycling of α-tocopherol by ascorbate during peroxidation of phosphatidylcholine liposomes. An electron paramagnetic resonance study, *Biochim. Biophys. Acta* **801**(2), 215-219 (1984).
101. F. Hruba, V. Novakova, and E. Ginter, The effect of chronic marginal vitamin C deficiency on the α-tocopherol content of the organs and plasma of guinea pigs, *Experientia* **38**(12),1454-1455 (1982).
102. A. Bendich, L. J. Machlin, O. Scandurra, G. W. Burton, and D. N. Wayner, The antioxidant role of vitamin C, *Adv. Free Radic. Biol. Med.* **2**, 419-444 (1986).
103. J. Huang, D. B. Agus, C. J. Winfree, S. Kiss, W. J. Mack, R. A. McTaggart, T. F. Choudhri, L. J. Kim, J. Mocco, D. J. Pinsky, W. D. Fox, R. J. Israel, T. A. Boyd, D. W. Golde, and E. S. Connolly Jr., Dehydroascorbic acid, a blood-brain barrier transportable form of vitamin C, mediates potent cerebroprotection in experimental stroke, *Proc. Natl. Acad. Sci. U. S. A.* **98**(20), 11720-11724 (2001).
104. F. Fornai, S. Piaggi, M. Gesi, M. Saviozzi, P. Lenzi, A. Paparelli, and A. F. Casini, Subcellular localization of a glutathione-dependent dehydroascorbate reductase within specific rat brain regions, *Neuroscience* **104**(1), 15-31 (2001).
105. M. Maden, Heads or tails? Retinoic acid will decide, *Bioessays* **21**(10), 809-812 (1999).
106. G. Begemann, and A. Meyer, Hindbrain patterning revisited: timing and effects of retinoic acid signalling, *Bioessays* **23**(11), 981-986 (2001).
107. G. Wolf, Vitamin A functions in the regulation of the dopaminergic system in the brain and pituitary gland, *Nutr. Rev.* **56**(12), 354-355 (1998).
108. B. Ahlemeyer, R. Huhne, and J. Krieglstein, Retinoic acid potentiated the protective effect of NGF against staurosporine-induced apoptosis in cultured chick neurons by increasing the trk A protein expression, *J. Neurosci. Res.* **60**(6), 767-778 (2000).
109. S. T. Omaye, Safety of megavitamin therapy, *Adv. Exp. Med. Biol.* **177**, 169-203 (1984).
110. M. R. McCall, and B. Frei, Can antioxidant vitamins materially reduce damage in humans? *Free Radic. Biol. Med.* **26**(7/8), 1034-1053 (1999).

111. R. E. Beyer, The participation of CoQ10 in free radical production and antioxidation, *Free Radic. Biol. Med.* **8**(6), 545-565 (1990).
112. F. L. Crane, Development of concepts for the role of ubiquinones in biological membranes, in: *Highlights in Ubiquinone Research,* edited by G. Lenaz, O. Barnabei, A. Rabbi, M. Battino (Taylor & Francis, London, 1990) pp. 3-17.
113. T. Takahashi, T. Okamoto, K. Mori, H. Sayo, and T. Kishi, Distribution of ubiquinone and ubiquinol homologues in rat tissues and subcellular fractions, *Lipids* **28**(9), 803-809 (1993).
114. L. Ernster, P. Forsmark, and K. Nordenbrand, The mode of action of lipid-soluble antioxidants in biological membranes: Relationship between the effects of ubiquinol and vitamin E as inhibitors of lipid peroxidation in submitochondrial particles, *BioFactors* **3**(4), 241-248 (1992).
115. F. Aberg, E. L. Appelkvist, G. Dallner, and L. Ernster, Distribution and redox state of ubiquinones in rat and human tissues, *Arch. Biochem. Biophys.* **295**(2), 230-234 (1992).
116. M. E. Götz, A. Dirr, W. Gsell, R. Burger, B. Janetzky, A. Freyberger, H. Reichmann, W.-D. Rausch, and P. Riederer, Influence of N-methyl-4-phenyl-1,2,3,6-tetrahydropyridine, lipoic acid, and L-deprenyl on the interplay between cellular redox systems, *J. Neural Transm.* **43** (Suppl.), 145-162 (1994).
117. C. W. Shults, R.H. Haas, D. Passov, and F. Beal, Coenzyme Q10 levels correlate with the activities of complexes I and II/III in mitochondria from parkinsonian and nonparkinsonian subjects, *Ann. Neurol.* **42**(2), 261-264 (1997).
118. J. P. Sheehan, R.H. Swerdlow, W.D. Parker, S.W. Miller, R.E. Davis, and J. B. Tuttle, Altered calcium homeostasis in cells transformed by mitochondria from individuals with Parkinson's disease, *J. Neurochem.* **68**(3), 1221-1233 (1997).
119. M. E. Götz, A. Dirr, R. Burger, B. Janetzky, M. Weinmüller, W. W. Chan, S. C. Chen, H. Reichmann, W.-D. Rausch, and P. Riederer, Effect of lipoic acid on redox state of coenzyme Q in mice treated with 1-methyl-4-phenyl-1,2,3,6-tetrahydropyridine and diethyldithiocarbamate, *Eur. J. Pharmacol. Mol. Pharmacol. Sect.* **266**(3), 291-300 (1994).
120. M. F. Beal, D. R. Henshaw, B. G. Jenkins, B. R. Rosen, and J. B. Schulz, Coenzyme Q10 and nicotinamide block striatal lesions produced by the mitochondrial toxin malonate, *Ann. Neurol.* **36**(6), 882-888 (1994).
121. J. B. Schulz, D. R. Henshaw, R. T. Matthews, and M. F. Beal, Coenzyme Q10 and nicotinamide and a free radical spin trap protect against MPTP neurotoxicity, *Exp. Neurol.* **132**(2), 279-283 (1995).
122. R. T. Matthews, L. Yang, S. Browne, M. Baik, and M. F. Beal, Coenzyme Q10 administration increases brain mitochondrial concentrations and exerts neuroprotective effects, *Proc. Natl. Acad. Sci. U.S.A.* **95**(15), 8892-8897 (1998).
123. J. Fallon, R. T. Matthews, B. T. Hyman, and M. F. Beal, MPP^+ produces progressive neuronal degeneration which is mediated by oxidative stress, *Exp. Neurol.* **144**(1), 193-198 (1997).
124. S. E. Stephans, T. S. Whittingham, A. J. Douglas, W. D. Lust, and B. K. Yamamoto, Substrates of energy metabolism attenuate methamphetamine-induced neurotoxicity in striatum, *J. Neuochem.* **71**(2), 613-621 (1998).
125. M. F. Beal, Coenzyme Q10 as a possible treatment for neurodegenerative diseases, *Free Radic. Res.* **36**(4), 455-60 (2002).
126. R. J. Ferrante, O. A. Andreassen, A. Dedeoglu, K. L. Ferrante, B. G. Jenkins, S. M. Hersch, and M. F. Beal, Therapeutic effects of coenzyme Q10 and remacide in transgenic mouse models of Huntington's disease, *J. Neurosci.* **22**(5), 1592-1599 (2002).
127. R. P. Ostrowski, Effect of coenzyme Q(10) on biochemical and morphological changes in experimental ischemia in the rat brain, *Brain Res. Bull.* **53**(4), 399-407 (2000).
128. H. Li, G. Klein, P. Sun, and A. M. Buchan, CoQ10 fails to protect brain against focal and global ischemia in rats, *Brain Res.* **877**(1), 7-11 (2000).
129. M. E. Götz, A. Gerstner, R. Harth, A. Dirr, B. Janetzky, W. Kuhn, P. Riederer, and M. Gerlach, Altered redox state of platelet coenzyme Q10 in Parkinson's disease, *J. Neural Transm.* **107**, 41-48 (2000).
130. E. Strijks, H. P. Kremer, and M. W. Horstink, Q10 therapy in patients with idiopathic Parkinson's disease, *Mol. Aspects Med.* **18**(Suppl), S 237-240 (1997).
131. K. Lonnrot, T. Metsa-Ketela, G. Molnar, J. P. Ahonen, M. Latvala, J. Peltola, T. Pietila, and H. Alho, The effect of ascorbate and ubiquinone supplementation on plasma and CSF total antioxidant capacity, *Free Radic. Biol. Med.* **21**(2), 211-217 (1996).
132. L. Packer, H. J. Tritschler, and K. Wessel, Neuroprotection by the metabolic antioxidant α-lipoic acid, *Free* Radic. Biol. Med. 22(1-2), 359-378 (1997).
133. A. Bast, and G. R. M. M. Haenen, Interplay between lipoic acid and glutathione in the protection against microsomal lipid peroxidation, *Biochem. Biophys. Acta* **963**(3), 558-561 (1988).
134. H. Scholich, M. E. Murphy, and H. Sies, Antioxidant activity of dihydrolipoate against microsomal lipid peroxidation and its dependence on α-tocopherol, *Biochem. Biophys. Acta* **1001**(3), 256-261 (1989).

135. M. Panigrahi, Y. Sadguna, B. R. Shivakumar, S. V. Kolluri, S. Roy, L. Packer, and V. Ravindranath, α-Lipoic acid protects against reperfusion injury following cerebral ischemia in rats, *Brain Res.* **717**(1-2), 184-188 (1996).
136. P. Wolz, and J. Krieglstein, Neuroprotective effects of α-lipoic acid and its enantiomers demonstrated in rodent models of focal cerebral ischemia. *Neuropharmacology* **35**(3), 369-375 (1996).
137. N. Aguirre, M. Barrionuevo, M. J. Ramirez, J. Del Rio, and B. Lasheras, α-Lipoic acid prevents 3,4-methylenedioxy-methamphetamine (MDMA)-induced neurotoxicity, *Neuroreport* **10**(17), 3675-3680 (1999).
138. O. A. Andreassen, R. J. Ferrante, A. Dedeoglu, and M. F. Beal, Lipoic acid improves survival in transgenic mouse models of Huntington's disease, *Neuroreport* **12**(15), 3371-3373 (2001).
139. O. A. Andreassen, A. Dedeoglu, A. Friedlich, K. L. Ferrante, D. Hughes, C. Szabo, and M.F. Beal, Effects of an inhibitor of poly(ADP-ribose)polymerase, desmethylselegiline, trientine, and lipoic acid in transgenic ALS mice, *Exp. Neurol.* **168**(2), 419-424 (2001).
140. M. F. McCarty, Versatile cytoprotective activity of lipoic acid may reflect its ability to activate signalling intermediates that trigger the heat-shock and phase II responses, *Med. Hypotheses* **57**(3), 313-317 (2001).
141. L. Zhang, G. Q. Xing, J. L. Barker, Y. Chang, D. Maric, W. Ma, B. S. Li, and Rubinow, α-Lipoic acid protects rat cortical neurons against cell death induced by amyloid and hydrogen peroxide through the Akt signalling pathway, *Neurosci. Lett.* **312**(3), 125-128 (2001).
142. B. Drukarch, and F. L. van Muiswinkel, Neuroprotection for Parkinson's disease: a new approach for a new millennium, *Expert Opinion on Investigational Drugs* **10**(10), 1855-1868 (2001).
143. J. Flier, F. L. Van Muiswinkel, C. A. Jongenelen, and B. Drukarch, The neuroprotective antioxidant α-lipoic acid induces detoxication enzymes in cultured astroglial cells, *Free Radic. Res.* **36**(6), 695-699 (2002).
144. K. Hager, A. Marahrens, M. Kenklies, P. Riederer, and G. Münch, α-Lipoic acid as a new treatment option for Alzheimer type dementia, *Arch. Gerontol. Geriatr.* **32**(3), 275-282 (2001).
145. J. C. Watkins, P. Krogsgaard-Larsen, and T. Honore', Structure-activity relationships in the development of excitatory amino acid receptor agonists and competitive antagonists, in: *Trends in Pharmacological Sciences, The Pharmacology of Excitatory Amino Acids, Special Report*, edited by D. Lodge, and G. L. Collingridge, (Elsevier, Amsterdam, 1991). pp. 4-12.
146. D.T. Monaghan, R. J. Bridges, and C. W. Cotman, The excitatory amino acid receptors: their classes, pharmacology, and distinct properties in the function of the central nervous system, *Ann. Rev. Pharmacol. Toxicol.* **29**, 365-402 (1989).
147. M. DiFiglia, M. Excitotoxic injury of the neostriatum: a model for Huntington's disease, *Trends Neurosci.* **13**(7), 286-289 (1990).
148. M. H. M. Bakker, and A. C. Foster, An investigation of the mechanism of delayed neurodegeneration caused by direct injection of quinolinate into the rat striatum *in vivo*. *Neuroscience* **42**(2), 387-395 (1991).
149. H. S. Chen, J. W. Pellegrini, S. K. Aggarwal, S. Z. Lei, S. Warach, F. E. Jensen, and S. A. Lipton, Open-channel block of N-methyl-D-aspartate (NMDA) responses by memantine: therapeutic advantage against NMDA receptor-mediated neurotoxicity. *J. Neurosci.* **12**(11), 4427-4436 (1992).
150. H. S. Lustig, K. V. Ahern and D. A. Greenberg, Antiparkinsonian drugs and *in vitro* excitotoxicity, *Brain Res.* **597**(1), 148-150(1992).
151. D. L. Small, and A. M. Buchan, NMDA antagonists: their role in neuroprotection, *Int. Rev. Neurobiol.* **40**, 137-171 (1997).
152. W. Danielczyk, Therapy of akinetic crises, *Med. Welt* **24**, 1278-1282 (1973).
153. R. J. Uitti, A. H. Rajput, J. E. Ahlskog, K. P. Offord, D. R. Schroeder, M. M. Ho, M. Prasad, A. Rajput, and P. Basran, Amantadine treatment is an independent predictor of improved survival in Parkinson's disease, *Neurology* **46**(6), 1551-1556 (1996).
154. R. J. Uitti, More recent lessons from amantadine, *Neurology* **52**(3), 676 (1999).
155. K. L. R. Jansen, R. L. M. Faull, M. Dragunow, and B. L. Synek, Alzheimer's disease: changes in hippocampal N-methyl-D-aspartate, quisqualate, neurotensine, adenosine, benzodiazepine, serotonin and opioid receptors, an autoradiographic study. *Neuroscience* **39**(3), 613-627 (1990).
156. J. W. Newcomer, and J. H. Krystal, NMDA receptor regulation of memory and behavior in humans, *Hippocampus* **11**(5), 529-542 (2001).
157. K. A. Haberny, M. G. Paule, A. C. Scallet, F. D. Sistare, D. S. Lester, J. P. Hanig, and W. Slikker Jr., Ontogeny of the N-methyl-D-aspartate (NMDA) receptor system and susceptibility to neurotoxicity, *Toxicol. Sci.* **68**(1), 9-17 (2002).
158. B. K. Siesjö, H. Memezawa, and M. L. Smith, Neurotoxicity: pharmacological implications, *Fundam. Clin. Pharmacol.* **5**(9), 755-767 (1991).

159. G. G. C. Hwa, and M. Avoli, The involvement of excitatory amino acids in neocortical epileptogenesis: NMDA and non-NMDA receptors. *Exp. Brain Res.* **86**(2), 248-256 (1991).
160. B. K. Kohl, and G. Dannhardt, The NMDA receptor complex: a promising target for novel antiepileptic strategies, *Curr. Med. Chem.* **8**(11), 1275-1289 (2001).
161. K. Williams, Ifenprodil, a novel NMDA receptor antagonist: site and mechanism of action, *Curr. Drug Targets* **2**(3), 285-298 (2001).
162. J. Ruel, M. J. Guitton, and J. L. Puell, Negative allosteric modulation of AMPA-preferring receptors by the selective isomer GYKI 53784 (LY303070), a specific non-competitive AMPA antagonist, *CNS Drug Rev.* **8**(3), 235-254 (2002).
163. D. M. Turetsky, L. M. T. Canzoniero, and D. W. Choi, Kainate-induced toxicity in cultured neocortical neurons is reduced by the AMPA receptor selective antagonist SYM2206, *Soc. Neurosci. Abstr.* **24**, 578(1998).
164. J. Cartmell, and D. D. Schoepp, Regulation of neurotransmitter release by metabotropic glutamate receptors, *J. Neurochem.* **75**(3), 889-907 (2000).
165. A. Stefani, A. Pisani, N. B. Mercuri, and P. Calabresi, The modulation of calcium currents by the activation of mGluRs, functional implications, *Mol. Neurobiol.* **13**(1), 81-95 (1996).
166. V. Bruno, G. Battaglia, I. Ksiazek, H. van der Putten, M. V. Catania, R. Giuffrida, S. Lukic, T. Leonhardt, W. Inderbitzin, F. Gasparini, R. Kuhn, D. R. Hampson, F. Nicoletti, and P. J. Flor, Selective activation of mGlu4 metabotropic glutamate receptors is protective against excitotoxic neuronal death. *J. Neurosci.* **20**(17), 6413-6420 (2000).
167. J. R. Brorson, P. A. Manzolillo, and R. J. Miller, Calcium entry via AMPA/KA receptors and excitotoxicity in cultured cerebellar Purkinje cells, *J. Neurosci.* **14**(1), 187-197 (1994).
168. K. S. Lee, S. Frank, P. Vanderklish, A. Arai, and G. Lynch, Inhibition of proteolysis protects hippocampal neurons from ischemia, *Proc. Natl. Acad. Sci. U. S. A.* **88**(16), 7233-7237 (1991).
169. C. G. Markgraf, N. L. Velajo, M. P. Johnson, D. R. McCarty, S. Medhi, J. R. Koehl, P. A. Chmielewski, and M. D. Linnik, Six-hour window of opportunity for calpain inhibition in focal cerebral ischemia in rats. *Stroke* **29**(1), 152-158 (1998).
170. M. Miyamoto, and J. T. Coyle, Idebenone atttenuates neuronal degeneration induced by intrastriatal injection of excitotoxins, *Exp. Neurol.* **108**(1), 38-45 (1990).
171. H. Monyer, D. M. Hartley, and D. W. Choi, 21-Aminosteroids attenuate excitotoxic neuronal injury in cortical cell cultures, *Neuron* **5**(2), 121-126 (1990).
172. J. S. Beckman, and W. H. Koppenol, Nitric oxide, superoxide, and peroxynitrite: the good, the bad, and ugly. *Am. J. Physiol.* **271**(5 Pt 1), C1424-1437 (1996).
173. L. L. Dugan, S. L. Sensi, L. M. Canzoniero, S. D. Handran, S. M. Rothman, T. S. Lin, M. P.Goldberg, and D. W. Choi, Mitochondrial production of reactive oxygen species in cortical neurons following exposure to N-methyl-D-aspartate, *J. Neurosci.* **15**(10), 6377-6388 (1995).
174. A. F. Schinder, E. C. Olson, N. C. Spitzer, and M. Montal, Mitochondrial dysfunction is a primary event in glutamate neurotoxicity, *J. Neurosci.* **16**(19), 6125-6133 (1996).
175. E. P. Wei, M. D. Ellison, H. A. Kontos, and J. T. Povlishock, O2 radicals in arachidonate-induced increased blood-brain barrier permeability to proteins, *Am. J. Physiol.* **251**(4 Pt 2), H693-699 (1986).
176. S. J. Hewett, T. F. Uliasz, A. S. Vidwans, and J. A. Hewett, Cyclooxygenase-2 contributes to N-methyl-D-aspartate-mediated neuronal cell death in primary cortical cell culture, *J. Pharmacol. Exp. Ther.* **293**(3), 417-425 (2000).
177. M. P. Mattson, Stabilizing calcium homeostasis, in: *Handbook of Experimental Pharmacology, CNS Neuroprotection*, edited by F. W. Marcoux, and D. W. Choi (Springer-Verlag, Berlin, New York, 2002), pp. 115-153.
178. J. W. Phillis, Neuroprotection by free radical scavengers and other antioxidants, in: *Handbook of Experimental Pharmacology, CNS Neuroprotection,* edited by F. W. Marcoux, and D. W. Choi (Springer-Verlag, Berlin, New York, 2002), pp. 245-280.

NEUROTRANSMITTER RELEASE IN EXPERIMENTAL STROKE MODELS: THE ROLE OF GLUTAMATE-GABA INTERACTION

Laszlo G. Harsing, Jr., Gabor Gigler, Mihaly Albert, Gabor Szenasi, Annamaria Simo, Krisztina Moricz, Attila Varga, Istvan Ling, Erzsebet Bagdy, Istvan Kiraly, Sandor Solyom, and Zsolt Juranyi*

1. INTRODUCTION

Stroke or cerebrovascular accident reduces blood flow and decreases oxygen supply (ischemia) in brain tissue. This may be resulted from vascular obstruction when a blood vessel is blocked or by hemorrhage when bleeding occurs into the brain tissue. Decrease in oxygen supply shifts pH to acidosis and increases extracellular K^+ concentration, which depolarizes neural cell membrane. Anoxic depolarization leads to excessive release of glutamate, which then activates various glutamate receptors in the synapse or the extrasynaptic space. Opening of ionotropic glutamate receptors (NMDA, AMPA and kainate receptors) causes influx of Na^+ through the activated glutamate-gated ion channels. In response to anoxia, Ca^{2+} also enters the cells in excessive amounts via activated NMDA receptors and Ca^{2+}-permeable AMPA receptors. This will lead to activation of several Ca^{2+}-dependent intracellular signal transduction pathways (proteases, kinases, endonucleases, lipoxygeneses and nitric oxide synthase), which ultimately leads to neural death (Vizi et al., 1996; Parsons et al., 1998).

The excessive release of glutamate induced by ischemic insults is a key element in the ischemic processes. In ischemia, the mechanism by which glutamate is released is changed from Ca^{2+}-dependent vesicular release to Ca^{2+}-independent reverse-mode operation of glutamate transporter (Szatkowski and Attwell, 1994). The direction of glutamate transport is largely depends on the concentrations of the ions cotransported by the carrier molecule and elevation of intracellular Na^+ concentration which occurs in hypoxia induces reverse-mode operation of the transporters. Following ischemia, extra- and intracellular Na^+, K^+, and Ca^{2+} concentrations are shifted to the normal direction and glutamate release occurs by exocytosis from the vesicular compartments. An important

* Laszlo G. Harsing, Jr, Gabor Gigler, Mihaly Albert, Gabor Szenasi, Annamaria Simo, Krisztina Moricz, Attila Varga, Istvan Ling, I., and Zsolt Juranyi, EGIS Pharmaceuticals Ltd., Budapest, Hungary, Erzsebet Bagdy, Istvan Kiraly, and Sandor Solyom, IVAX Drug Research Institute, Budapest, Hungary.

difference between the normoxic and postischemic vesicular release of glutamate is that the latter elicits overstimulation of postsynaptic NMDA receptors. NMDA receptors possesses several binding sites for allosteric modulators and after ischemia, some of these allosteric modulators will lead to postischemic NMDA receptor overactivation (Szatkowski and Attwell, 1994).

Excessive efflux of glutamate in ischemia and overactivation of postsynaptic NMDA receptors following ischemic insult will result in increase of the release of several other neurotransmitters in the brain. Of these neurotransmitters, GABA might particularly be important as enhanced GABA release inhibits further release of glutamate from nerve endings (Nelson et al., 2000). Therefore, the shift in balance of the neurotransmitters glutamate and GABA in ischemia is extensively studied (Globus et al., 1988; Lyden, 1997; Seif-El-Nasr and Khattab, 2002). In this paper, we investigated the changes in glutamate and GABA contents in ischemic brain samples, ex vivo and in vitro release of GABA from tissues exposed to ischemia and the role of the corticostriatal glutamatergic pathway in activation of GABAergic spiny neurons in the striatum. Changes in glutamatergic and GABAergic neurotransmission occurred in close parallel in all ischemic models we used in this study.

2. EFFECT OF TRANSIENT OCCLUSION OF THE MIDDLE CEREBRAL ARTERY ON BRAIN TISSUE VIABILITY

2.1. Transient Occlusion of the Middle Cerebral Artery as a Cerebral Ischemia Model

To test the viability of brain tissue after transient occlusion of the middle cerebral artery, male Sprague-Dawley rats weighing 360-400 g were used (Warner et al., 1995; Longa et al., 1989; Sopala et al., 2002). The rats were anesthetised with a gas mixture of 70% N_2O and 2% halothane in oxygen. During surgery, the body temperature of the animals was kept at the preoperative levels (36-38 °C) with the help of a heating pad and a heating lamp. The animals were placed in supine position; a median incision was made on the neck skin. The right common carotid artery and the cervical carotid bifurcation were exposed and dissected free from the vagal nerve. The external carotid artery and common carotid artery were occluded proximally by silk ligatures and an atraumatic aneurysm clip was placed on the distal end of the common carotid artery. A 3-0-nylon monofilament with a heat-blunted tip was coated with poly-L-lysine and was used as an embolus. The monofilament was introduced from the bifurcation of the internal carotid artery and the ipsilateral common and external carotid arteries were ligated. After introduction of the monofilament, the internal carotid artery was ligated just distal to the insertion. The filament was extended from the bifurcation of the internal carotid artery to the proximal portion of the anterior cerebral artery. As the filament was passed 21-22 mm in the lumen of the internal carotid artery, the middle and the posterior cerebral arteries were occluded. Following surgery, the anesthesia was discontinued for 60 min. The animals were reanesthetised and the thread on the internal carotid artery was loosen, the monofilament was pulled out and a ligature permanently closed the internal carotid artery. The ischemic brain area was reperfused via the cerebral arterial circle, i.e. the contralateral carotid and basilar arteries. After surgery, the animals were kept in a temperature-controlled room and water and food were given ad libitum.

2.2. Staining the Brain Tissue with 2,3,5-Triphenyltetrazolium Chloride

The viability of brain tissue after ischemic injury of cerebral tissue was evaluated by using 2,3,5-triphenyltetrazolium chloride (TTC) as described by Bederson et al. (1986). TTC is a water soluble salt which is enzymatically reduced to a fat soluble, light sensitive compound in living tissue and colors normal areas to deep red. The conversion of TTC depends on tissue dehydrogenase activity, which is markedly depleted after ischemic damage. Since enzymic reduction of TTC does not occur in infarcted brain tissue normal and pathological areas are stained with marked differences in intensity.

Twenty-four hours after the operation, the animals were deeply anesthetized with pentobarbital (100 mg/kg ip) and perfused through the heart with 4% TTC. One hour later, the animals were decapitated; the brains were removed and placed in saline containing 8% formalin for at least 24 hours. Each brain was cut into 1 mm coronal sections. In some experiments, the removed brain was sectioned into coronal slices in a brain matrix then they were immersed into a 2% solution of TTC in normal saline at 37 °C for 30 min and the sections were fixed in 10% phosphate-buffered formalin. The area of the ischemic damage was measured and expressed in mm^2 in each slice using a morphometric software program (DigiCell for Windows 4.0) and the hemispheric extent of the infarct areas was estimated from the infarct areas.

Figure 1. shows ischemic areas on coronal section of rat brain demonstrated by transcardiac perfusion with TTC 24 hours after the right middle cerebral artery occlusion. TTC stained normal areas of the brain deep red but it did not stain infarcted tissue. The colorless area was considered to correspond to the territory supplied by the occluded middle cerebral artery. The colorless area was extended to the frontoparietal (somatosensory) cortex as well as the lateral and medial segments of the striatum. Cortex could clearly be distinguished from caudate nucleus after TTC staining within the infarcted area. This made possible to dissect tissue samples from the frontoparietal cortex and the lateral caudate nucleus of the ischemic and contralateral normal sides and these samples were then used for histological analysis, determination of glutamate and GABA tissue concentrations and for measurements of [^{3}H]GABA release (Figure 2).

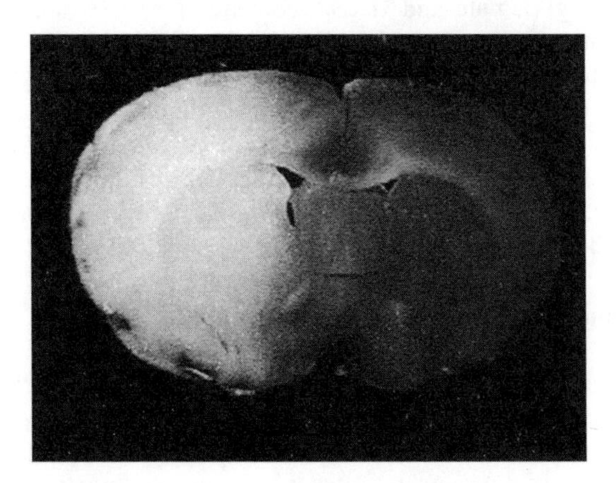

Figure 1. Coronal section of rat brain in which transcardiac perfusion with TCC was performed after transient right middle cerebral artery occlusion. Deep red areas: normal blood supply, colorless regions indicate ischemic brain damage. Swelling of brain tissue occurs as a consequence of anoxic insult.

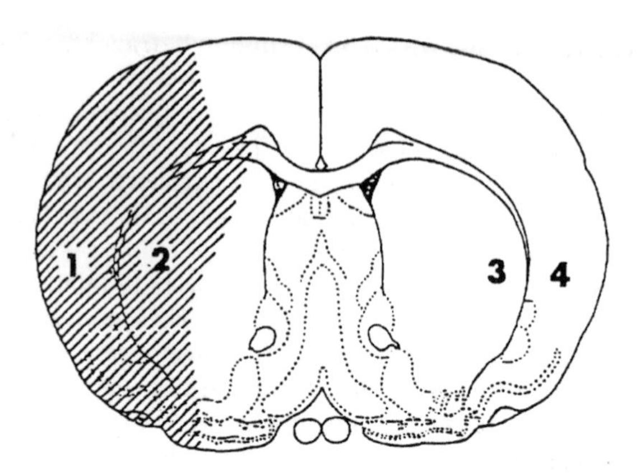

Figure 2. Coronal section of rat brain showing anatomical regions dissected for histological and neurochemical analysis after transient occlusion of the middle cerebral artery in the rat. Shaded area indicates ischemic tissues. Samples 1 and 4 represent frontoparietal cortices and samples 2 and 3 are the lateral segment of caudate-putamen from hypoxic and contralateral sides.

3. CHANGES IN GLUTAMATE AND GABA CONTENT IN THE CORTEX AND STRIATUM AFTER TRANSIENT OCCLUSION OF THE MIDDLE CEREBRAL ARTERY

3.1. Preparation of Brain Tissue for Amino Acid Determination

Tissue samples from the frontoparietal cortex and the striatum were dissected from coronal slices of rat brain 24 hours after a 60-min transient occlusion of the middle cerebral artery. Cortical and striatal tissue weighing 25-35 mg from the right side exposed to hypoxic insults and those obtained from the normoxic left side were collected and used for determination of glutamate and GABA contents. For determination of endogenous amino acid concentrations, ischemic and normoxic cortical and striatal tissues were homogenized by ultrasonication in 1 ml of 80% ethanol and centrifuged. The supernatant of the samples was diluted 10-times with water and aliquots of the samples or standards were derivatized by addition of 20 µl derivatization mixture (Rowley et al., 1989). The derivatization reagent was prepared by adding the following order: 22 mg o-phthaldialdehyde, 0.5 ml 1 M sodium sulfite (Na_2S0_3), 0.5 ml absolute ethanol and 9 ml 0.1 M sodium tetraborate buffer, pH adjusted to 10.4 with 5 M NaOH.

3.2. Determination of Glutamate and GABA Contents

For amino acid determination in brain tissue samples the HPLC-electrochemical detector system consisted of a Knauer high pressure pump type 64 with pulse dampener, a Rheodyne 7125 injector with 20 µl sample loop and a 150 x 4 mm Nucleosil C_{18} reversed-phase column, 5 µm particles with guard column (20 x 4 mm). A Bioanalytical Systems CC-5 cross-flow thin layer electrochemical flow cell equipped with a BAS LC-4C Amperometric Controller interfaced with a Shimadzu C-R6A Chromatopac integrator was used for electrochemical detection and data analysis. Mobile phase consisted of 0.1 M NaH_2PO_4/Na_2HPO_4, pH 4.9 containing 0.5 mM Na_2EDTA and 10% methanol for glutamate and 20% methanol for GABA determination. The mobile phase was filtered then degassed in an ultrasonic generator. Flow rate was kept at 0.7 ml/min.

3.3. Changes in Glutamate Contents of the Frontoparietal Cortex and Striatum after Transient Occlusion of Rat Middle Cerebral Artery

As shown in Figure 3, assay of glutamate concentrations in the frontoparietal cortex and the striatum showed significant reduction in both areas after transient occlusion of the right middle cerebral artery in the rat. Glutamate in the cerebral cortex is concentrated in afferent pathways originated from the thalamus and also in efferent pathways originated from neurons located in the layer of pyramidal cells (Aghajanian and Marek, 2000). The striatum, however, contains almost exclusively afferent glutamatergic projections only which originated from the cerebral cortex and the thalamus (Smith and Bolam, 1990). The reduction in glutamate content observed in the cerebral cortex and striatum after occlusion of the middle cerebral artery may indicate neural damage in the thalamocortical, thalamostriatal and corticostriatal glutamatergic projections.

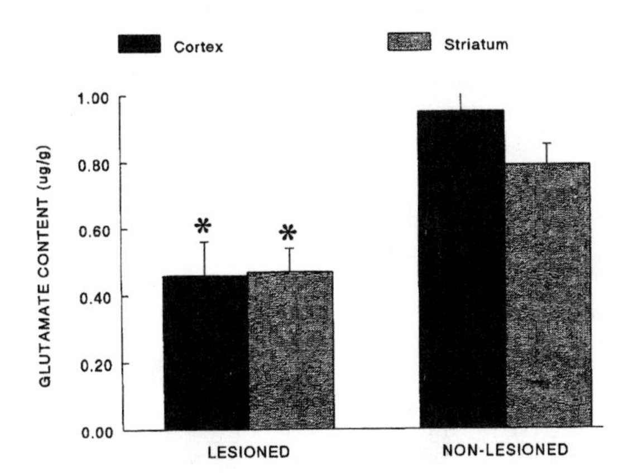

Figure 3. Decrease in glutamate concentrations in cerebral cortex and striatum after 1-hour occlusion of the right middle cerebral artery in the rat. Rats were killed 24 hours after ischemic insult and 2 mm thick coronal sections were prepared from the brain and samples were punched out from the frontoparietal cortex and the striatum of the damaged and intact sides as shown in Figure 2. Glutamate content was measured by using HPLC-electrochemistry. Mean±S.E.M., n=8, Student t-test for two-means, *$P<0.05$.

3.4. Changes in GABA Content of the Frontoparietal Cortex and Striatum after Transient Occlusion of Rat Middle Cerebral Artery

Measurement of GABA contents in the ischemic cerebral cortex and striatum indicates that hypoxic neural damage alters not only glutamatergic neurotransmission but also is extended to other neural systems. Figure 4 shows that GABA content in the striatum was markedly decreased in the damaged side after temporal occlusion of rat middle cerebral artery. GABA content in the cerebral cortex also tended to be lowered after ischemia, albeit this change was statistically not significant. The decrease in GABA content of different brain areas may be explained by injury of GABAergic neurons that takes place in ischemia. GABA in the cerebral cortex is mostly concentrated in interneurons. These neurons are involved in neural interactions established between raphe-cortical afferent and cortical glutamatergic efferent pathways (Carlsson et al., 1997). GABA interneurons are believed to be more resistant to cell death compared to principal neurons (Nitsch et al., 1989; Schwartz-Bloom and Sah, 2001). The high number of GABAergic interneurons in the cerebral cortex, which may be more resistant to hypoxia, may explain the smaller rate of reduction in GABA content observed in this brain area in response to transient ischemia. The striatum, on the other hand, contains both GABAergic interneurons and projection neurons. GABAergic interneurons in the striatum are interfaced between corticostriatal glutamatergic projection and GABAergic efferent neurons establishing feed-forward inhibition (Kita, 1996) whereas GABAergic projections neurons form the indirect and direct efferent pathways for the globus pallidus or the substantia nigra (Bolam and Bennett, 1995). In striatum, the medium-size GABAergic spiny neurons are highly vulnerable to transient forebrain ischemia whereas different groups of interneurons are much more resistant (Ren et al., 1997; Zoli et al., 1997). Gonzales and coworkers (1992) also found that GAD-immunopositive interneurons in the striatum are relatively resistant to cerebral ischemia. It is likely the activities of the ischemia-vulnerable projection neurons are more severely damaged by ischemic insult than the ischemia-resistant interneurons. The higher reduction in striatal GABA content may be explained by the fact that at least 90% of all striatal neurons are GABAergic projection neurons sensitive to hypoxic insults.

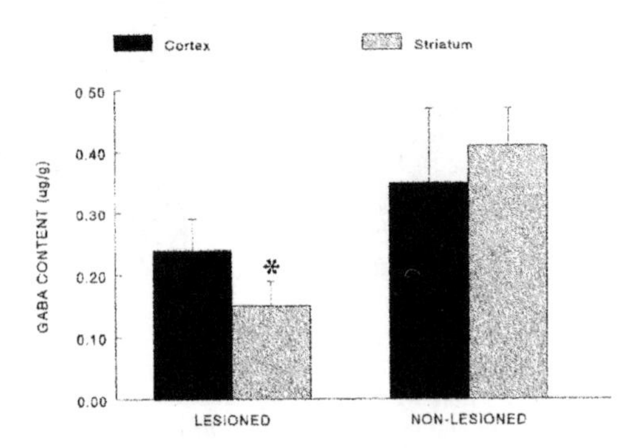

Figure 4. Decrease in GABA concentrations in cerebral cortex and striatum after 1-hour occlusion of the right middle cerebral artery in the rat. Rats were killed 24 hours after ischemic insult and 2 mm thick coronal sections were prepared from the brain and samples were punched out from the frontoparietal cortex and the striatum of the damaged and intact sides as shown in Figure 2. GABA content was measured by using HPLC-electrochemistry. Mean±S.E.M., n=4-5, Student t-test for two-means, *P<0.05.

Others also found that GABAergic neurotransmission is altered in cerebral ischemia (Gajkowska and Mossakowski, 1994; Kawai et al., 1995; Leyden, 1997). Thus, induction of cerebral ischemia followed by reperfusion reduced GABA content in rat cerebral cortex and hippocampus (Katsura et al., 1992; Seif-El-Nasr and Khattab, 2002). In addition, microsphere embolism also resulted in decrease of GABA content in the cortex, striatum, and hippocampus (Taguchi et al., 1993). Our findings that GABA content in striatum is altered by transient ischemia further support the view that GABAergic systems may be particularly important in mediation of ischemic events. This importance may be explained by the fact that GABAergic neurotransmission functions in opposition to that of glutamate.

4. MORPHOLOGICAL CHANGES IN THE CORTEX AND PUTAMEN AFTER TRANSIENT OCCLUSION OF THE MIDDLE CEREBRAL ARTERY

Tissue samples from the cerebral cortex and putamen from the right side exposed to hypoxic insults and those obtained from the intact left side were dissected to observe histological changes. Brain tissues were fixed in 10% buffered formalin then dehydrated through graded alcohols and after a xylene step, embedded in Histoplast (Shandon, UK). Brain sections 6 μm were stained with hematoxylin and eosin and Luxol fast blue. The sections were examined under a light microscope and ischemic neuronal damage was graded according to the morphological changes of the cells.

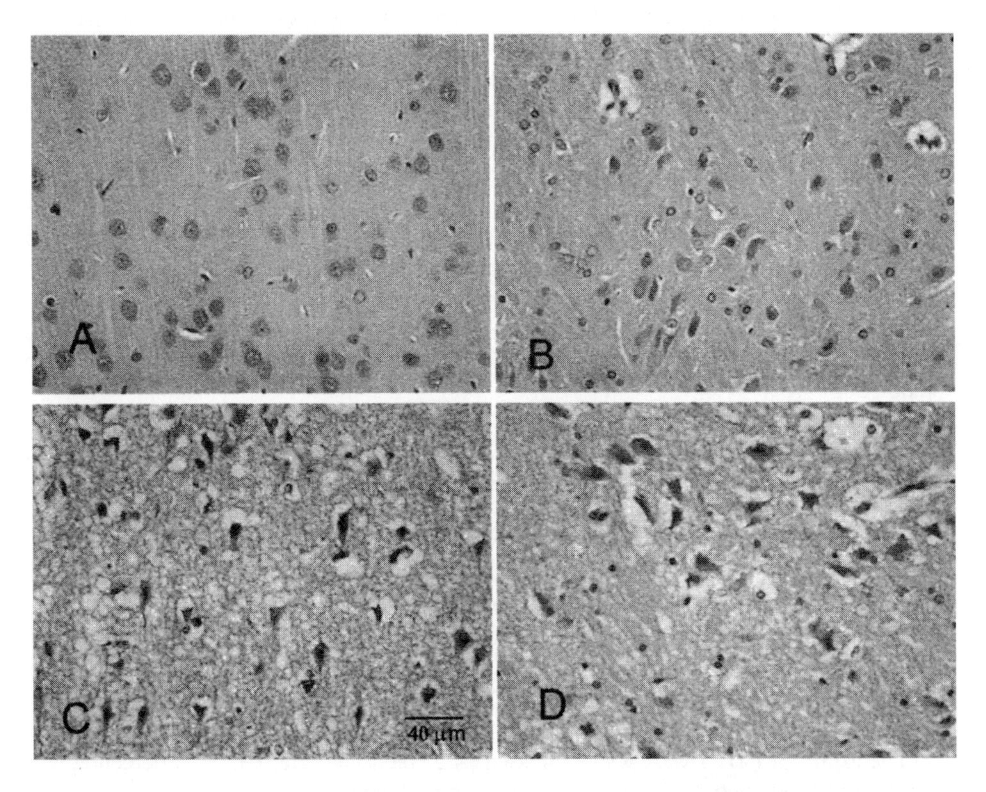

Figure 5. Histologic pictures of cortex and putamen obtained from rat brain after ischemic insult induced by transient occlusion of the middle cerebral artery. Intact cortex (A) and putamen (B), filament-induced necrosis in the cortex (C) and putamen (D), hematoxylin eosin staining. Note that besides edematous substances, vacuolated glial cells can also be seen. In addition, some neurons showed preserved structures as well.

Histologic examination of the cortex and putamen from the normoxic left side of the brain showed no pathologic alterations (Figure 5). On the contrary large necrosis was found in the cortex and putamen of the ischemic injured right side of the brain. These necrotic areas contained debris, remnants of neurocytes, lytic neurophile and glial cells. There were several shrunken neurocytes within the necrotic area and damage observed suggested irreversible changes. Light microscopy also revealed some relatively mild ischemic changes in striatal neurons, some of them showed relatively preserved structure (Figure 5). Edema of neuroglia was also seen around the necrotic areas.

5. [^{3}H]GABA RELEASE FROM CORTEX AND STRIATUM OBTAINED FROM RATS WITH TRANSIENT OCCLUSION OF THE MIDDLE CEREBRAL ARTERY

5.1. Measurement of [^{3}H]GABA Release

Cortical and striatal tissues were collected from the right side exposed to hypoxic insults and from the normoxic left side of the brain and used for measurement of [^{3}H]GABA release (Harsing and Zigmond, 1997). The slices were incubated with [^{3}H]GABA (2.5 μCi/ml) in oxygenated Krebs-bicarbonate buffer for 30 min at 37 °C. β-Alanine (1 mM), an inhibitor of GABA uptake in glial cells (Iversen and Kelly, 1975), was present in the incubation buffer. The tissues were then transferred into low-volume (300 μl) superfusion chambers (Experimetria, Inc., Budapest, Hungary) and superfused with aerated and preheated (37 °C) Krebs-bicarbonate buffer that contained the aminotransferase inhibitor aminooxyacetic acid (0.1 mM). The GABA uptake inhibitor nipecotic acid was present in Krebs-bicarbonate buffer in a concentration of 0.1 mM. The flow rate was kept at 1 ml/min during superfusion. The superfusate was discarded in the first 60-min period of the experiment then twenty-two 3-min fractions were collected. When used, electrical stimulation (20 V, 20 Hz, 2 msec for 2 min) was carried out in fraction 10.

At the end of superfusion, the tissues were collected from the superfusion chambers and homogenized in Soluene-100 and an aliquot was processed for determination of tissue content of radioactivity. To determine the radioactivity released from the tissue, a sample of the superfusate was mixed with liquid scintillation reagent and subjected to liquid scintillation spectrometry. Previously we separated the released radioactivity into [^{3}H]GABA and [^{3}H]metabolites on thin layer chromatography and found that, in these experimental conditions, 92-108% of the radioactivity released from striatal slices is in the form of authentic [^{3}H]GABA (Harsing and Zigmond, 1998).

5.2. Determination of [^{3}H]GABA Efflux from Brain Slices

The efflux of [^{3}H]GABA was expressed as a fractional rate, i.e., as a percentage of the amount of radioactivity in the tissue at the time the release was determined (Harsing et al., 1992). A computer program (Quattro Pro V6.0) was used to estimate the fractional rate of [^{3}H]GABA efflux. To estimate the electrical stimulation-induced [^{3}H]GABA overflow, the mean of the basal release determined before and after stimulation was subtracted from each sample and the evoked release represents the sum of the release in the fractions collected.

5.3. [^{3}H]GABA Release from the Frontoparietal Cortex and Striatum after Transient Occlusion of Rat Middle Cerebral Artery

Histological studies revealed that some neurons exhibited relatively mild morphological changes and their structures were well preserved after ischemic insult. Whether or not these neurons remained functionally intact, release of [^{3}H]GABA was measured in tissue samples obtained from the frontoparietal cortex and the caudate nucleus of the ischemic and contralateral sides (Figure 2). As shown in Figure 6, the electrical stimulation-induced [^{3}H]GABA release was significantly increased in the striatum but was without changes in the frontoparietal cortex of the ischemia-lesioned side. The increase in [^{3}H]GABA release observed in the ischemic caudate nucleus was associated with reduction in GABA content. These data may indicate that while striatal GABAergic neurons are damaged and thereby GABA content is reduced after transient ischemic insult, the release of GABA was actually increased as a compensation of neuronal loss. Neuronal damage is frequently associated with compensatory increase of transmitter release as was also demonstrated in experimental parkinsonism (Snyder et al., 1990). Administration of the neurotoxin 6-hydroxydopamine not only reduces dopamine content but also increases the release of striatal dopamine as a compensation to maintain dopaminergic tone in the basal ganglia (Zigmond et al., 1990).

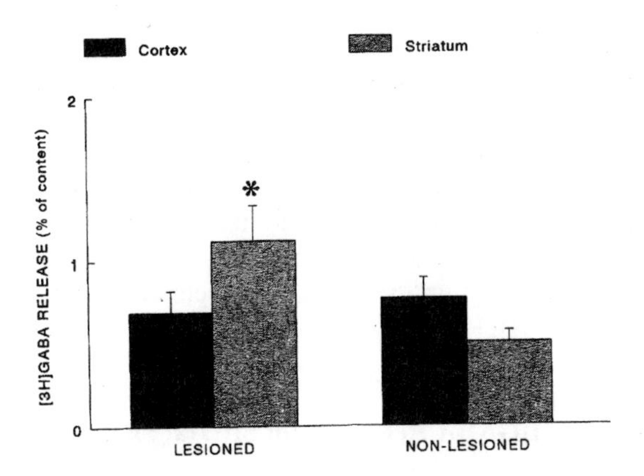

Figure 6. [^{3}H]GABA release in cerebral cortex and striatum after 1-hour occlusion of the right middle cerebral artery in the rat. Rats were killed 24 hours after ischemic insult and 2 mm thick coronal sections were prepared from the brain. Samples were punched out from the frontoparietal cortex and the striatum of the damaged and intact sides as shown in Figure 2. The tissue samples were loaded with [^{3}H]GABA and superfused. The release of [^{3}H]GABA evoked by electrical stimulation and was expressed as a fractional rate. Mean±S.E.M., n=8-10, Student t-test for two-means, *$P<0.05$.

Other studies also suggest an increased GABA turnover in response to transient ischemic insults. Globus and coworkers (1991) reported that GABA accumulates in the extracellular space during focal cerebral ischemia and it returns to normal levels within an hour of onset of reperfusion. In addition, increase in extracellular GABA concentrations after cerebral ischemia has been demonstrated in striatum, cerebral cortex and hippocampus using in vivo microdialysis or cortical cup technique (Schwartz-Bloom and Sah, 2001). Moreover, one can speculate that the enhanced GABAergic inhibitory neural activity can oppose the spread of excessive impulse activity after ischemic insult and this finding may correlate with the observation that GABA inhibits glutamate release by activation of $GABA_A$ receptors (Nelson et al., 2000).

6. IN VITRO ANOXIA MODEL IN STRIATAL SLICES

6.1. Measurement of [^{3}H]GABA and [^{3}H]Glutamate Release from Striatal Slices

For experiments when brain slices were exposed to in vitro hypoxia, male Sprague-Dawley rats weighing 200-250 g were killed by decapitation and the brain was removed from the skull. Coronal slices approximately 350 μm thick were cut from the striatum using a McIlwain tissue chopper. The slices were collected into ice-cold Krebs-bicarbonate buffer, pH 7.4 with the following composition in mM: NaCl 118, KCl, 4.7; $CaCl_2$ 1.25, NaH_2PO_4 1.2, $MgCl_2$ 1.2, $NaHCO_3$ 25, glucose 11.5. The Krebs-bicarbonate buffer used throughout the experiments was continuously gassed with 5% CO_2 in O_2.

The release of [^{3}H]GABA was determined from striatal slices of the rat as described above. In another set of experiments, striatal slices were loaded with [^{3}H]D-aspartate (5 μCi/ml), an indicator for glutamatergic neurotransmission. After a 30-min incubation at 37 °C, the slices were superfused as described above and the efflux of [^{3}H]D-aspartate was determined. To investigate the effect of anoxia on [^{3}H]GABA or [^{3}H]glutamate release in vitro, Krebs-bicarbonate buffer saturated with a gas mixture of 95% N_2 in oxygen with 11.5 mM mannitol instead of glucose was used to introduce anoxic conditions for superfused striatal slices. Striatal tissues were superfused with oxygen/glucose-free buffer from fraction 10 and anoxia was maintained throughout the rest of the experiment. Drugs were added to the superfusion buffer 9 min before introduction of anoxia. At the end of the experiments, the tissue content of radioactivity and the radioactivity released from the tissue were determined as described above. The efflux of [^{3}H]GABA or [^{3}H]D-aspartate was expressed as a fractional rate as described above.

6.2. Anoxia-Induced [^{3}H]GABA Release from Striatal Slices of the Rat

Based upon our neurochemical observations carried out on brain tissue obtained from rats exposed to transient focal ischemia, the question rose whether anoxia also induces [^{3}H]GABA release from striatal slices in in vitro conditions. When striatal slices preloaded with [^{3}H]GABA and superfused were deprived of oxygen and glucose, [^{3}H]GABA levels increased rapidly in the superfusate (Figure 7). This release was completely reversible as readmission of oxygen and glucose normalized the efflux of [^{3}H]GABA from the slices. The mechanism how [^{3}H]GABA may be released in ischemia is not clear. We found that the anoxia-induced [^{3}H]GABA release was partially Ca^{2+}-dependent indicating that depolarization-induced Ca^{2+}-dependent vesicular release as well as depolarization-induced Ca^{2+}-independent reversal of GABA transporters may be involved. It is generally believed that the reverse-mode operation of amino acid transporters may be at least in part responsible for the excessive release observed in anoxic conditions (Szatkowski and Attwell, 1994). The reverse-mode operation of GABA transporter in anoxic conditions may be insensitive for nipecotic acid, a substrate-type inhibitor of GABA transporter (Szerb, 1982).

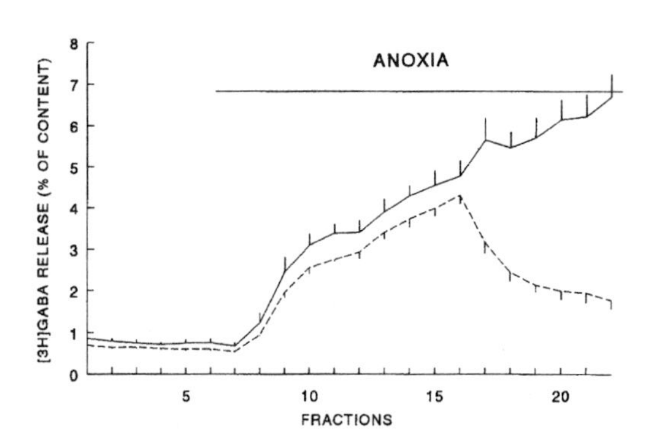

Figure 7. Anoxia-induced [^{3}H]GABA release from striatal slices of the rat. Readmission of oxygen and glucose normalized the efflux of [^{3}H]GABA from the slices. Striatal slices were prepared, loaded with [^{3}H]GABA and superfused. Glucose was omitted from the superfusion buffer and it was saturated with 95% N_2/5% CO_2 instead of 95% O_2/5% CO_2 (anoxia) from fraction 6. Mean±S.E.M., n=4.

Since a close connection has been demonstrated between glutamatergic and GABAergic neurotransmission in the striatum (Gerfen, 1992; Chen et al., 1998; Harsing et al., 2000) the question arose whether anoxia influences GABAergic neurotransmission directly or the enhanced [^{3}H]GABA release observed is an indirect effect and is a consequence of anoxia-evoked excessive glutamate release. In fact, hypoxia and glucose deprivation was shown to induce glutamate release from striatum (Miluseva et al., 1992). To investigate this hypothesis, a number of NMDA and AMPA receptor antagonists were tested on anoxia-induced [^{3}H]GABA release in rat striatal slices. As shown in Table 1, the NMDA receptor antagonist MK-801 as well as GYKI-52466 and GYKI-53655, two AMPA receptor negative modulators (Phillis et al., 1993; Tarnawa and Vizi, 1998; Abraham et al., 2000), inhibited anoxia-evoked [^{3}H]GABA release from striatal slices of the rat. These findings indicate that in anoxia [^{3}H]GABA release is evoked at least in part indirectly by stimulation of NMDA and AMPA types of glutamate receptors.

6.3. Anoxia-induced [^{3}H]D-aspartate Release from Striatal Slices of the Rat

To further study the role of glutamate in stimulation of [^{3}H]GABA efflux in anoxia, the release of [^{3}H]D-aspartate was also determined from striatal slices of the rat (Figure 8). D-aspartate is assumed as an indicator of endogenous glutamate, however it is not subject for transport mediated by glutamate carrier (Beani et al., 1991). The increase of [^{3}H]D-aspartate in response to anoxia indicates that glutamate and GABA are released simultaneously from striatal slice preparation and the increase of GABA efflux may be at least in part consequence of the release of the excitatory amino acid glutamate.

Table 1. Effect of NMDA and AMPA receptor antagonists on anoxia-induced [^{3}H]GABA release from striatal slices of the rat

Compounds		Concentration (μM)	[^{3}H]GABA release (Per cent of content)
Control		-	50.39±2.50
MK-801	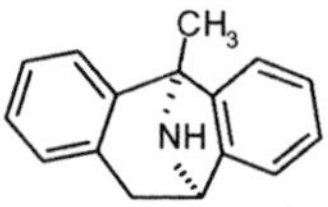	5	37.85±1.62*
NBQX	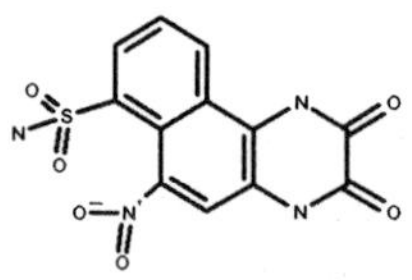	100	28.68±2.65*
GYKI-52 466	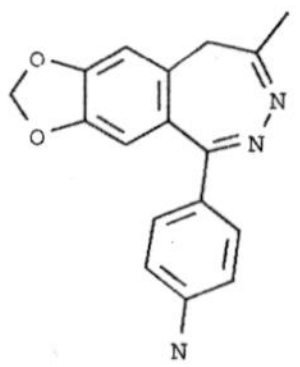	100	33.04±2.88*
GYKI-53 655	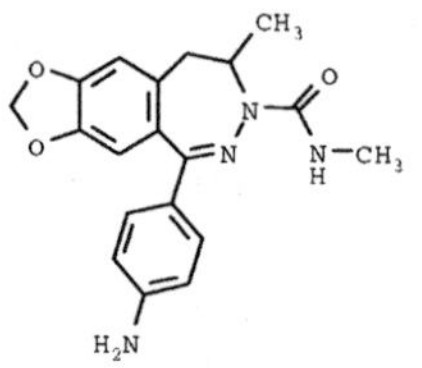	100	28.02±3.03*

Striatal slices of the rats were prepared, loaded with [^{3}H]GABA and superfused. [^{3}H]GABA release was induced by superfusion of the slices with glucose-free Krebs-bicarbonate buffer saturated with 95% N_2/5% CO_2 (anoxia) from fractions 5 to 22.
GYKI-52 466: 5-(4-aminophenyl)-8-methyl-9H-1,3-dioxolo[4,5-n][2,3]-benzodiazepine,
GYKI-53 655: 5-(4-aminophenyl)-7-methylcarbamoyl-8(R,S)-methyl-8,9-dihydro-7H-1,3-dioxolo[4,5-n]-[2,3]-benzodiazepine. Mean±S.E.M., ANOVA followed by the Dunnett's test, F(4,17)=14.366, P<0.01, mean±S.E.M., *P<0.05, n=4-6.

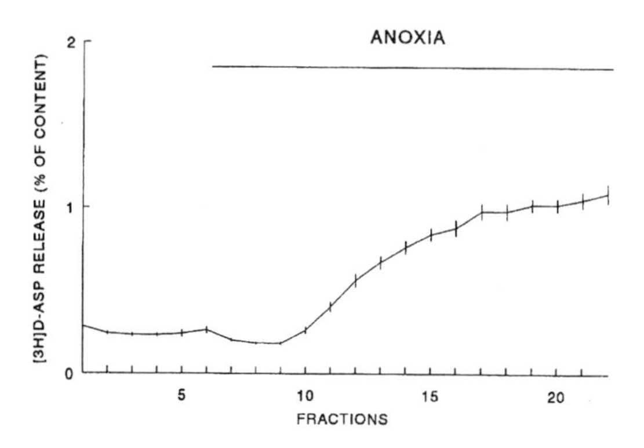

Figure 8. Anoxia-induced [^{3}H]D-aspartate release from striatal slices of the rat. Striatal slices were prepared, loaded with [^{3}H]D-aspartate and superfused. Glucose was omitted from the superfusion buffer and it was saturated with 95% N_2/5% CO_2 instead of 95% O_2/5% CO_2 (anoxia) from fraction 6. Mean±S.E.M., n=4.

There was, however, a time delay found in the onset of glutamate and GABA release evoked by anoxia (Baldwin et al., 1994). This finding also suggests the primarily role of glutamate in the mediation of anoxic neurochemical processes. The increase of glutamate release is due to the injury of energy producing processes within the cell and the decreased ATP levels leads to reduced activity of Na^+-K^+-ATPase increasing intracellular Na^+ concentration (Coyle and Puttfarcken, 1993). The excessive release of glutamate further increases intracellular Na^+ concentration by activating Na^+-permeable AMPA-gated ion channels and also by activation of Na^+-dependent neurotransmitter transporters. Further, the increase of intracellular Na^+ concentration induces reverse-mode operation of the transporters and extracellular neurotransmitter concentration reaches even higher levels. An additional factor in ischemic injury is the activation of Ca^{2+}-permeable NMDA receptors at the postsynaptic membranes increasing Ca^{2+} influx into the cells. This further may lead to increase the release of other neurotransmitters and activate Ca^{2+}-dependent signal transduction pathways (Martin et al., 1994).

7. INVESTIGATION THE ROLE OF CORTICOSTRIATAL PATHWAY

7.1. Measurement of [^{3}H]Glutamate Release in Dual Brain Slice Tissue Chamber

Experiments, which demonstrated the simultaneous release of [^{3}H]D-aspartate and [^{3}H]GABA evoked by anoxia from striatal slices on one hand, and the inhibitory influence of NMDA and AMPA receptor antagonists on anoxia-induced striatal [^{3}H]GABA release on the other hand, led us to the conclusion that the interaction between glutamatergic and GABAergic neurotransmission may have basic role in mediation of ischemic insults in the brain tissue. To further characterize this interaction in anoxia, corticostriatal complex slices were used (Juranyi et al., 2003). Horizontal slices (400-600 μm thick) containing the striatum and the adjacent prefrontal cortex from rat brain were cut with a Vibratome in a plane that maintains corticostriatal connections. After loading with [^{3}H]glutamate (2.5 μCi for 30 min), a slice was submerged into a two-compartment tissue bath so that the cortical part was contained entirely in one compartment, the corpus callosum passed through a silicone greased slot, and the striatal part was contained in the other compartment. The cortical and the striatal parts of the complex brain slice were superfused separately with Krebs-bicarbonate buffer at room temperature. D. C. potential was monitored between the cortical and the striatal parts of the slice by Ag/AgCl electrodes built in the two compartments. A cannula was placed over the striatal part of the submerged slice. Effluent was collected in 3-min fractions and [^{3}H]glutamate released from the tissue into the effluent was determined. At the end of the experiment, the cortical and striatal parts of the slice were separated and the [^{3}H]glutamate tissue content was also determined. To study the effect of anoxia on glutamate release, the cortical part of the slice was superfused with Krebs-bicarbonate buffer containing no glucose and saturated with 95% N_2/5% CO_2 instead of 95% O_2/5% CO_2 gas mixture. The release of [^{3}H]glutamate was expressed as a fractional rate.

7.2. Activation by Anoxia of the Corticostriatal Neural Pathway in a Complex Brain Slice Preparation

The corticostriatal pathway, which is originated from pyramidal cells of the cerebral cortex and makes synaptic contact with medium-size spiny GABAergic projection neurons may mediate the glutamatergic-GABAergic interaction in striatum. Exposure of the cortical tissue part to anoxia resulted in depolarization of the corticostriatal complex slice preparation (Figure 9). An anoxia-induced depolarization of neurons in the corticostriatal pathway increases cortical excitatory influence to striatal neurons, which then leads to enhancement of [^{3}H]glutamate release from the striatal part (Figure 10). The changes in cortically evoked synaptic transmission following ischemia therefore may provide further information regarding the importance of the glutamatergic-GABAergic interaction in the pathology of forebrain ischemia.

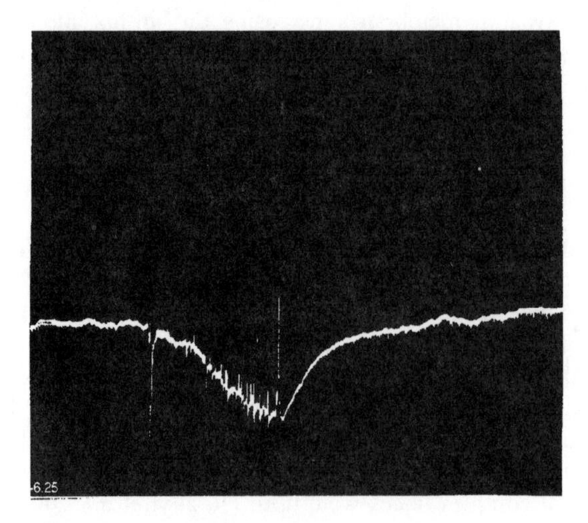

Figure 9. D. C. potential between cerebral cortex and striatum registered from a dual-compartment tissue bath for complex brain slices. Note that superfusion of the cortical part with glucose-free Krebs-bicarbonate buffer saturated with 95% N_2/5% CO_2 (anoxia) evoked changes in D. C. potential.

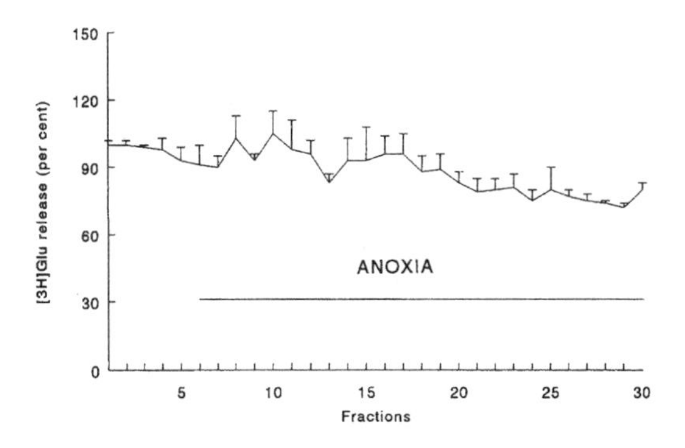

Figure 10. Striatal [^{3}H]glutamate release induced by anoxia from a complex cortex-striatum slice prepared from rat brain. The cortical part of the complex slice was superfused with glucose-free Krebs-bicarbonate buffer saturated with 95% N_2/5% CO_2 (anoxia) from fraction 6. The increase of striatal [^{3}H]glutamate release by exposition of the cortex to anoxic buffer indicates an activation of the corticostriatal pathway.

8. CONCLUSION

Various experimental stroke models were used to characterize glutamatergic-GABAergic interaction in ischemia in this study. We found that glutamate content in the frontoparietal cortex and striatum was decreased in a focal ischemia model (transient occlusion of the middle cerebral artery in the rat). GABA concentration was also reduced in striatum after occlusion of the middle cerebral artery and this decrease was associated with an enhanced compensatory GABA release. Histological analysis provided evidence that although sever neural loss occurred in the cortex and the basal ganglia after ischemia, several neurons survived ischemic insult. These neurons may actually function in an overactivated fashion ensuring an increased release of GABA after ischemic injury of the brain. Hypoxia also induces striatal GABA release from slice preparation, and this enhanced neurotransmitter efflux may be a consequence of activated glutamatergic neurotransmission. These was evidenced by the facts, that [^{3}H]GABA and [^{3}H]D-aspartate, a marker for glutamate, are released simultaneously in anoxic conditions and furthermore, NMDA and AMPA receptor antagonists suspended at least in part of anoxia-induced GABA release. The corticostriatal glutamatergic pathway, which synapses to medium-size spiny GABA projection neurons in the striatum, may be the neural substrate for the glutamatergic-GABAergic interaction in the caudate nucleus.

9. ACKNOWLEDGMENTS

A preliminary report has been made at the European Society of Clinical and Neuropharmacology (ESCNP) Conference, Budapest, 2002. This research was supported in part by the Research Council for Health Sciences, Hungarian Ministry of Health and Welfare (ETT-107/2000) and the Hungarian Science Research Fund (OTKA) grant No. T-32983. The authors acknowledge the excellent technical assistance of Mrs. Erika Hajdune-Gosi and Zsuzsa Major.

10. REFERENCES

Abraham, G., Solyom, S., Csuzdi, E., Berzsenyi, P., Ling, I., Tarnawa, I., Hamori, T., Pallagi, I., Horvath, K., Andrasi, F., Kapus, G., Harsing, L. G., Jr., and Kiraly, I., 2000, New non competitive AMPA antagonists, *Bioorg. Med. Chem.* **8**:2127.

Aghajanian, G. K., and Marek, G. J., 2000, Serotonin model of schizophrenia: emerging role of glutamate mechanisms, *Brain Res. Rev.* **31**:302.

Baldwin, H. A., Williams, J. L., Snares, M., Ferriera, T., Cross, A. J., Green, A. R., 1994, Attenuation by chlormethiazole of the rise in extracellular amino acids following focal ischemia in the cerebral cortex of the rat, *Br. J. Pharmacol.* **112**:188.

Beani, L., Tanganelli, S., Antonelli, T., Ferraro L., Morari, M., Spalluto, P., Nordberg, A., and Bianchi, C., 1991, Effect of acute and subchronic nicotine treatment on cortical efflux of [^{3}H]-D-aspartate and endogenous GABA in freely moving guinea-pigs, *Br. J. Pharmacol.* **104**:15.

Bederson, J. B., Pitts, L. H., Gremano, S. M., Nishimura, M. C., Davis, R. L., and Bartkowski, H. M., 1986, Evaluation of 2,3,5-triphenyltetrazolium chloride as a stain for detection and quantification of experimental cerebral infarction in rats, *Stroke* **17**:1304.

Bolam, J.P., and Bennett, B. D., 1995, Microcircuitry of the neostriatum, *in: Molecular and Cellular Mechanisms of Neurostriatal Function,* M. A. Ariano, and D. J. Surmeier, EDS., Landes Company, Austin, TX, pp. 1-19.

Carlsson, A., Hansson, L. O., Waters, N., and Carlsson, M. L., 1997, Neurotransmitter aberration in schizophrenia: new perspectives and therapeutic implications, *Life Sci.* **61**:75.

Chen, Q., Veenman, I., Knopp, K., Yan, Z., Medina, L., Song, W.-J., Surmeier, D. J., and Reiner, A., 1998, Evidence for the preferential location of glutamate receptor-1 subunits of AMPA receptors to the dendritic spines of medium spiny neurons in rat striatum, *Neuroscience* **83**:749.

Coyle, J. T., Puttfarcken, P., 1993, Oxidative stress, glutamate and neurodegenerative disorders, *Science* **262**:689.

Gajkowska, B., Mossakowski, M. J., 1994, Ischemia inhibits GABAergic neurons of the rat thalamic reticular nucleus. An immunocytochemical study, *Folia Neuropathol,* **32**:139.

Gerfen, C. R., The neostriatal mosaic: multiple levels of compartmental organization, *TINS* **15**:133.

Globus, M. Y.-Y., Busto, R., Martinez, E., Valdes, I., Dietrich, W. D., and Ginsberg, M. D., 1991, Comparative effect of transient global ischemia on extracellular levels of glutamate, glycine, and γ-aminobutyric acid in vulnerable and nonvulnerable brain regions in the rat, *J. Neurochem.* **57**:470.

Gonzales, C., Lin, R. C.-S., and Chesselet, M.-F., 1992, Relative sparing of GABAergic interneurons in the striatum of gerbils with ischemia-induced lesions, *Neurosci. Lett.* **135**:53.

Harsing, L. G., Jr., Sershen, H., Lajtha, A., 1992, Dopamine efflux from striatum after chronic nicotine: evidence for autoreceptor desensitization, *J. Neurochem.* **59**:48.

Harsing, L. G., Jr., Zigmond, M. J., 1997, Influence of dopamine on GABA release in striatum: evidence for D1-D2 interactions and non-synaptic influences, *Neuroscience* **77**:419.

Harsing, L. G., Jr., Zigmond, M. J., 1998, Postsynaptic integration of cholinergic and dopaminergic signals on medium size GABAergic projection neurons in the neostriatum. *Brain Res. Bull.* **45**:607.

Harsing, L. G., Jr., Csillik-Perczel, V., Ling, I., and Solyom, S., 2000, Negative allosteric modulators of AMPA-preferring receptors inhibit [^{3}H]GABA release in rat striatum, *Neurochem. Internat.* **37**:33.

Iversen, L. L., and Kelly, J. S., 1975, Uptake and metabolism of γ-aminobutyric acid by neurones and glial cells, *Biochem. Pharmac.* **24**:933.

Juranyi, Zs., Harsing, L. G., Jr., Zigmond, M. J., 2003, A new method for the investigation of transmitter release in complex corticostriatal slice preparation in vitro, *J. Neurosci. Methods* in press.

Katsura, M., Lino, T., and Kuriyama, K., 1992, Changes in content of neuroactive amino acids and acetylcholine in the rat hippocampus following transient forebrain ischemia, *Neurochem. Int.* **21**:243.

Kawai, K., Penix, L. P., and Kawara, N., 1995, Development of susceptibility to audiogenic seizures following cardiac arrest cerebral ischemia, *J. Cereb. Blood Flow Metab.* **15**:248.

Kita, H., 1996, Glutamatergic and GABAergic postsynaptic responses of striatal spiny neurons to intrastriatal and cortical stimulation recorded in slice preparation. *Neuroscience* **70**:925.

Leyden, P.D., 1997, GABA and neuroprotection, *Int. Rev. Neurobiol.* **40**:233.

Longa, E. Z., Weinstein, P. R., Carlson S., and Cummins, R., 1989, Reversible middle cerebral artery occlusion craniectomy in rats, *Stroke* **28**:84.

Martin, R. L., Lloyd, H. G.E., and Cowan, A. I., 1994, The early events of oxygen and glucose deprivation: setting the scene for neural death? *TINS* **17**:251.

Miluseva, E., Doda, M., Pasztor, E., Lajtha, A., Sershen, H., and Vizi, E.S., 1992, Regulatory interactions among axon terminals affecting the release of different transmitters from rat striatal slices under hypoxic and hypoglycemic conditions, *J. Neurochem.* **69**:946.

Nelson, R. M., Green, A. R., Lambert, D. G., and Hainsworth, A. H., 2000, On the regulation of ischaemia-induced glutamate efflux from rat cortex by GABA; in vitro studies with GABA, clomethiazole and pentobarbitone, *Br. J. Pharmacol.* **130**:1124.

Nitsch, C., Goping, G., and Klatzo, I., 1989, Preservation of GABAergic perykaria and boutons after transient ischemia in the gerbil hippocampal CA1 field, *Brain Res.* **495**:243.

Parsons, C. G., Danysz, W., and Quack, G., 1998, Glutamate in CNS disorders as a target for drug development: An update, *Drug News Perspect.* **11**:523.

Phillis, J. W., Smith-Barbour, M., Perkins, L. M., O'Regan M. H., 1993, GYKI 52466 and ischemia-evoked neurotransmitter amino acid release from rat cerebral cortex, *NeuroReport* **4**:109.

Ren, Y., Li, X., and Xu, Z. C., 1997, Asymmetrical protection of neostriatal neurons against transient forebrain ischemia by unilateral dopamine depletion, *Exp. Neurol.* **146**:250.

Rowley, H. L., Martin, K. F., Marsden, C. A., 1995, Determination of in vivo amino acid neurotransmitters y high-performance liquid chromatography with o-phthalaldehyde-sulphite derivatisation, *J. Neurosci. Methods,* **54**:93.

Schwartz-Bloom, R. D., and Sah, R., 2001, γ-Aminobutyric $acid_A$ neurotransmission and cerebral ischemia, *J. Neurochem.* **77**:353.

Seif-El-Nasr, M., and Khattab, M., 2002, Influence of inhibition of adenosine uptake on the γ-aminobutyric acid level of the ischemic rat brain, *Arzneim. Forsch./Drug Res.* **52**:353.

Smith A. D., Bolam, J. P., 1990, The neural network of the basal ganglia as revealed by the study of synaptic connections of identified neurones, *TINS* **13**: 259.

Snyder, G., Keller, R. W., Zigmond, M. J., 1990, Dopamine efflux from striatal slices after intracerebral 6-hydroxydopamine: evidence for compensatory hyperactivity of residual terminals, *J. Pharm. Exp. Ther.* **253**:867.

Sopala, M., Schweizer, S., Schaffer, N., Nurnberg, E., Kreuter, J., Seiller, E., and Danysz, W., 2002, Neuroprotective activity of a nanoparticulate formulation of the $glycine_B$ site antagonist MRZ 2/576 in transient focal ischaemia in rats, *Arzneim.-Forsch/Drug Res.* **52**:168.

Szatkowski, M., Attwell, D., 1994, Triggering and execution of neural death in brain ischaemia: two phases of glutamate release by different mechanisms, *TINS* **17**:359.

Szerb, J. C., 1982, Effect of nipecotic acid, a γ-aminobutyric acid transport inhibitor, on the turnover and release of γ-aminobutyric acid in the rat cerebral cortex, *J. Neurochem.* **39**:850.

Taguchi, T., Miyake, K., and Tanonnaka, K., 1993, Sustained changes in acetylcholine and amino acid contents of brain regions following microsphere embolism in rats, *Jpn. J. Pharmacol.* **62**:269.

Tarnawa, I., Vizi, E. S., 1998, 2,3-Benzodiazepine AMPA antagonists, *Rest. Neurol. Neurosci.* **13**:41.

Vizi, E. S., Mike, A., and Tarnawa, I., 1996 2,3-Benzodiazepines (GYKI 52466 and analogs): negative allosteric modulators of AMPA receptors, *CNS Drug Reviews* **2**:91.

Warner, D. S., Martin, H., Ludwig, P., McAllister, A., Keane, J. F. W., and Weber, E., 1995, In vivo models of cerebral ischemia: effects of parenterally administered NMDA receptor glycine site antagonists, *J. Cereb. Blood Flow and Metab.* **15**:188.

Zigmond, M. J., Abercrombie, E. D., Berger, T. W., Grace A. A., and Stricker, E. M., 1990, Compensations after lesions of central dopaminergic neurons: some clinical and basic implication, *TINS* **13**:290.

Zoli, M., Grimaldi, R., Ferrari, R., Zini I., and Agnati, L. F., 1997, Short- and long-term changes in striatal neurons and astroglia after transient forebrain ischemia in rats, *Stroke* **28**:1049.

NEUROPROTECTION IN ISCHEMIC/HYPOXIC DISORDERS

From the Preclinical to the Clinical Testing

Zoltán Nagy and László Simon*

1. INTRODUCTION

In the brain, the most common lesion is ischemic and/or hypoxic by origin. The insult results in neuronal loss. The primary aim of therapeutic interventions to reduce the volume of brain damage and thus, to lessen the neurological impairment, disability and handicap among stroke survivors. Reduction in infarct size may also reduce the risk of early death, particularly due to ischemic cerebral edema and transtentorial herniation.

The pathomechanism of neuronal loss in stroke is complex, many potential treatments have more than one mechanism of action. For some, the precise mechanism is far from being well elucidated. Blockage of a nutritive artery, sequential events in the hemostasis, with an active participation of the vascular endothelium, energy crisis, cell necrosis in the infarct's core and adjacent to the core (in the penumbra region), reperfusion injury, delayed neuronal death and/or necrosis are modeling the infarct's volume. The spatial and sequential complexity of these events are a challenge to all potential therapeutic strategies.

A fall in cerebral blood flow to a certain critical level causes rapid depletion in energy associated with changes in cerebral function. This initiates a complex cascade of biochemical changes that cause cell death. These changes has fundamentally two mechanisms: the rapid development of acidosis and the entry of ionic calcium into the cell. The high calcium concentration present in the cells activates the phospholipases, proteases, NOS, and turns on the immediate early gene expression.

Energy crisis, oxygen and glucose deprivation induce serial events of various cascade in the biochemical machinery of the cells of the central nervous system (CNS). Neuronal damage induced by brain ischemia is mediated by prolonged elevation of intracellular calcium levels and accumulation of free radicals. Ischemia induced glutamate release causes calcium overload in the cells via the operation of voltage-sensitive or agonist-operated calcium channels. Calcium is an activator of enzymes which degrade phospholipids, proteins and DNA, or alter the state of phosphorylation of

*Semmelweis University, Department of Vascular Neurology, Budapest, Hungary

proteins. By a rise in the intracellular Ca^{2+}, the production of reactive oxygen species (ROS) is augmented.

Neurons communicate with the environment through ligands (neurotransmitters, peptides, proteins, matrix molecules, or gaseous molecules like NO). The ligands activate intracellular signals by binding to receptors and, in this way, initiate different intracellular cascade mechanisms with the assistance of second messenger molecules (diglycerids, cAMP, cGMP, NO, H_2O_2). Ligand binding activates kinases or phosphatases, induces redox changes in proteins or turn a proteolytic process on.

In ischemic/hypoxic conditions the energy crisis results in ATP loss and subsequently in the loss of the activity of ion-pumps of membranes. Increased amounts of intracellular Na and Ca^{2+} enhance neurotransmitter release and decrease uptake. Overload of intracellular Ca^{2+} activates proteases and lipases. In the lesion, the cells become catabolic: degradation of macromolecules is not counteracted by resynthesis.

The energy crisis leads either to cell necrosis, or programmed cell death (apoptosis) and/or may activate repaire mechanisms.

2. CELL NECROSIS, APOPTOSIS, CELL RESCUE

Apoptosis and necrosis are two distinct forms of cell death with different morphological and molecular features. Around the infarct's core a rim corresponding to the penumbra region contains a great number of TUNEL stained neurons with fragmented DNA, which represent apoptosis, or programmed cell death.[1,2] Apoptosis is an active process of cellular selfdestruction both in physiological and pathological conditions. Cell necrosis is always pathological. The susceptibility to apoptosis is tightly regulated. Plasma membranes are intact until late onset cellular enzymes are activated (chromatin condensation, nuclear fragmentation, DNA fragmentation, selective protein degradation by means of specific proteases/caspases). Cytolysis could occur secondary to apoptosis. Cell necrosis is unregulated or poorly regulated, the plasma membrane is destroyed early: leakage of cell content, inflammation, cytoplasmic swelling, and mitochondrial swelling are well documented. Although apoptosis and necrosis have long been viewed as antagonistic by nature, it is now generally assumed that both forms of cell death constitute two extremes of a continuum. During apoptosis, cells may undergo secondary necrosis. Many pathologies labeled necrotic involve apoptosis. This concept is substantiated by the observations, that oncoprotein Bcl-2 can inhibit both apoptosis and in some models, primary necrosis. Overexpression of Bax and Bak causes apoptosis and in the presence of caspase inhibitors, non-apoptotic cytolysis. It is important to note, that there are several rescue pathways in the course of apoptosis by means of upregulation of cellular repair pathways. The delicate balance of heterodimerisation of anti- and proapoptotic proteins/genes represents this regulatory mechanism.

Following the ischemic/hypoxic injury, there is a severe DNA damage and repair genes are expressed within the core and around it, in the penumbra region. Cerebral ischemia induces genomic instability in the brain. Mitosis of postmitotic cells is associated with genomic instability.[3] It was demonstrated, that pro-apoptotic genes (Bax) in the infarct core and adjacent to the core are upregulated and in the same area TUNEL positive cells evidenced apoptosis.[4] In contrast, the upregulation of anti-apoptotic genes, Bcl-2 and Bclx, could be related to cell survival.[5] The "terminator" protease, caspase-3 is associated with programmed cell death and its presence has been documented in the core and adjacent to the infarcted area.[6,7]

Mitochondria play a major role in apoptosis triggered by ligand binding related pathways or by calcium activated pathways. The different signals converge onto mitochondria and are mediated through members of the Bcl-2 protein family (Bid, Bad). Following a death signal Fas-ligand, Fas receptor and FADD/DISC activation leads caspase-8 activation, which cleaves Bid, and thus tBid translocates to mitochondria. It activates Bax and bax-like proteins and this results in cytochrome-c release. Cytochrome-c activates caspase-9 by binding to Apaf-1 and a dATP and finally caspase-3 is activated. By the upregulation of "survival factors", Bad could be phosphorylated by several kinases (Akt, MAPK, Erk, PKA, PAK). Beside this rescue mechanism, due to calcium influx, Bad could be dephosphorylated by the Ca^{2+} sensitive phosphatase calcineurin, or by the protein phosphatase 1α (PP-1α) which translocates to the outer membrane of the mitochondria and binds to the antiapoptotic Bcl-XL.

TNFα mediates injury by initiating apoptotic cell death,[8] but it could also be neuroprotective.[9] TNF in neutrophils and in vascular endothelial cells could mediate injury, whereas TNF induction in neurons is protective.[10]

NFκB, upregulates both survival factors (MnSOD, Bcl-2, calbindin, TNFR, IAPs) and factors resulting in cell injury or death (iNOS, MMP-9, Cox-2, IL-1, ICAM-1, IL-8, CINC, gro and Bcl-x).

Aspirin and IL-10 downregulate NFκB, and TNFα[11] in part by inhibiting ikKs.[12] Detailed decription of the signaling pathways of pro-apoptosis and anti-apoptosis cascades could open new perspectives in the therapeutic interventions in the future.

3. CALPAIN AND CASPASE

Neurotoxic challenges (hypoxia, excitotoxicity, or metabolic inhibition) result in the activation of both cystein proteases.[13] Calpain is activated in various necrotic and apoptotic conditions, while caspase-3 is activated only in neuronal apoptosis.

Calpains could be overactivated in extreme situations, that result in sustained elevation of cytosolic Ca^{2+} levels, which is generaly associated with cell necrosis. Calpain substrates include cytoskeletal proteins, plasma-membrane associated proteins (such as PDGFr), signal transduction, calmodulin-dependent proteins and transcription factors.

Among caspases, caspase-3 (CPP32) is a common downstream apoptosis effector. The proenzyme form (pro-caspase) exists in most of the cells including neurons. It is processed and activated by caspase-9 or caspase-8 to the heterodimeric form by two different pathways. The first is the mitochondria dependent pathway: complexing of cytochrome-c and apoptotic protease-activating factor 1 (APAF 1) to caspase-9. After the dimerisation and autolytic activation steps, the processing of caspase-3 follows. The second pathway is related to the activation of death-domain containing receptors (TNF-αr1, FAS). The receptors recruit adapter proteins (FADD, TRADD) and following the activation of caspase-8, caspase-8 in turn activates caspase-3 proteolitically. Inhibitors of apoptosis proteins (IAP1, IAP2, and neuronal apoptosis inhibitor protein NAIP) supress caspase-3 activation in resting cells. Caspase-3 activation does not require Ca^{2+} influx.

4. SELECTIVE NEURONAL VULNERABILITY AND ISCHEMIC STROKE

Focal brain ischemia is usually caused by thromboembolic or atherothrombotic occlusion of an extracranial or intracranial cerebral blood vessel. Such an occlusion rarely if ever produces a complete loss of blood flow to its vascular territory. The rate of blood flow gradually decreases in space from a mild reduction in areas proximal to the collateral channels to more marked blood flow reduction of regions distal to the collateral supply, that is, the area located at the center of the ischemic infarct. The histopathological outcome is directly influenced by both the degree and duration of cerebral ischemia. Focal brain ischemia typically causes infarction with a central area of necrosis (all cell types are necrotized) surrounded by a thin rim of neuronal injury (penumbra). Selective ischemic necrosis limited to a specific population of neurons have been empirically identified. Examples of such selectively vulnerable populations include the CA1 sector of the hippocampus, medium size striate neurons, and cerebellar Purkinje cells. At least three distinct time-related properties of selective ischemic necrosis have been identified; the ischemic interval required to trigger irreversible injury, the time after cerebral recirculation before irreversible injury begins (delay interval), and the time it takes for irreversible injury to evolve once it has been initiated.

5. CALCIUM INFLUX

Cell signaling induced by Ca^{2+} initiates protein kinase C (PKC) and calcium calmodulin protein kinase II (CaMK II) activity. These enzymes phosphorylate receptor residues such as a glutamate receptor subunit, or transcription factors and eventually regulatory proteins. Both enzymes translocate from the cytosol into the membrane in the course of ischemia and reperfusion period. The translocation of these enzymes correlates with cell death.[14] Phosphorylation of receptors, ion-channels or a Na/Ca antiporter of the mitochondrial inner membrane enhances Na^{+} and Ca^{2+} permeability. Tyrosine phosphorylation of the NR2 subunite of NMDA receptor complex could be detrimental by augmenting calcium influx.[15]

Because Ca^{2+} influx has a pivotal role in the neuronal necrosis and apoptosis, great effort has been concentrated on the development of a Ca^{2+} antagonist for blocking the voltage-sensitive or ligand activated Ca^{2+} channels as a therapy of ischemic stroke. In the last 15 years, most of the clinical studies have been working with a voltage sensitive Ca^{2+} channel blocker, a dihydropyridine derivate, Nimodipine, though no conclusive results could be obtained.[16,17]

NMDA receptor activation by glutamate is another basic mechanism by which Ca^{2+}/Na^{+} ions accumulate in the cytosol. Non-competitive antagonists like phenylcyclidine (PCP), ketamine, disolcipine (MK-801), dextrometorphan and dexorphan bind to the PCP recognition site in the NMDA-gated ion channels. Different side effects in human use terminated the clinical trials with the exception of Cerestat. Mg^{2+} ion is the other voltage dependent blocker of the ion channels of NMDA receptors. Mg^{2+} as a neuroprotective ion in ischemic stroke is still under evaluation. Competitive NMDA antagonists, Selfotel, Eliprodil or a glutamate release inhibitor Lubeluzole have not been proved to be useful in stroke therapy.[18]

Sodium and potassium channel modulators are the other candidates in maintaining ionic homeostasis in the cells during the course of brain ischemia. The Na^+ channel blockers, Lamotrigine and Riluzole have been introduced into the clinical practice but not in stroke. A K+ channel opener with down-regulation of voltage-gated Na^+ channels decreases membrane excitability and stimulus-coupled transmitter release while promoting neuronal survival.[19] The preclinical data have not yet been confirmed with clinical trial because of different side-effects.

AMPA and kainate antagonists appear to be on their way towards clinical trials, but there are no data yet on clinical application.

6. NITRIC OXIDE

Nitric oxide (NO) has well recognized roles in the vasculature such as controlling cerebral blood flow, reducing thrombosis, and reducing adherence of inflammatory cells. NO has a special, but not completely elucidated role in the brain, where neurons express as much as 20 times more NOS activity than all of the endothelium in the body (Beckman 1997). NO is closely linked to the Ca^{2+} influx mediated by NMDA receptors in neurons, it has a role in local modulation of neuronal activity, in certain types of learning behaviors (in spatially oriented tasks), it helps to regulate synaptic remodeling by acting as a retrograde messenger. In humans there are at least two different promoters that control the transcription of NOS mRNA by alternative splicing, even though finally the expressed protein is the same.[20]

The close association between the NMDA receptor and NO production suggests that NO might be involved in neurodegenerative processes.[21]

Immediately after focal ischemia occurs, NO is derived mainly from neuronal NOS (nNOS or NOS-1) and endothelial NOS (eNOS or NOS-3) in the core and rim of the infarct. Inducible NOS (iNOS or NOS-2) is produced by neutrophils, macrophages, microglia, astroglia or blood vessels after an ischemic period.[22] Studies using an NO-sensitive microelectrode revealed that NO concentration in the hemispheres with permanent or transient MCA occlusion present was remarkably increased 15-45 min and 1.5-4 h after MCA occlusion.

NOS inhibitors were reported to reduce, or even to enhance brain injury. In a stroke model, the protective effects were due to the inhibition of vascular NO (controlling thrombosis), while neuronal NO seems to be more detrimental.[23] Neuronal NO caused an excessive activation of the NMDA receptor. Enhanced NO production within the cerebral microvessels protects brain tissue during focal ischemia via hemodynamic mechanisms, whereas neuronal NO overproduction may facilitate or mediate neurotoxicity. Using transgenic animals lacking NOS activity support these observations.[24] Repeated intraperitoneal administration of N(G)-nitro-L-arginine (L-NNA), a NOS inhibitor, diminished the increments in NO production during ischemia and reperfusion leading to a remarkable reduction in infarct volume.[23]

Recent data documented a close relationship between prostaglandin synthesis and the pathomechanisms of NO.[25] COX-2, the prostaglandin synthesizing enzyme cyclooxygenase –2 interacts with the inducible NO synthase (iNOS).

7. DECREASED PROTEIN SYNTHESIS

A decrease in or block of protein synthesis is one of the first biochemical changes to occur after focal brain injury. Ribosomal protein synthesis appears to be a step sensitive to reduction in CBF, because of interaction of the initiation factor 2 (eIF2). The guanine nucleotide exchange factors, GTP and the eukariotic elongation factor (eEF2).[26] Glutamate dependent phosphorylation of eEF2 provides a direct link between increased extracellular glutamate and initiation of protein synthesis. In the core, however, at least part of the protein synthesis (hsp70, eNOS) could be maintained mainly in the microvascular endothelium.[27]

Intracellular adhesion molecule-1 (ICAM-1) is expressed by the endothelial cells in both the core and in the penumbra region.[28] Membrane-bound ICAM-1 and VCAM-1 expression in human brain microvessel endothelial culture have been documented as early as 4 hours following cytokine (IL-1β, and TNFα) treatment.[29] Cytokine induced neutrophil chemo-attractant protein (CINC) is also induced in the infarct and in its immediate surroundings.[30] Reduction in the number of inflammatory cells or inhibition of adhesion molecules reduced the infarct size in experimental stroke models.[31] A recently completed trial with anti-ICAM-1 antibody in humans failed to show any benefit.[32] The metalloproteinase MMP-9 (gelatinase B) is increased in the core within 2 hours. MMP-9mRNA could be mediated by an NFκB site in the MMP promoter.[33] Activation of MMP-9 correlates with blood-brain barrier breakdown[34] and with hemorrhagic transformation.[35]

8. IMMEDIATE EARLY GENE EXPRESSION

Fos and Juns are members of a class of genes called immediate early genes (IEGs) or early response genes.[36] After focal ischemia IEGs (C-fos) are upregulated in the entire affected hemisphere.[37] It has been suggested, that spreading depression is related to C-fos induction in the diseased hemisphere. IEGs are thought to be the third messengers in a cellular cascade of stimulus-transcription coupling, which convert extracellular signals into alteration of cellular function by regulating late response or target genes.[38] Fos/Juns complexes are capable of regulating genes such as NGF in the nervous system following experimental global ischemia in gerbils and rats.

BDNF, bFGF, GFAP have AP-1 sites in their promoters, so as members of the Fos and Jun families, they could induce these genes. The expression of IEGs' is induced throughout the infarcted hemisphere after spreading depression. Based on these data it could be assumed, that IEGs have a neuroprotective role.[39] Other genes with AP-1 sites in their promoters that could be induced by fos-jun family members include dynorphin, encephalin, iNOS, APP, tyrosine hydroxilase, GAP-43, NGF.[40,41]

C-jun can form a homodimer as well as a complex with c-fos family members. Therefore c-jun expression has been associated with cell survival as well as cell death (Fig 3) Phosphorilated c-jun is expressed in cells that are going to die by apoptosis.[42]

Glutamate-receptor mediated induction of IEG encoded mRNAs can be observed in hippocampal neurons. Due to the irreversible translation block in vulnerable CA1 pyramidal neurons, transcription is not followed by translation into proteins. In contrast, expression of jun and fos proteins is restricted to neurons less vulnerable or resistant to ischemia.[43] Selective and delayed neuronal death of CA1 pyramidal cells after transient global ischemia can be prevented by conditioning the brain with a short period of ischemia 1-4 days prior to the subsequent normally lethal ischemic period. Endogenous neuroprotection is associated with a genetic program which involves the selective expression of c-jun. C-jun is a potential rescue gene during excitotoxic injury.[44]

9. HEAT SHOCK PROTEINS

Heat shock protein (hsp) expression has been documented in different global or focal ischemia models. From the class of heat shock transcription factors (HSF) a single factor, HSF1 appears to be responsible for the regulation of hsp genes in hyperthermia or ischemia.

Hsp72 is not constitutively present in the brain and it seems to be an optimal marker-protein involved in the trafficking of proteins between and within intracellular compartments after insult. Hsp72 mRNA induction after global ischemia correlates with the relative vulnerability of neuron population such as in the hippocampus CA1 segment, on the other hand the expression of hsp72mRNA and the accumulation of the encoded protein were found to be dissociated from one another. The functional translation deficit could be related to the compromised protein synthesis in stressful conditions.[45] Hsp72 mRNA expression in the focal ischemia model delineates the penumbra zone as well as indicates the presence of vascular endothelium in the ischemic core. This localisation indicates surviving cells in the area of compromised circulation. Hsp72 could be the mediator of induced tolerance following ischemic insults, so this stress protein could play a beneficial role. The defense mechanism related to the hsp72 in the ischemic brain, however, is still incompletely elucidated.

Hsp70, the major inducible hsp is constitutively expressed in all cells. Injuries responsible for protein denaturation, activate transcription of hsp70. Denatured proteins appear to be the major stimulus for hsp induction. Hsfs binds to hsp90 in the inactivated cells. Denaturing proteins bind to hsp90 and HSFs could be released.[46] Hsp70 induction represents the zone of protein denaturation and the zone of attempted protein renaturation. Most of the hsp70 mRNA is expressed in vessels within the infarct's core, and in some glia and neurons at some distance from the core in the penumbra. The zone of protein denaturation extends beyond the zone of selective neuronal death. Hsp70 expression seems to protect cells against injury,[47] but may not protect against apoptosis or severe cell injury.[48]

Hsp27 is expressed almost entirely exclusively in astrocytes and can be induced by spreading depression.[49] Hsp70 is an other constitutive heat shock protein, and it probably chaperons cell proteins in order to prevent abnormal folding during protein synthesis. Modest ischemia induced tolerance can upregulate hsp70.[50]

Hsp 32 (hemoxigenase-1, HO-1) is an inducible hsp. It plays an important role in metabolizing heme, released from hemoglobin (in hemorrhagic stroke), or heme proteins from mitochondria after ischemic cell injury. Hsp32 is expressed in microglia and macrophages for a very long time period after stroke.[51] It has a role in inflammation as well.

10. HYPOXIA INDUCIBLE FACTORS (HIFS)

HIF-1 is a recently recognized transcription factor that is induced by changes in molecular oxygen levels in tissue.[52,53] Once induced HIF-1α protein binds to HIF-1β, which is constitutively expressed in most cells. HIF-1 dimerization is stabilizing both proteins and leads to binding to hypoxia response sequences in various target genes: erythropoietin, tyrosine hydroxylase, iNOS, VEGF, GLUT-1, HSP 32, transferrin, and glycolytic enzymes. During reoxygenation, the HIF-1 protein complex has a very short half-life.[52] HIF is induced in the brain after focal ischemia.[54]

Other hypoxia inducible factors are HIF-2 and EPAS-1.[55] EPAS-1 is expressed in capillary endothelial cells and plays a role in inducing vascular target genes (VEGF).

The role of HIF-1 in acute focal cerebral ischemia is unclear. It increases NO production by iNOS, dopamine production by tyrosine hydroxilase, and lactate production by lactate dehydrogenase. These substances may be detrimental in ischemic conditions. On the other hand HIF-1 increases the expression of glucose transporter or erythropoietin that might be protective in brain following ischemia.

11. CYTOKINE EXPRESSION (TNFα, IL-1, IL-6, IL-10)

TNFα, including CD40, CD27 and CD30 ligands, expression is induced about 10-500-fold following ischemic insult. TNF damages oligodendrocytes, but could protect hippocampal cells against excitotoxic, metabolic or oxidative stress. TNF acts on the receptors p55 and p75. The p55 receptor activation through the ceramide release upregulates NFκB.

TNFα produced in the brain would appear to be critical in mediating tissue injury as well as orchestrating the production of other key inflammatory and procoagulant mediators. TNFα is expressed at very low levels in the normal brain tissue. Different injuries augment the expression rapidly (2-6 hr) and intensively, about 10-500-fold. TNF damages oligodendrocytes, but can protect hippocampal or cortical neurons against ischemic (excitotoxic, metabolic, oxidative) insults. TNF through the p55 receptor activates the sphingomyelinase that cleaves sphingomyelin and releases ceramide which activates a transcription factor NFκB, which results in more calbindin and manganese-superoxide dismutase. TNFα has been shown to act synergistically with IL-1β to stimulate the production of NGF.

In the *in vitro* hippocampal neuron culture model NMDA neurotoxicity has been blocked by a short-term (60 sec) pretreatment of TNFα or TGFβ. The mechanisms of the signal transduction pathway of TGFβ is still not elucidated.

IL-1 is markedly induced after focal ischemia, mostly in the ipsilateral, but also in the contralateral non-ischemic cortex.[56] In both hemispheres, IL-1 expression is documented mostly in the endothelial cells and in microglia. Spreading depression and the the IEG expression could explain the bilateral IL-1 expression. IL-1 appears to worsen ischemic injury and could induce apoptosis and brain edema.

IL-6 is expressed in neurons and microglia cells. It is expressed following ischemic insult.

Administration of IL-6 protects against ischemic injury.[57]

IL-10 is induced only in the ischemic hemisphere. The expression is maintained for days following stroke.[58]

12. COX-2

COX-2 initiates ischemic injury by release of superoxides and toxic prostanoids. Phospholipids are metabolized to arachidonic acid by phospholipase A2. Arachidonic acid is metabolized to prostaglandins by COX-2 and metabolized to leukotriens by 5-lipoxigenase. COX-2 is induced by focal ischemia.[59] Spreading depression throughout the affected hemisphere is likely to be related to COX-2. COX-2 inhibitors decrease stroke volume in some but not all of the studies. Cell protection produced by COX-2 inhibitors appears to be linked to iNOS mediated injury.[60] The failure of NS3988 to reduce infarct volume in iNOS null mice suggests that iNOS derived NO is required for the deleterious effects of COX-2 to occur.

13. GROWTH FACTORS AND SIGNAL TRANSDUCTION PATHWAYS

The endogenous neuroprotective factors (fibroblast growth factor-FGF, nerve growth factor-NGF, brain-derived neuroprotective factor-BDNF, insuline-like growth factors-IGFs, platelet-derived growth factors-PDGFs, tumor necrosis factors-TNFs, transforming growth factor-β–TGF-β, secreted forms of β–amyloid precursor protein-sAPPs and protease nexin-1) have neuroprotective functions, partly by stabilizing Ca^{2+} homeostasis, partly inducing signaling pathways resulting in expression of antioxidant enzymes.

Activated by tyrosine kinase, gene expression of neurotrophic factors is upregulated. bFGF suppresses expression of NMDA receptors, furthermore bFGF and BDNF increase expression of calbindin, a neuroprotective Ca^{2+} binding protein. NGF increases levels of catalase, BDNF increases levels of gluthatione peroxidase.

Different growth factors produced by neurons and/or glia in response to cerebral ischemia have been demonstrated to protect neurons against metabolic, excitotoxic and oxidative insults in animal models of ischemic stroke. The neuroprotective factors include fibroblast growth factors (FGFs), nerve growth factor (NGF), brain-derived neurotrophic factor (BDNF), insulin-like growth factors (IGFs), platelet-derived growth factors (PDGFs), tumor necrosis factors (TNFs), transforming growth factorβ (TGFβ), secreted forms of β-amyloid precursore protein (sAPPs) and nexin-1.

Growth factors result in changes in neuroprotective, antiapoptotic and antioxidant gene expression, or modulate proteins involved in the regulation of ion homeostasis (regulation of Ca^{2+}, K^{+} channels). In different *in vivo* or *ex vivo* model systems growth factors have been demonstrated to be neuroprotective by activating receptors with intrinsic protein tyrosine kinase activity, cascade of phosphorylation and changes of gene expression. Growth factors suppress the accumulation of Ca^{2+} and free radicals under ischemia-like conditions.

The proteinase nexin-1 is the product of glial cells and it is upregulated by different brain injuries. Nexin-1, by suppressing the detrimental thrombin induced Ca^{2+} and accumulation of free radicals in the neurons, is neuroprotective.

VEGF is a potential HIF target gene,[61] that is induced by focal ischemia. VEGF could be detected in the core and in the penumbra as well.[62] VEGF mRNA is located in microglia, macrophages and in the endothelial cells. VEGF receptors Flt-1 and Flk-1 are induced after ischemia, Flt-1 on neurons, glial and endothelial cells, while Flk-1 mainly on glia and endothelium. Induction of VEGF receptors and other VEGF target genes could be mediated by Ets-1, a vascular related transcription factor.[63] VEGF induced by HIF-1 or EPAS-1 could mediate new vessel formation after stroke.[64] VEGF could influence furthermore the permeability of endothelial cells and post-ischemic brain edema.[65]

14. GENOMIC INSTABILITY INDUCED BY BRAIN ISCHEMIA

By means of the neurotrophin receptors tyrosine kinases activate mitogen activated protein kinases (MAPKs),[66] and initiate gene transcription by which apoptosis or cell rescue cascade is initiated, furthermore activate protective and regenerative processes, related to synaptic plasticity. C-jun phosphorylation by stress activated protein kinase (SAPK) prompts gene transcription, which could be related to cell rescue. On the other hand, cytokine induced SAPK activation resulted in apoptosis. TNFα, Fas ligand by receptor binding and through the docking proteins and death domains activate NFκB, the transcription factor which regulates both the cell suicide or survival. Other NGF receptor, p75 lacks the death domain which induces cell death by activation of NFκB.[67]

Protein expression related to DNA damage such as Gadd 45, p53, Bax, MDM2 have been detected in severely damaged neurons. On the other hand, DNA repaire protein expression decreased in the core.[68,69] Following ischemia, protein expression related to the brain plasticity has been demonstrated in ischemic brain. Cell-cycle proteins, a cytoskeleton protein (MAP-2), a growth-associated protein (GAP-43) and a neurofilament protein (nestin) could be visualised mainly in the penumbra region.[70] It is not yet clear, how the expression of plasticity proteins orchestrate the remodeling of synapses in the peri-infarct rim. The concept of genomic instability in the ischemic brain have been substantiated by the data of the diverse, detrimental and beneficial consequences of TNF and NFκB upregulation.

Cerebral ischemia induces genomic instability in the brain. DNA damage and repair ensue, and these processes tend to be localised to neurons after stroke. Stroke activates protein synthesis machinery that both kills cells and fosters conditions that promote cellular proliferation and repair. Wild type p53 is a tumor-supressor protein and a potent transcription factor, decreases the treshold for genomic activity, both for repair and damage. Among p53 response proteins expressed are DNA repair proteins such as GADD45 and proliferating cell nuclear antigen (PCNA). Proteins associated with DNA damage, such as Bax family of proteins are also expressed and are primarily localized to apoptotic cells. The genetic instability following the ischemic episode has detrimental consequences with upregulation of proapoptotic genes and delayed or programmed neuronal death. On the other hand it has been demonstrated that upregulated genes are responsible for expression of developmental proteins such as cell-cycle regulating protein (cyclin D1), cytosceletal protein (MAP-2), growth-associated protein (GAP-43), neurofilament protein (nestin) and vascular endothelial growth factor (VEGF). Mostly in the penumbra region these developmental proteins could signal a cell repair mechanism. Brain plasticity could be explained by this mechanism.

Adult cortex is capable of neuronal regeneration in response to neuronal injury or degeneration under specific condition. It has been recently demonstrated, that synchronous targeted apoptosis could induce neurogenesis by releasing a repressed program of neural plasticity.[71,72] The relationships between different escape mechanisms in the apoptosis cascade and late genomic changes including the upregulation of developmental proteins are still speculative.

15. NO CLINICAL TRIAL ON NEUROPROTECTION HAD A SIGNIFICANT EFFECT ON ISCHEMIC STROKE PATIENTS

As much as 50 various neuroprotective agents were found in preclinical studies to be effective and there were more than 100 stroke trials involving the drugs and thousends of patients and no success at all.[73]

A great number of pittfalls could be mentioned: the therapeutic window differences in experimental versus clinical studies, that the preclinical studies target only the penumbra, that the preclinical studies are concentrated on exclusively on the grey matter, that the optimal duration of the neuroprotective action is not known, that the endpoints are different: infarct size versus behavioural outcome (early outcome versus late outcome), that stroke models are homogenous in contrast with human stroke cases which are heterogenous (risk factor profile, age, genetic inheritance), that the otcome measure, the endpoint is critical in clinical trials, and that small trials are trying to answer questions that only large trials can answer. These challenges must be answered by appropriate trial planning and management.

However, there are new therapeutic frontiers in the area of neuroprotetion. Delayed neuronal death/apoptosis could be prevented or attenuated by anti-apoptosis gene therapy (Bcl-2, Bcl-XL) by high capacity adenovirus constructs, and by the induction of anti-apoptotic genes via selegiline administration.

Furthermore, not only neuroprotection is the important issue, but the exploitation, the promotion of the repair mechanisms, the plasticity processes, that is the interrelationships between the anti-apoptosis strategies and brain plasticity processes.

The new strategy may involve multiaction drugs which are not only neuroprotectants but are also the promoters of the post-stroke plasticity processes.

Vinpocetine is proved to be a multi-action drug effective in cases of vascular dementia.

Meta-analysis of clinical trials on vascular dementia has documented the efficacy of this compound,[74] as well as a pilot study by Feigin et al., 2001[75] shows that a full-scale randomized double-blind, placebo-controlled trial of vinpocetine treatment in acute ischaemic stroke is feasible and warranted.

16. VINPOCETINE, THE DRUG OF CHOICE

Vinpocetine (14-ethoxycarbonyl-(3alpha, 16 alpha-ethyl)14,15-eburnamine, or Cavinton), a vasoactive vinca alcaloid has been used in clinical practice for the treatment of various cerebrovascular syndromes for nearly three decades. Several studies have demonstrated that indole derivatives such as vinpocetine enhance survival rate under various experimental anoxic/hypoxic conditions.[76] The most of the beneficial effects of Vinpocetine is due to cerebral vasodilation[77] resulting from the inhibition of Ca2+/calmodulin dependent phosphodiesterase (PDE1) and an increase in the level of cyclic nucleotides in blood vessel myocytes.[78] Vinpocetine protects neurons against Ca^{2+}-induced death;[79,80] and increases the magnitude of long term potentiation in hippcampal slices.[81] A neuroprotective effect is due to a restricted inward Ca^{2+} current,[79,80] while the enhancement of synaptic transmission is related to an increase in the Ca^{2+} influx. The mechanisms of increase in the amount of intracellular Ca^{2+} includes a blockade of inward Na^{+} and Ca^{2+} currents.[82-84] The effect of vinpocetine on $[Ca^{2+}]i$ changes in single pyramidal neurons in the vulnerable CA1 region of rat hippocampal slices was investigated by using a cooled CCD camera-based ratio imaging system and cell loading with fura 2/AM, and vinpocetine, at a pharmacologically relevant concentration, proved to be able to decrease the pathologically high $[Ca^{2+}]i$ levels in individual rat hippocampal CA1 pyramidal neurons.[85] The antioxidant effect of vinpocetine was examined by Santos and his colleagues.[86] Vinpocetine prevented the formation of free radicals and lipid peroxidation in rat brain synaptosomes. They have also found that vinpocetine inhibited ascorbate/Fe^{2+} stimulated oxygen consumption and thiobarbituric acid reactive substances (ROS) formation, an indicator of lipid peroxidation. The ROS formation was also prevented by Vinpocetine. They concluded that the antioxidant effect of vinpocetine might contribute to its protective role in pathological conditions. Vinpocetine is proved to be protective against global anoxia in mice[87,88] and in global ischemia induced by decapitation and hypoxia.[89-91] Vinpocetine also proved to be effective in a carotid occlusion model in mice and rats.[92,93]

Vinpocetine is also able to lessen memory and learning deficits induced by scopolamine and hypoxia in rodents.[94-97]

In a model of focal ischemia the extent of the necrotic area decreased by nearly 25% following vinpocetine introduction.[98,99] Vinpocetine pretreatment in a bilateral carotid artery occlusion model led to 28% neuronal necrosis in the CA1 region of rat hippocampus, while the neuronal loss in the control group reached 60%. Rieschke and Krieglstein reported 50% inhibition of the neuronal loss by vinpocetine.[100-102] Vinpocetine pretreatment in a carotid occlusion model prevented the long-term increase in glucose utilisation and reduction of the blood flow in the CA1 region. Following vinpocetine treatment, 40% inhibition of neuronal loss in the hippocampal CA1 region was observed in gerbils after bilateral carotid artery occlusion[103-104] and 40% inhibition of neuronal loss in the hippocampal CA1 region was observed in gerbils after bilateral carotid artery occlusion.[103,104]

REFERENCES

1. Y. Li, M. Chopp, N. Jiang, C. Zaloga, In situ detection of DNA fragmentation after focal cerebral ischemia in mice, Brain Res Mol Brain Res **28**, 164–168 (1995).
2. Y.Li, M. Chopp, N. Jiang, Z.H. Zang, C. Zaloga, Induction of DNA fragmentation after 10 to 120 minutes of focal cerebral ischemia in rats, *Stroke* **26**, 1252–1258 (1995).
3. M. Chopp, Y. Li, and Z.G. Zhang, Protein Expression and brain plasticity after transient middle Cerebral Artery Occlusion in the rat, In: Maturation Phenomenon in Cerebral Ischemia II./Springer (1998).
4. K. Matsushita, T. Matsuyama, K. Kitagawa, M. Matsumoto, T. Yanagihara, M. Sugita, Alternations of Bcl-2

family proteins precede cytoskeletal proteolysis in the penumbra, but not in infarct centers following focal celebral ischemia in mice, *Neuroscience* **83**, 439–448 (1998).
5. J. Chen, S.H. Graham, P.H. Chan, J. Lan, R.L. Zhou, R.P. Simon, Bcl-2 in expressed in neurons that survive focal ischemia in the rat, *Neuroreport* **6**, 394–398 (1995).
6. M. Asahi, M. Hoshimaru, Y. Uemura, T. Tokime, M. Kojima, T. Ohtsuka, N. Matsura, T. Aoki, K. Shibahara, H. Kikuchi, Expression of interleukin-1 beta converting enzyme gene family and bcl-2 gene family in the rat brain following permanent occlusion of the middle cerebral artery, *J. Cereb Blood Flow Metab* **17**, 11–18 (1997).
7. S. Namura, J. Zhu, K. Fink, M. Endres, A. Srinivasan, K.J. Tomaselli, J. Yuan, M.A. Moskowith, Activation and cleavage of caspase-3 in apoptosis induced by experimental cerebral ischemia, *J Neurosci* **18**, 3659–3668 (1998).
8. G.Y. Yang, C. Gong, Z. Qin, W. Ye, Y. Mao, A.L. Bertz, Tumor necrosis factor alpha expression produces increased blood-brain barrier permeability following temporary focal cerebral ischemia in mice, *Brain Res Mol Brain Res* **69**,135–143 (1999).
9. H. Nawashiro, D. Martin, J.M. Hallenbeck, Inhibition of tumor necrosis factor and amelioration of brain infarction in mice, *J Cereb Blood Flow Metab* **17**, 229–232 (1997).
10. M.P. Mattson, Neuroprotective signal transduction: relevance to stroke, *Neurosci Biobehav Rev* **21**, 193–206 (1997).
11. X. Shi, M. Ding, Z. Dong, F. Chen, J. Ye, S. Wang, S.S. Leonard, V. Castranova, V. Vallyathan, Antioxidant properties of aspirin: characterization of the ability of aspirin to inhibit silica-induced lipid peroxidation, DNA damage, NF- kappaB activation, and TNF-alpha production, *Mol Cell Biochem* **199**, 93–102. (1999).
12. M.A. Stevenson, M.J. Zhao, A. Asea, C.N. Coleman, S.K. Calderwood, Salicylic acid and aspirin inhibit the activity of RSK2 kinase and repress RSK2-dependent transcription of cyclic AMP response element binding protein- and NF-kappa B-responsive genes, *J. Immunol* **163**, 5608-5616 (1999).
13. K.K.W. Wang, Calpain and caspase: can you tell the difference? *TINS* No1, **23**, 20–26 (2000).
14. J.S. Kim, S.C. Gautam, M. Chopp, C. Zaloga, M.L. Jones, P.A. Ward, K.M. Welch, Expression of monocyte chemoattractabt protein-1 and macrophage3 inflammatory protein-1 after focal cerebral ischemia in the rat, *J Neuroimmunol* **56**, 127–134 (1995).
15. I.S. Moon, M.L. Apperson, and M.B. Kennedy, The major tyrosine-phosphorylated protein in the postsynaptic density fraction is N-methyl D-aspartate receptor subunit 2B, *Proc.Natl.Acad.Sci.* **91**, 3954–3958 (1994).
16. J.P. Mohr, J.M. Orgogozo, M.J.G. Harrison, N.G. Wahlgren, J.H. Gelmers, E. Martinez-Vila, J. Dycka, D. Tettenborn, Meta-analysis of oral nimodipine trials in acute ischaemic stroke *Cerebrovasc Dis.* **4**, 197–203 (1994).
17. N.G. Wahlgren, Cytoprotective therapy for acute ischaemic stroke, In: M. Fisher, (ed.) Stroke Therapy, Butterworth and Heinemann, Boston, pp. 315–350. (1995).
18. N.G. Wahlgren, A Review of earlier clinical studies on neuroprotective agents and current approaches. Neuroprotective agents and cerebral ischeamia, *International Review of Neurobiology* **40**, 337–353. (1997).
19. X. Xie, B. Lancaster, T. Peakman, and J. Garthwaite, Interaction of the antiepileptic drug lamotrigine with recombinant rat brain type IIA Na+ channels and with native Na+ channels in rat hyppocampal neurones, *Pflügers Arch* **430**, 437–446 (1995).
20. J.S. Beckman, Nitric oxide, superoxide and perosynitrite in CNS injury, *Cerebrovascular Diseases* pp:209–210. (1997).
21. S.A Lipton, and J.S. Stamler, Actions of redoxrelated congeners of nitric oxide at the NMDA receptor, *Neuropharmacology* **33**, 1229–1233 (1994).
22. C. Iadecola, Bright and dark sides of nitric oxide in ischemic brain injury, *Trends Neurosci* **20**, 132–139 (1997).
23. U. Dirnagl, C. Iadecola, M.A. Moskowitz, Pathobiology of ischemic stroke: an integrated view, *Trends Neurosci* **22**, 391–397. (1999).
24. Z. Huang, P.L. Huang, N. Panahian, T. Dalkara, M.C. Fishman, M.A. Moskowitz, Effects of cerebral ischemia in mice deficient in neuronal nitric oxide synthase, *Science* **265**, 1883–1885 (1994).
25. M. Nagayama, K. Niwa, T. Nagayama, M.E. Ross, C.Iadecola, The cyclooxygenase-2 inhibitor NS-389 ameliorates ischemic brain injury in wild-type mice but not in mice with deletion of the inducible nitric oxide synthase gene, *J Cereb Blood Flow Metab* **19**, 1213–1219 (1999).
26. P. Marin, K.L. Nastiuk, N. Daniel, J.A. Girault, A.J. Czernik, J. Glowinski, A.C. Nairn, J. Premont, Glutamate-dependent phosphorylation of elongation factor-2 and inhibition of protein synthesis in neurons, *J Neurosci* **17**, 3445–3454 (1997).
27. C. Iadecola, F. Zhang, R. Casey, H.B. Clark, M.E. Ross, Inducible nitric oxide synthase gene expression in vascular cells after transient focal cerebral ischemia *Stroke* **27**, 1371–1380 (1996).

28. G.Y. Yang, C. Gong, Z. Qin, W. Ye, Y. Mao, A.L. Bertz, Tumor necrosis factor alpha expression produces increased blood-brain barrier permeability following temporary focal cerebral ischemia in mice, *Brain Res Mol Brain Res* **69**, 135–143 (1999).
29. M. Vastag, J. Skopál, Z. Voko, É. Csonka, Z. Nagy, Expression of Membrane-bound and soluble cell adhesion molecules by human brain microvessel endothelial cells. *Microvascular Research* **57**, 52–60. (1999).
30. Y. Li, M. Chopp, Z.G. Zhang, R.L. Zhang, J.H. Garcia, Neuronal survival is associated with 72-kDA heat shock protein expression after transient middle cerebral artery occlusion in the rat, *J. Neurol Sci* **120**, 187–194 (1993).
31. M. Chopp, Y. Li, N. Jiang, R.L. Zhang, J. Prostak, Antibodies against adhesion molecules reduce apoptosis after transient middle cerebral artery occlusion in rat brain, *J Cereb Blood Flow Metab* **16**, 578–587 (1996).
32. Zhang, R.L., Zhang, Z.G., Chopp, M., Zivin, J.A. (1999) Thrombolysis weith tissue plasminogen activator alters adhesion molecule expression in the ischemic rat brain. Stroke, 30,624–629.
33. S. Mun-Bryce, G.A. Rosenberg, Matrix melloproteinases in cerebrovascular disease, *J Cereb Blood Flow Metab* **18**, 1163–1172 (1998).
34. Y. Gasche, M. Fujimura, Y. Morita-Fujimura, J.C. Copin, M. Kawase, J. Massengale, P.H. Chan, Early appearance of activated matrix metalloproteinase-9 after focal cerebral ischemia in mice: a possible role in blood-brain barrier dysfunction, *J Cereb Blood Flow Metab* **19**, 1020–1028 (1999).
35. J.H. Heo, J. Lucero, T. Abumiya, J.A. Koziol, B.R. Copeland, G.J. del Zoppo, Matrix metalloproteinases increase very early during experimental focal cerebral ischemia, *J Cereb Blood Metab* **19**, 624–633 (1999).
36. R. Bravo Growth factor responsive genes in fibroblasts, *Cell Growth Diff.* **1**, 305–309 (1990).
37. P.J. Lindsberg, K.U. Frerichs, A.L. Siren, J.M. Hallenbeck, T.S. Nowak, Jr. Heat-shock protein and C-fos expression in focal microvascular brain damage, *J Cereb Blood Flow Metab* **16**, 82–91 (1996).
38. J.I. Morgan, and T. Curran, Stimulus-transcription coulping in the nervous system: Involvement of the inducible proto-oncogenes fos an jun, *Annu. Rev. Neurosci* **14**, 421–451 (1991).
39. K. Matsushima, R. Schmidt-Kastner, M.J. Hogan, A.M. Hakim, Cortical spreading depression activates tropic factor expression in neurons and astrocytes and protects against subsequent focal brain ischemia, *Brain Res* **807**, 47–60 (1998).
40. T.S. Nowak, M. Kiessling, Reprogramming of gene expression aafter ischemia. In: W. Walz, N.J. Totowa, (eds.) Cerebral Ischemia: Molecular and Cellular Pathophysiology Humana Press, pp. 145–216. (1999).
41. J.I. Morgan, T. Curran, Immediate-early genes : ten years on, *Trends Neurosci* **18**, 66–67 (1995).
42. M. Walton, B. Conner, P. Lawlor, D. Young, E. Sirimanne, P. Gluckman, G. Cole, M. Dragunow, Neuronal death and survival in two models of hypoxic-ischaemic brain damage, *Brain Res Brain Res Rev* **29**, 137–168 (1999).
43. M. Kiessling, G. Stumm, Y. Xie, T. Herdegen, A. Aguzzi, R. Bravo, and P. Gass, Differential transcription and translation of immediate early genes in the gerbil hippocampus after transient global ischemia, *J Cereb Blood Flow Metab* **13**, 914–924 (1993).
44. C. Sommer, P. Gass, M. Kiessling, Selective c-JUN expression in CA1 neurons of the gerbil hippocampus during and after acquisition of an ischemia-tolerant state, *Brain Pathol* **5**, 135–144. (1995).
45. T.S.Nowak, Jr. Localization of 70 kDa heat shock protein mRNA following transient focal cerebral ischemia in the rat. *J Cereb Blood Flow Metab* **11**, 432–439 (1991).
46. J. Zou, Y. Guo, T. Guettouche, D.F. Smith, R.Voellmy, Repressing of heat shock transcription factor HSF1 activation by HSP90 (HSP90 complex) that forms a stress-sensitive complex with HSF1, *Cell* **94**, 471–480 (1998).
47. M.A. Yenari, S.L. Fink, G.H. Sun, L.K. Chang, M.K. Patel, D.M. Kunis, D. Onley, D.Y. Ho, R.M. Sapolsky, G.K. Steinberg, Gene therapy with HSP72 is neuroprotective in rat models of stroke and epilepsy, *Ann Neurol* **44**, 584–591. (1998).
48. M.J. Wagstaff, J. Smith, Y. Collaco-Moares, J.S. De Belleroche, R. Voellmy, R.S. Coffin, D.S. Latchman, Delivery of a constitutively active form of the heat shock factor using a virus vector protects neural cells from thermal or ischemic stress but not from apoptosis, *Suppl Eur J Neurosci* **10**, 3343–3350. (1998).
49. J.C. Plumier, J.N. Armstrong, N.I. Wood, J.M. Babity, T.C. Hamilton, A.J. Huntrer, H.A. Robertson, R.W. Currie, Differential expression of c-fos, Hsp70 and Hsp27 after photothrombic injury in rat brain, *Brain Res Mol Brain Res* **45**, 239–246 (1997).
50. J. Chen, R. Simon, Ischemic tolerance in the brain, *Neurology* **48**, 306–311. (1997).
51. M. Bergeron, D.M. Ferriero, H.J. Vreman, D.K. Stevenson, F.R. Sharp, Hypoxia-ischemia, but not hypoxia alone, induces the expression of heme oxygenase-1 (HSP32) in newborn rat brain. J Cereb Blood Flow Metab, **17**, 647–658 (1997).
52. G.L. Semenza, Perspectives on oxygen sensing, *Cell* **98**, 281–284 (1999).
53. P. Ratcliffe, J. Rourke, P. Maxwell, Oxygen sensing, hypoxia-inducible factor-1 and the regulation of mammalian gene expression, *J Exp Biol* **201**, 1153–1162 (1998).

54. M. Bergeron, A.Z. Yu, K.E. Solway, G.L. Semenza, F.R. Sharp, Induction of hypoxia-inducible factor-1 (HIF-1) and its target genes following focal ischemia in rat brain, *Eur J Neurosci* **11**, 4159–4170 (1999).
55. I. Flamme, T. Frohlich, M. von Reutern, A. Kappel, A. Damert, W. Risau, HRF, a putative basic helix-loop-helix-PAS-domain transcription factor is closely related to hypoxia-inducible factor-1 alpha and developmentally expressed in blood vessels, *Mech Dev* **63**, 51–60 (1997).
56. Q.H. Zhai, N. Fuell, F.J. Che, Gene expression of IL-10 in relationship to TNF-alpha, IL-1beta and IL-2 in the rat brain- following middle cerebral artery occlusion, *J Neurol Sci* **152**, 119–124 (1997).
57. S.A. Loddick, A.V. Turnbull, N.J. Rothwell, Cerebral interleukin-6 is neuroprotective during permanent focal cerebral ischemia in the rat, *J Cereb Blood Flow Metab* **18**, 176–179 (1998).
58. X. Wang, J.A. Ellison, A.L. Siren, P.G. Lysko, T.L. Yue, F.C. Barone, A. Shatzman, G.Z. Feuerstein, Prolonged expression of interferoninducible protein –10 in ischemic cortex after permanent occlusion of the middle cerebral artery in rat, *J Neurochem* **71**, 1194–1204. (1998).
59. H. Tomimoto, M. Shibata, M. Ihara, I. Akiguchi, R. Ohtani, H. Budka, A comparative study on the expression of cyclooxygenase and 5-lipoxygenase during cerebral ischemia in humans, *Acta Neuropathol* **104**(6), 601–7 (2002).
60. M. Nagayama, K. Niwa, T. Nagayama, M.E. Ross, C. Iadecola, The cyclooxygenase-2 inhibitor NS-389 ameliorates ischemic brain injury in wild-type mice but not in mice with deletion of the inducible nitric oxide synthase gene. *J Cereb Blood Flow Metab* **19**, 1213–1219 (1999).
61. N.S. Levy, M.A. Goldberg, A.P. Levy, Sequencing of the human vascular endothelial growth factor (VEGF) 3' untranslated region (UTR): conversation of five ypoxia-inducible RNA-protein binding sites, *Biochim Biophys Acta* **1351**, 167–173 (1997).
62. C.S. Cobbs, J. Shen, D.A. Greenberg, S.H. Graham, Vascular endothelial growth factor expression in transient focal cerebral ischemia in the rat, *Neurosci Lett* **249**, 79–82 (1998).
63. M.M. Valter, A. Hugel, H.J. Huang, W.K. Cavenee, O.D. Wiestler, T. Pietsch, N. Wernert, Expression of the Ets-1 transciption factor in human astrocytomas is associated with Fms-like tyrosine kinase-1 (Flt-1)/vascular endothelial growth factor receptor-1 synthesis and neoangiogenesis. *Cancer Res* **59**, 5608–5614 (1999).
64. J.C. LaMamma, N.T. Kuo, W.D. Lust, Hypoxia-induced brain angiogenesis. Signals and consequences, *Adv. Exp Med Biol* **454**, 287–293 (1998).
65. N. van Bruggen, H. Thibodeaux, J.T. Palmer, W.P.P. Lee, LFu, B. Cairns, DTumas, R. Gerlai, S.P. Williams, M. van Lookeren Campagne, N. Ferrara, VEGF antagonism reduces edema formation and tissue damage after ischemia/reperfusion injuri in the mouse brain, *J Clin Invest* **104**, 1613–1620 (1999).
66. M. Verheij, R. Bose, X.H. Lin, B. Yao, W.D. Jarvis, S. Grant, M.J. Birrer, E. Szabo, L.I. Zon, J.M. Kyriakis, A. Haimovitz-Friedman, Z. Fuks, and R.N. Kolesnick, Requirement for ceramide-initiated SAPK/JNK signalling in stress induced apoptosis *Nature* **380**, 75–79 (1996).
67. W.H. Lee, G.M. Wang, L.B. Seaman, S.J. Vannucci, Coordinate IGF-I and IGFBP5 gene expression in perinatal rat brain after hypoxia-ischemia *J Cereb Blood Flow Metab* **16**, 227–236 (1996).
68. M. Fujimura, Y. Gasche, Y. Morita-Fujimura, M. Kawase, P.H. Chan, Early decrease of apurinic/apyrimidinic endonuclease expression after transient focal cerebral ischemia in mice, *J Cereb Blood Flow Metab* **19**, 495–501 (1999).
69. M. Fujimura, T. Morita-Fujimura Ysugawara, P.H. Chan, Early decrease of XRCCI, a DNA base excision repair protein, may contribute to DNA fragmentation after transient focal cerebral ischemia in mice *Stroke* **30**, 2456–2462 (1999).
70. M. Chopp, Y. Li, and Z.G. Zhang, Protein Expression and brain plasticity after transient middle Cerebral Artery Occlusion in the rat. In: Maturation Phenomenon in Cerebral Ischemia II./Springer(1998).
71. S.S. Magavi, B.R. Leavitt, J.D. Macklis, Induction of neurogenesis in the neocortex of adult mice, *Nature* **405**(6789), 951–5 (2000).
72. S.S. Magavi, J.D. Macklis, Induction of neuronal type-specific neurogenesis in the cerebral cortex of adult mice: manipulation of neural precursors in situ, *Brain Res Dev Brain Res* **134**(1-2), 57–76 (2002).
73. D.J. Gladstone, S.E. Black, A.M. Hakim, Toward wisdom from failure: lessons from neuroprotective stroke trials and new therapeutic directions, *Stroke* **33**(8), 2123–36 (2002).
74. Z. Nagy, P. Vargha, L. Kovács, P. Bönöczk, Meta-analysis of Cavinton, *Praxis* **49**, 420–4 (1998).
75. V.L. Feigin, B.M. Doronin, T.F. Popova, E.V. Gribatcheva, D.V.Tchervov, Vinpocetine treatment in acute ischaemic stroke: a pilot single-blind randomized clinical trial, *Eur J Neurol* **8**(1), 81-5 (2001).
76. C. Stolc, Indole derivativesas neuroprotecrants, *Life Sci.* **65**, 1943–1950 (1999).
77. M. Miyazaki, The effect of a cerebral vasodilator, vinpocetine, on cerebral vascular resistance evaluated by the Doppler ultrasonic technique in patients with cerebrovascular diseases, *Angiology* **46**(1), 53–8 (1995).
78. N. Miyata, H. Yamaura, M. Tanaka, M. Muramatsu, K. Tsuchida, S. Okuyama, S. Otomo, Effects of VA-045, a novel apovincaminic acid derivative, on isolated blood vessels: cerebroarterial selectivity, *Life Sci* **52**(18), PL181-6 (1993).

79. V. Lakics, P. Molnar, S.L. Erdo, Protection against veratridine toxicity in rat cortical cultures: relationship to sodium channel blockade, *Neuroreport* **7**(1), 89–92 (1995).
80. V. Lakics, M.G. Sebestyen, S.L. Erdo, Vinpocetine is a highly potent neuroprotectant against veratridine-induced cell death in primary cultures of rat cerebral cortex, *Neurosci Lett* 185(2), 127–30 (1995).
81. K. Ishihara, H. Katsuki, M. Sugimura, M. Satoh, Idebenone and vinpocetine augment long-term potentiation in hippocampal slices in the guinea pig, *Neuropharmacology* **28**(6), 569–73 (1989).
82. S. Kaneko, H. Takahashi, M. Satoh, The use of Xenopus oocytes to evaluate drugs affecting brain Ca2+ channels: effects of bifemelane and several nootropic agents, *Eur J Pharmacol* **189**(1), 51–8 (1990).
83. P. Molnar, S.L. Erdo, Vinpocetine is as potent as phenytoin to block voltage-gated Na+ channels in rat cortical neurons, *Eur J Pharmacol* **273**(3), 303–6 (1995).
84. L. Tretter, V. Adam-Vizi, The neuroprotective drug vinpocetine prevents veratridine-induced [Na+]i and [Ca2+]i rise in synaptosomes, *Neuroreport* **9**(8), 1849–53 (1998).
85. T. Zelles, L. Franklin, I. Koncz, B. Lendvai, G. Zsilla, The nootropic drug vinpocetine inhibits veratridine-induced [Ca2+]i increase in rat hippocampal CA1 pyramidal cells, *Neurochem Res* **26**(8-9):1095–100 (2001).
86. M.S. Santos, A.I. Duarte, P.I. Moreira, C.R. Oliveira, Synaptosomal response to oxidative stress: effect of vinpocetine, *Free Radic. Res.* **32**(1):57–66 (2000).
87. G.A. King, Protective effects of vinpocetine and structurally related drugs on the lethal consequences of hypoxia in mice, *Arch.Int.Pharmacodyn.Ther.* **286**, 299–307 (1987).
88. M. Yamamoto, M. Shimizu, S. Kawabata, Cerebral vasodilators potentiate the anti-anoxic activity of indeloxazine hydrochloride, a new cerebral activator, *Neuropharmacology* **28**,313–317 (1989).
89. H. Hara, A. Ozaki, M.S.T. Yoshidomi, Protective effect of KB-2796, a new calcium antagonist, in cerebral hypoxia and ischemia, *Arch. Int. Pharmacodyn. Ther.* **304**,206–218 (1990).
90. J.C. Lamar, M. Beaughard, C. Bromont, H. Pignet, Effects of vinpocetine in four pharmacological models of cerebral ischemia, In: Krieglstein J ed. Pharmacology of cerebral ischemia. Amsterdam: Elsevier, pp 334–339 (1986).
91. K. Yamaguchi, S. Yamada, M. Yoshida, K. Kyuki, S. Okuyama, Anti-anoxic effects of VA-045, *J. Pharmacol.* **61** (suppl.1),184 (1993).
92. S. Takeo, K. Tanonaka, T. Hirano, T. Miyake, J. Okamoto, Cerebroprotective action of naftidrofuryl oxalate I:prolongation of survival time and protection of cerebral energy metabolism in bilateral carotid artery ligated mice, *Folia Pharmacol. Japon.* **91**,267–273 (1988).
93. V. Vaizov, T.M. Plotnikova, T. Yakimova, O. Vaizova, A. Saratikov, Ammonium succinate: An effective corrector of circulatory cerebral hypoxia, *Byull. Eksp. Biol.Med.* **118**,276–278 (1995).
94. V.J. DeNoble, Vinpocetine enhances retrieval of a step-through passive avoidance response in rats, *Pharmacol. Biochem. Behav.* **26**,183–186 (1987).
95. D. Groó, E. Pálosi, L. Szporny, Effect of vinpocetine in scopolamine-induced learning and memory impairments, *Drug Dev. Res.* **11**,29–36 (1987).
96. D. Groó, E. Pálosi, L. Szporny, Comparison of the effects of vinpocetine, vincamine, and nicergoline on the normal and hypoxia-damaged learning process in spontaneously hypertensive rats, *Drug Dev.Res.* **15**,75–85 (1988).
97. D. Groó, E. Pálosi, L. Szporny, Effect of vinpocetine in memory disturbances induced by different damaging agents, In: Krieglstein J, ed.Pharmacology of cerebral ischemia. Stuttgart: Wissenschaftliche Verlagsgesellschaft mbH;229–305.
98. C. Backhaus, C Karkoutly, M. Welsch, J. Krieglstein, A mouse model of focal cerebral ischemia for screening neuroprotective drug effects, *J. Pharmacol. Toxicol.Meth.* **27**,27–32 (1992).
99. G. Bielenberg, Effects of vincamine and vinpocetine on infarct size in focal cerebral ischemia, *Arch Pharmacol* **354** (suppl. 1):R122, (1992).
100. R. Rischke, J. Krieglstein, Increased LCGU and decreased LCBF in rat hippocampus 7 days after ischemia, *J Neurochem.* **52**(supplS56), (1989).
101. R. Rischke, J. Krieglstein, Effects of vinpocetine on local cerebral blood flow and glucose utilization seven days after forebrain ischemia in the rat, *Pharmacology* **41**,153–160 (1990).
102. R. Rischke, J. Krieglstein, Protective effect of vinpocetine against brain damage caused by ischemia, *Jpn. J. Pharmacol.* **56**,349–356 (1991).
103. T. Araki, K. Kogure, K. Nishioka, Comparative neuroprotective effects of pentobarbital, vinpocetine, flunarizine and ifenprodil on ischemic neuronal damage in the gerbil hippocampus, *Res. Exp.Med.* (Berl.) **190**, 19–23 (1990).
104. K. Kogure, H. Kato, Pharmacological modification of post-ischemic brain cell injury, *Clin. Neuropharmacol.* **13**(suppl. 2), 154–155 (1990).

NEUROPROTECTION AND DOPAMINE AGONISTS

Zvezdan Pirtošek and Dušan Flisar*

1. INTRODUCTION

Parkinson's disease is a neurodegenerative disorder characterized by a progressive loss of the dopaminergic neurons in the substantia nigra pars compacta. Accumulating evidence indicates that apoptosis contributes to the cell death in Parkinson's disease patients' brain. Excitotoxicity, oxidative stress, and mitochondrial respiratory failure are thought to be the key inducers of the apoptotic cascade. The chapter will review the evidence suggesting that some agents – and dopamine agonists in particular – are neuroprotective and the possible mechanisms whereby these effects might occur.

2. ETIOLOGY OF PARKINSON'S DISEASE

Despite prodigious investigative effort over the years, the etiology of idiopathic Parkinson's disease remains unknown. However, a tantalizingly large selection of biochemical pointers has emerged along the way and it now seems likely that there are a number of predisposing causes.

Both, genetic and environmental factors interact and play an important role in Parkinson's disease which should be considered more of a syndrome than a disorder with a single cause.

The first unequivocal description of idiopathic Parkinson's disease was that of James Parkinson (Parkinson, 1819). If it did appear for the first time in James Parkinson's time, it coincided with the build-up of the industrial revolution. However, no precise toxic component of environmental industrial effluent as such has so far been identified, although some environmental factors may increase the risk of Parkinson's disease in the population and others appear to be protective against the development of Parkinson's disease. Associated with a lower risk of developing Parkinson's disease are cigarette smoking and coffee drinking (Hernan et al., 2002). Associated with a slight increase in

* Zvezdan Pirtošek, Centre for Extrapyramidal Disorders, Neurology Hospital, University Medical Centre, SI-1525 Ljubljana, Slovenia. Dušan Flisar, Neurology Department, General Hospital, SI-2000 Maribor, Slovenia.

risk of developing Parkinson's disease are pesticide use, rural living (Semchuk et al., 1992), the use of well water and certain chemicals (1-methyl-4-phenyl-1,2,3,6-tetrahydropyridine; MPTP). The MPTP story gave an important dimension to our understanding of the neurotoxic mechanism likely to be involved. A trace component, MPTP, was produced accidentally during the chemical synthesis of an illicit opiate drug. When administered, it induced an acute parkinsonian state which responded to antiparkinsonian drugs. MPTP is a prodrug, converted by glial MAO B to a neurotoxic substance, MPP+ which is taken up by the dopamine re-uptake system and accumulates in dopaminergic neurons where it enters mitochondria and inhibits Complex 1 of the mitochondrial respiratory chain. Many analogues of MPTP have been described, some of which are substrates for MAO B. It seems quite possible that an MPTP-like compound could be elaborated endogenously and, indeed, Naoi and his colleagues (1993) have pinpointed just such a possible pathway, including the further metabolism of tetrahydroisoquinoline to the neurotoxin, N-methyl tetrahydroisoquinolinium. This compound, like MPP+, seems capable of causing a deficit of Complex 1.

Recent studies clarified a genetic component of Parkinson's disease. This seems to be more significant in younger-onset patients. The first gene mutation to be identified in familial Parkinson's disease came from a study into a large Italian-American family with autosomal dominant inheritance (Golbe et al., 1990). The gene locus was mapped to chromosome 4 (Polymeropoulos et al., 1996) and characterized as a single alanine to threonine substitution in the gene for alpha-synuclein (Polymeropoulos et al., 1997). Alpha synuclein is a small protein known to be part of the non-amyloid component of Alzheimer's disease plaques. It is expressed at the nerve synapse (Iwai et al., 1995) but its physiological role is not yet understood. The second Parkinson's disease associated mutation was identified in autosomal recessive juvenile onset Parkinson's disease (Ishikawa and Tsuji, 1996). The gene responsible was mapped to chromosome 6 (Matsumine et al., 1997) and subsequently mutations were found in the parkin gene (Kitada et al., 1998). Parkin is a protein that functions as an E3-ubiquitin ligase and participates in the metabolism of proteins via the ubiquitin proteosomal system. Several other loci on chromosomes 1, 2, and 4 have been described in association with hereditary Parkinson's disease and other mutated genes responsible for parkinsonism in some families were found (ubiquitin C-terminal hydrolase-L1 and DJ-1).

3. PATHOGENESIS OF PARKINSON'S DISEASE

Pathogenetic cascade in Parkinson's disease has been illuminated in recent years, but it still remains unexplained. Studies have identified several biochemical abnormalities which add to the mosaic of the disease mechanism:

- Free radicals and oxidative stress
- Excitotoxic damage
- Mitochondrial dysfunction
- Inflammatory change
- Non-availability of trophic factors.

Free radicals form during normal respiration and oxidation. During oxidation, a molecule transfers one or more electrons to another. Stable molecules usually have matched pairs of protons and electrons, whereas free radicals have unpaired electrons and tend to be highly reactive, oxidizing agents. Damage to cells caused by free radicals includes protein oxidation, DNA destruction, an increase of intracellular calcium, activation of damaging proteases and nucleases, peroxidation of cellular membrane lipids and leads to the formation of tissue-damaging mediators. The damaging effect of free radicals is controlled to some extent by the antioxidant defense system and by cellular repair mechanisms, which include enzymes such as superoxide dismutase, catalase, glutathione peroxidase, and vitamins such as tocopherol, ascorbate and beta carotene. Dopaminergic system is a fertile potential source of free radicals and oxidative stress. The oxidative deamination of dopamine itself generates hydrogen peroxide, as does the dismutation of superoxide and this can lead to the formation of the highly toxic hydroxyl radical. In the brain, any hydrogen peroxide that is generated is primarily scavenged by glutathione peroxidase. Jenner and colleagues (1992) confirmed that a deficit of glutathione peroxidase appears early in the disease, before florid clinical signs become evident.

Parameters indicative of oxidative stress and damage in symptomatic Parkinson's disease include a decrease in reduced glutathione (Sian et al., 1994), increased levels of malondialdehyde and hydro-peroxide (Jenner, 1991), signs for damage to DNA (increased levels of 8-hydroxydeoxy guanosine; Sanchez-Ramos et al., 1994) and increased iron concentrations demonstrated in the substantia nigra and in the basal ganglia of Parkinson's disease patients (Dexter et al., 1987). Iron is capable of catalyzing oxidative reactions that will generate hydrogen peroxide and the hydroxyl iron, inducing a considerable oxidative stress. However, increased iron is not specific for Parkinson's disease and has been found also in other parkinsonian syndromes.

The cellular and molecular mechanisms of excitotoxicity are still not fully understood. Excessive stimulation with the release of excitatory amino acid transmitters causes overactivation of glutamate receptors. Consequently, a dramatic increase of intracellular calcium occurs, which can then activate a self destructive cellular cascade involving many calcium-dependent enzymes, such as phosphatases, proteases and lipases and which leads to the formation of free radicals, such as nitric oxide. The excitatory amino acids, N-methyl-D-aspartate and glutamic acid, believed to cause many neurodegenerative diseases, induce apoptosis in neurons. The activation of 'death' genes characteristic of programmed cell death (apoptosis) may be involved in exitotoxic damage of neurons. In Parkinson's disease, there is glutaminergic input from the subthalamic nucleus to the globus pallidus pars-interna and the substantia nigra pars-reticularis.

A mitochondrial Complex 1 defect was identified in the substantia nigra of patients with idiopathic Parkinson's disease (Schapira et al., 1989) and in patients with MPTP-induced parkinsonism. MPTP is converted to active compound 1-methyl-4-phenylpyridinum (MPP+) by monoaminooxydase-B. MPP+ is then concentrated within dopaminergic neurons into mitochondria, where it specifically inhibits complex 1, causing a dramatic fall in ATP production. In addition, MPP+ generates free radicals and thus mediates its toxicity with both mitochondrial inhibition and oxidative stress (Tipton and Singer, 1993). Pesticide rotenone, a specific complex 1 inhibitor, has been shown to produce nigral dopaminergic cell loss (Betarbet et al., 2000).

There is evidence of inflammation in Parkinson's disease (McGeer et al., 1988). Activated microglia, which may release proteinases and cytokines, was found in Parkinson's disease brains. Brains demonstrated an increase in levels of interleukin 1 beta, interferon gamma and tumor necrosis factor alpha (Hirsch et al., 1998).

Trophic factors are substances, usually proteins, which enhance differentiation as well as survival of neurons. Several factors have been shown to be neurotrophic for developing dopaminergic neurons in culture. They include brain derived neurotrophic factor (BDNF), the mitogens epidermal growth factor (EGF), acidic fibroblast growth factor (FGF-1) and basic fibroblast growth factor (FGF-2), insulin-like growth factor 1 and 2 (IGF 1 and 2), muscle derived differentiation factor (MDF) and the glial cell line derived factor (GDNF). These substances support survival and/or differentiation of embryonic dopaminergic mesencephalic neurons in tissue culture. Most of these factors lack specificity for dopaminergic neurons, as they also promote survival of non-dopaminergic cells. GDNF, however, is considered to be specific for developing dopaminergic neurons.

Many of these abnormalities and processes – oxidative stress, exitotoxicity, non-availability of trophic factors – may lead to *apoptosis.* This hypothesis was first put forward by Appel (1981). Apoptosis is mostly inherently programmed cell death, distinct from cell necrosis. Apoptosis mediates cell deletion in tissue homeostasis, embryological development and in pathological conditions, such as neurodegenerative diseases. In apoptosis, the cascade of events occurs which in Parkinson's disease finally causes death of dopaminergic nigral cells.

The sequence of biochemical abnormalities remains unclear. It seems that the Complex 1 defect results in increased free radical production. This sequence is more likely than the reverse because free radical damage to mitochondria results in deficiences of complexes 1-3, not in the specific defect of complex 1 as seen in Parkinson's disease. Inflammatory changes are likely to be secondary (Schapira, 2002b).

4. LEVODOPA IN PARKINSON'S DISEASE: BENEFITS AND PROBLEMS

Current therapies for Parkinson's disease are aimed mainly at symptomatic relief. Levodopa is the most potent drug for controlling Parkinson's disease symptoms. However, although extremely effective in reducing symptoms, levodopa has substantial limitations in that it

- does little, if anything, to slow the underlying rate of progression of the disease;
- does not alleviate all symptoms of Parkinson's disease equally;
- following the long-term therapy, levodopa is associated with the emergence of several problems, A major problem are motor fluctuations typified by a shortening of the time for which the drug exerts its actions (the wearing off phenomenon) and a sudden, unpredictable lack of efficacy (on-off fluctuations). Long-term levodopa treatment is further accompanied by the emergence of psychosis and by the emergence of debilitating involuntary movements (dyskinesia) affecting over 75% of patients within 5 years of starting treatment with levodopa (Rinne, 1983; Stocchi et al., 1997).
- after more than 40 years of clinical experience, there is still doubt whether levodopa accelerates nigral cell degeneration or not. Levodopa can promote the formation of free radicals and reactive oxidant species by way of auto-oxidation or by conversion to dopamine, as shown by an increase in oxidized glutathione, formation of hydroxyl radicals, and increased lipid peroxidation, DNA oxidation and mitochondrial damage (Spina and Cohen, 1988; Tanaka et al., 1992; Przedborski et al., 1993; Spencer et al.,

1994; Spencer-Smith et al., 1994). Levodopa can be toxic to cultured dopaminergic and other neurons in high concentrations in vitro studies (Michel et al.1990; Tanaka et al., 1991; Pardo et al., 1993; Pardo et al., 1995a; Melamed et al., 1998). The mechanisms underlying this toxic effects are not fully understood. The enzymatic oxidation of levodopa to dopamine and its further metabolism by monoamine oxydase B (MAO-B) could give rise to the formation of hydrogen peroxide (H_2O_2), which would be normally inactivated by glutathion peroxidase. Administration of levodopa to the culture of neuronal cells actually produces an increase in oxidized glutathione (GSSG), demonstrating the occurrence of oxidative stress (Spina et al., 1988). Excess of H_2O_2 could than be also converted by iron mediated Fenton reaction to produce very toxic hydroxyl radical, which could further induce oxidative damage. Experimental results, however, do not support completely the above explanation, since MAO-B inhibitors do not prevent dopamine toxicity in cell cultures. It appears that levodopa and dopamine toxicity result from autoxidation rather than enzymatic metabolism of levodopa (Mytilineou et al., 1993). Levodopa and dopamine can be autoxidated to toxic quinones on their metabolic pathway to melanin (Graham, 1978), leading to free radical generation. Dopamine toxicity could be prevented by antioxidants (Masserano et al., 1996; Offen et al., 1996) where iron-mediated oxidation seems to be involved for levodopa, as it can be prevented by iron chelator desferrioxamine (Tanaka et al., 1991).

In levodopa treated neuronal cultures, mitochondrial dysfunction may occur as a result of Complex I dysfunction. Levodopa can also have some excitotoxic effect by evoking glutamate release in rodent striatal neuronal cell cultures (Goshima et al., 1993), which could be prevented by glutamate antagonists (Cheng et al., 1996) and complex IV inhibition. Dopamine inhibits pyruvate- and succynate-dependent respiration in rat brain mitochondria. It seems that H_2O_2 formed in some way by dopamine metabolism increases GSSG levels, which can form protein-mixed disulfides suppressing mitochondrial electron transport (Pardo et al., 1995b; Ben-Shachar et al., 1995).

Most of these in vitro studies were performed on neuronal cultures containing only small number of glial cells or in the complete absence of glial cells. It was proven that glial cells can take up levodopa and dopamine and deactivate both substances by dopa-decarboxylase, MAO and catechol-O-methyltransferase (Pelton et al., 1981). Astrocytes can protect striatal neurons in the culture from toxic hydrogen peroxide. H_2O_2 is efficiently detoxified by catalase and glutathion peroxidase which are abundant in glial cells. It was also shown that glial cells release some soluble factors promoting survival of dopaminergic cells and making them more resistant to oxidative stress (Desagher et al., 1996).

Recent studies have even suggested the neuroprotective and neurotrophic activity of levodopa in low concentrations. Adding of levodopa to mixed neuronal mesencephalic and glial cell cultures increases the level of reduced glutathion (GSH), a substrate for glutathion peroxydase, the cell protector from toxic effects of oxidants (Han et al., 1996), however this can be interpreted as a secondary reaction to oxidative stress.

Levodopa might also have a neurotrophic effect on its own, not dependent on dopamine synthesis, because this effect is not blocked by decarboxylase inhibitor carbidopa. Addition of low concentration of levodopa to cell cultures, in a range normally observed in plasma of parkinsonian patients, promotes the survival of dopaminergic neurons of the rat fetal midbrain neurons mixed with astroglia (Mena et al., 1997).

However, there is no evidence to show that levodopa is toxic to normal mice, rats or humans (Cotzias et al., 1977; Hefti et al., 1981, Papavasiliou et al., 1981; Perry et al., 1984; Quinn et al., 1986). There is concern that this may not be the same in Parkinson's disease where oxidant defense mechanisms were shown to be compromised. Levodopa has been shown to promote cell death induced by 6-OHDA or MPTP (Blunt et al., 1993; Ogawa et al., 1994a), but this finding could not be replicated in another more recent study (Dziewczapolski et al., 1997), and, on the other hand, levels of superoxide dismutase are increased in Parkinson's disease and this suggests a protective response against active free radicals (Saggu et al., 1989). It was also demonstrated, that chronic administration of levodopa is not toxic for remaining dopamine neurons, but instead promotes their recovery, in rats with moderate nigrostriatal lesions (Murer et al., 1998).

There is also no compelling evidence from human clinical and pathological studies to support the toxicity of levodopa on nigral cells. Normal subjects mistakenly treated with levodopa for many years have not developed parkinsonism or alterations in PET scan, nor have been found to have reduced number of nigral cells on post-mortem histological examination (Maier Hoehn, 1983; Quinn et al., 1986; Rajput et al., 1997). At the moment we do not have any convincing evidence that levodopa is toxic for nigral dopamine cells, despite some indication by in vitro experiments of such an effect.

5. DOPAMINE AGONISTS

For all of the above reasons there has been a search for better therapies that can provide symptomatic benefit while avoiding the side effects and putative neurotoxicity that characterize levodopa therapy. Dopamine agonists, analogues of dopamine, have some of these features. Older dopamine agonists were mainly synthetic ergoline compounds derived from plant alkaloids. More recently non-ergoline synthetic drugs were developed. It is interesting to know that the first dopaminergic agonist was apomorphine, synthesized from morphine already in 19th century. The first reference of its use for parkinsonism already comes from Weill in 1884 (Weill, 1884), after obtaining beneficial results in a patient with Sydenham's chorea. This presumption was proven clinically almost 70 years later by Schwab et al. (1951) who noted short but marked improvement in Parkinson's disease patients.

Dopamine agonists offer several theoretical advantages as a treatment for Parkinson's disease:

- direct stimulation of dopamine receptors;
- no need to be converted to an active product;
- potential to stimulate subsets of dopamine receptors;
- relatively long half-life;
- do not undergo oxidative metabolism;
- reduced potential to induce motor complications and
- putative neuroprotective effects.

Dopamine receptors were classified into D1 and D2 families, according to the positive linkage of the former to adenylate cyclase and the lack of, or negative, coupling of the latter to adenylate cyclase (Kebabian and Calne, 1979). More recently, molecular biology studies have led to cloning and characterization of a least five dopamine receptor subtypes, D1, D2,

D3, D4 and D5 (Schwartz et al., 1992). Dopamine and dopamine agonists show differential selectivity for these receptor subtypes. The D1 family consists of two receptor subtypes, D1 and D5, which couple to the stimulatory G protein. This protein activates adenyl cyclase directly and increases cAMP production from ATP. The mRNA for D1 is located primarily in the caudate, putamen, nucleus accumbens and olfactory tubercle. The mRNA for D5 is located primarily in the hippocampus and hypothalamus.

The D2 family consists of D2, D3 and D4 receptor subtypes linked to inhibitory G protein. Activation of these receptors decreases the concentration of cAMP. D2 receptors are located mostly in the caudate, putamen, nucleus accumbens and olfactory tubercle. D3 receptors are highly expressed in the striatum but also in the limbic areas. D4 receptors are present in the frontal cortex, midbrain, amygdala, medulla and to much lesser extent in the basal ganglia. The relative antipsychotic effect of atypical neuroleptics may be due to their antagonism to D3, D4 and D5 receptors. Parkinsonian symptoms on the other hand may be caused by blocking of D2 receptors in the striatum by typical neuroleptics.

Dopamine agonists are effective in reversing the motor symptoms of Parkinson's disease and they have also been shown that they can delay or prevent the onset of motor complications associated with levodopa use. The side effect spectrum of dopamine agonists is similar to that of levodopa with a greater tendency of dopamine agonists to produce adverse mental reactions. Other common side-effects include nausea, orthostatic hypotension, dizziness, somnolence, daytime-sleepiness and insomnia. Ergot agonists may very rarely induce retroperitoneal and pulmonary fibrosis, but leg edema may occur with ergot and non-ergot agonists. Recently there have been reports of sudden sleep attacks (Frucht et al., 1999), especially with new dopamine agonists. It became clear in retrospect that this might have been a potential and non-specific propensity of all dopaminergic drugs, including levodopa (Homann et al., 2002).

6. NEUROPROTECTION AND PARKINSON'S DISEASE

In addition to the benefit gained from treatment of the symptoms of Parkinson's disease measures taken to delay the death of the remaining dopaminergic neurons may also be of benefit to slow the progress of the disease. Neuroprotection can be defined as the protection of neurons from cell death induced by the various biochemical abnormalities associated with aetiology and pathogenesis and it should be distinguished from neurorescue. Neurorescue implies the reversal of some or all of the biochemical abnormalities associated with aetiology and pathogenesis, and the restoration of normal neuronal function to damaged cells (Schapira, 2002b). Dead, abnormal and normal cells in normal ageing, in Parkinson's disease, in the neuroprotection and in the neurorescue condition are schematically represented in Fig.1.

6.1. Methods For Studying Neuroprotection

There are several methodological approaches to the study of neuroprotection.

- The basic mechanisms of neuroprotection are often studied in vitro. Unfortunately such experiments do not fatefully simulate the situation in humans for several reasons. The lesion inflicted to the cells in the culture is usually acute. Therefore only acute damage of the cells can be observed and not their progressive degeneration like in Parkinson's disease. If the culture cells are only neurons, there is no protective effect on neurons from the surrounding glial cells. This was shown as very important in the studies of levodopa toxicity. The neurons in the culture are also dispersed and functionally not organized like in animals or in humans. For example, it would be impossible in cell cultures to study and replicate the effect of some drugs, like levodopa or dopamine agonists, which they exert indirectly on the striatal glutamathergic neurons, by decreasing their potentially excitotoxic activity, as explained by the current model of basal ganglia organisation.

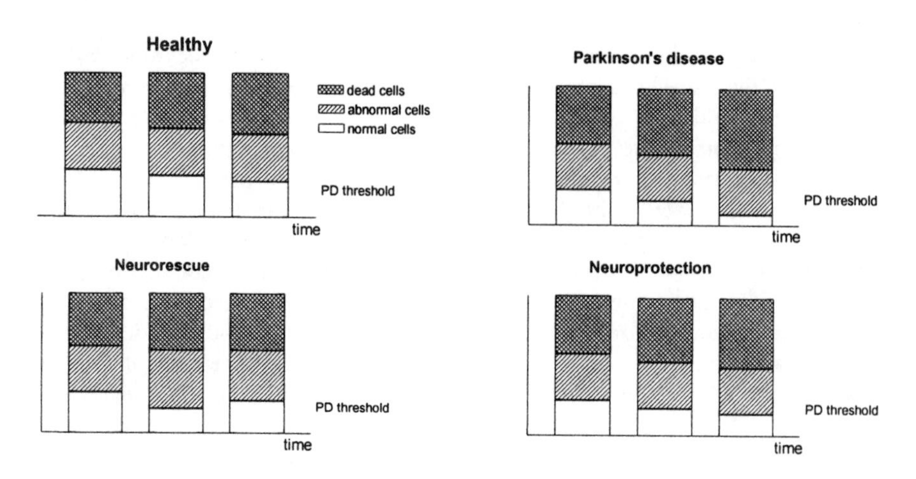

Figure 1. Proportions of dead cells, abnormal cells, and normal cells in 4 conditions: healthy, Parkinson's disease, neurorescue, neuroprotection.

- Several animal models have been developed for testing the neuroprotection in Parkinson's disease. 1-methyl-4-phenyl-1,2,3,6-tetrahydropyridine (MPTP) and 6-hydroxydopamine are two commonly used toxic substances for dopaminergic neurons in these models. We face the same problem in animal studies as in the cell culture studies. Relatively slow degeneration of dopaminergic neurons typical for Parkinson's disease can not be replicated in the animal model. This might be one of the reasons why drugs effectively protecting dopaminergic neurons from toxic substances in animal experiments, did not prove to be effective in clinical trials.With recent discoveries of mutated genes responsible for parkinsonism in some families (alpha- synuclein, parkin, ubiquitin C-terminal hydrolase-L1 and DJ-1), it is now possible to create transgenic animals carrying genes with these mutations. These animals do not replicate the biological situation of the diseased members of these families and especially not the situation of the majority of Parkinson's disease patients, who are not carriers of these genes. However, they provide a very useful tool for the research of the pathogenesis and probably of neuroprotection in the future.

- The progression of Parkinson's disease in humans can be measured by imaging radioligands which selectively bind to dopaminergic neurons. These imaging methods are nowadays used in all clinical trials looking for the potential neuroprotective effect of the drugs. The imaging has in this respect replaced the common clinical motor scales, like UPDRS and Webster scale, because the symptomatic effect of drugs under investigation can not be separated from their neuroprotective effect. Attempts to washout the medication to control for this are only partially successful. Current rating scales subjectively assess multiple symptoms and signs which are not equally weighted. The treatment effect on a particular symptom may produce unduly large or small effects on the overall outcomes. In this situation the imaging methods should play the decisive role in judging the neuroprotective effect of the drugs. In this respect an independent biomarker which provides an objective means of assesing disease progression is of value. The positron emission tomography (PET) and single photon emission computer tomography (SPECT) are being used for the quantification of dopaminergic neurons in the subtantia nigra. In the late 80s it was already possible to draw a conclusion on the number of dopaminergic neurons with the use of 18F-fluorodopa. With this ligand it was possible at the beginning of parkinsonian symptoms to discover the reduction of its concentration in dopaminergic neurons (Brooks, 1993). 18F-fluorodopa enters the dopaminergic neuron the same way as levodopa and is there further metabolized. It therefore reflects more the metabolic activity of dopaminergic neurons rather than their actual number. Moreower, 18F-dopa is transported or metabolized at multiple sites that are all regulated and influenced by levodopa and dopaminergic drugs. Chronic administration of carbidopa (together with levodopa) upregulates the activity of plasma L-aromatic amino acid decarboxylase (Boomsma et al., 1989). 18F-dopa and levodopa cross the blood-brain barrier via a large neutral amino acid transporter (LAT). Whether chronic levodopa treatment alters LAT activity remains open to speculation. 18F-dopa is transported into striatal neurons by a high affinity, saturable carrier, similar to blood-brain barrier LAT. The extent to which this carrier activity is regulated is unknown, but could be influenced by chronic levodopa treatment (Sampaio-Maia et al., 2001; Yee et al., 2001). Autoreceptor activation inhibits nigrostriatal neuron firing and dopamine release. Dopamine reduce firing rates (Kreiss et al., 1995) and dopamine efflux (Pothos et al.1998; Sotnikova et al., 2001). Reduced efflux should result in increased intraneuronal 18F-dopamine from administered 18F-dopa. How long term agonist or levodopa administration affects regulation of this function is unknown. Binding of specific ligands on striatal dopamine transporters (DAT) is another way of presenting presynaptic dopaminergic neurons, since DAT are located only on neuronal terminals. Several cocaine derivatives with high specificity and affinity for DAT have been discovered in recent years and can be detected by SPECT. There are some major problems of interpretation with this method as well (Ahlskog, 2003). These ligands bind to DAT on dopaminergic cells. DAT is responsible for removing released dopamine from synaptic regions and is critical for proper modulation of dopaminergic neurotransmission. It is a highly regulated protein with a high turnover rate. Its expression varies with factors influencing dopaminergic neurotransmission. Animal studies suggest that levodopa (Rioux et al., 1997) and DAT (Kimmel et al., 2001) administration may influence DAT regulation, but not neccessarily in the same direction. In Parkinson's disease patients, the reduced brain dopamine levels are also compensated by downregulation of DAT (Lee et al., 2000). With this potential for regulation of DAT expression, it follows that changes in DAT imaging could simply reflect a pharmacologic effect on this regulation, instead of reflecting the change in the actual number of dopaminergic neurons.

6.2. Neuroprotective Approaches In Parkinson's Disease

Recent understanding of the molecular and biochemical pathophysiology of Parkinson's disease provides the possibility of neuroprotective therapy to halt or reverse the degenerative process. Important reasons for considering neuroprotective therapy in Parkinson's disease are (i) the slow evolution of the neurodegenerative process, which begins years before the symptomatic phase and (ii) a possibility that levodopa may add to the degenerative process by causing lipid peroxidation, membrane disruption and damage to the mitochondrial respiratory mechanisms.

Research on Parkinson's disease has discovered impaired mitochondrial function, altered iron metabolism and increased lipid peroxidation in the substantia nigra. All this emphasized the importance of oxidative stress and free radical formation in the pathogenesis of Parkinson's disease. Environmental toxins could play an important role in the generation of this oxidative stress. It was shown for example, that the toxicity of 6-hydroxydopamine (Kumar et al., 1995) and MPTP (Przedborski et Jackson-Lewis, 1998) could also result from the oxidative stress and free radical formation.
However, we must stay aware of the importance of the genetic factors, which can effectively attenuate this environmental effect. Apoptosis is the most important form of neuronal death in Parkinson's disease. It is known that an increase in the expression of anti-apoptotic proteins inhibits neuronal apoptosis during normal development as well as that induced by cytotoxic compounds. For example, in transgenic neuronal cells, the level of expression of Bcl-2 determines the number of surviving neurons exposed to MPTP or 6-hydroxydopamine (Offen et al., 1998).

Several approaches to neuroprotection, aimed at various steps in the cascade of events leading to death of dopaminergic neurons have been tried in laboratories – oxidative stress, mytochondrial dysfunction, exitotoxicity, but translation of this research into clinical practice remains largely unfilled. The most important categories of neuroprotective therapies in Parkinson's disease are:

- dopamine agonists (see below);
- antioxidants: deprenyl, rasageline, vitamin C, beta carotene, iron chelators (such as desferrioxamine which maintain iron in a relatively unreactive form), and agents that may promote glutathione-like activity such as thioctic acid;
- antiexcitotoxic agents: remacemide, riluzole;
- neurotrophic factors: GCDNF, BDNF;
- immunosupresants : NIL-A;
- cell transplants: e.g. fetal pig dopaminergic transplant can take over the function of damaged dopaminergic neurons;
- surgical interventions: inhibit neuronal firing in the subthalamic nucleus and possibly glutamate-induced excitotoxicity. Lesions of the subthalamic nucleus have been shown to protect against dopamine neuronal damage and degeneration in the substantia nigra (Saji et al., 1996);
- anti-apoptotic agents: therapeutic approaches to block apoptosis may theoretically include: immunosupresants; increasing the localized expression of bcl-2 protein in

the brain by gene therapy; neurotrophic factors; modulating p53 expression by antisense therapy; using apoptose inhibitors such as caspase inhibitors, calpain inhibitors and poly (ADP-ribose) polymerase inhibitors;

- trophic factors: The use of neurotrophic factors is made difficult, largely through problems of delivery. In respect to Parkinson's disease it is as yet unknown whether the degeneration of nigrostriatal dopaminergic neurons is in part due to a reduced trophic support. It likewise remains to be demonstrated whether additional trophic support may prevent, slow down, or even reverse the progressive degeneration of dopaminergic neurons in parkinsonism;
- gene therapy: gene therapy can be broadly defined as the transfer of defined genetic material to specific target cells of a patient for the ultimate purpose of preventing or altering a particular disease state. Carriers, or delivery vehicles, for therapeutic genetic material are called vectors which are usually viral, but several nonviral techniques are being used as well. It is feasible to protect dopaminergic neurons using a genomic virus-based vector that expresses either neurotrophic factors or anti-apoptotic peptides (Fink et al., 2000). Glial cell-line derived neurotrophic factor delivered by a lentiviral vector system, can prevent nigrostriatal degeneration and induce regeneration in primate models of Parkinson's disease (Kordower et al., 2000).

7. DOPAMINE AGONISTS AS NEUROPROTECTIVE AGENTS

When it was demonstrated in in vitro experiments that levodopa could be toxic to neuronal cells and able to generate free radicals, the potential significance of dopamine agonists, not only for the symptomatic treatment, but also as neuroprotective compounds, became suddenly very important. Preclinical studies demonstrated the potential of dopamine agonists to protect dopaminergic neurons from a variety of stresses. This has now been observed in tissue culture, rodents and non-human primates and recent clinical studies indicate that nigrostriatal function as measured by the striatal uptake of fluorodopa or ^{123}Iß-CIT on SPECT may be protected when initial Parkinson's disease treatment is begun with dopamine agonist as compared to levodopa.

There are various mechanisms whereby dopamine agonists might provide protection against cell degeneration:

- A cumulative reduction in lifetime exposure to levodopa

Dopamine agonists can reduce the lifetime exposure to levodopa by delaying its introduction (Montastruc et al., 1994) and permitting reduction of levodopa dosage (Olanow et al., 1994). As dopamine agonists do not undergo oxidative metabolism, their use reduces the formation of reactive oxygen species generated from levodopa metabolism and the likelihood that levodopa-generated free radicals will contribute to the neurodegeneration process.

- Stimulation of autoreceptors

Dopamine agonists might also prevent dopamine-related neurotoxicity by stimulation of D2 autoreceptors on presynaptic dopaminergic neurons. This will decrease dopamine release and metabolism, with consequent decreased free radical formation. In vivo studies have actually shown that D2 agonists decrease dopamine turnover (Carter et Muller.

1991) in rats and dopamine agonist-induced reduction of dopaminergic neuronal firing in rodents has been demonstrated (Piercey et al., 1995). It is however still not known, if this turnover reduction has any protective influence on nigral neurons.

- Direct anti-oxidopamine agonistsnt effects:
  The potential of dopamine agonists for neuroprotection has been shown in vitro and in vivo.
- Attenuation of STN-mediated excitotoxicity
- Receptor-mediated anti-apoptotic effect.

7.1. Basic Studies

Several in vitro and in vivo studies have shown that dopamine agonists might have some neuroprotective effects on their own.

In vitro studies have clearly demonstrated a common characteristic of dopamine agonists. They are all antioxidants or free radical scavengers including apomorphine (Gassen et al., 1996; Gassen et al., 1998; Gassen et al., 1999; Youdim et al., 2000), bromocriptine (Ogawa et al., 1994b), pergolide (Nishibayashi et al., 1996; Gomez-Vargas et al., 1998; Gille et al., 2002), pramipexole (Zou et al., 1999), ropinirole (Iida et al., 1999) and cabergoline (Yoshioka et al., 2002).

Apomorphine was shown to scavenge free radicals from isolated rat-brain mitochondria (Gassen et al., 1996) and to reduce iron (Ubeda et al., 1993). Pramipexole, the best studied dopamine agonist in terms of neuroprotection, protects against MPP+ and rotenone toxicity in SHSY-5Y cells (Schapira et al., 2002a). This effect was not blocked by dopamine antagonists, moreover, similar effect was also seen in JKAT cells, which do not have dopamine receptors. The protective action of pramipexole is therefore not entirely dependent on dopamine receptors, despite the fact that D3 antagonist attenuated its neuroprotective effect on mesencephalic cell cultures (Ling et al., 1999). Pramipexole protects also against mitochondrial swelling induced by MPP+ or calcium and increases Bcl2 expression, thereby protecting against apoptotic cell death (Kitamura et al., 1998).

In vivo studies were performed on different animal models. Animals were given dopamine agonists for neuroprotection first and later one of the neurotoxic compounds like 6-hydroxydopamine, MPTP or methamphetamine to induce lesions in dopaminergic neurons (Ogawa et al., 1994a; Hall et al., 1996; Sethy et al., 1997; Muralikrishnan et Mohanakumar, 1998; Grunblatt et al., 1999; Iida et al., 1999; Ferger et al., 2000; Grunblat et al., 2001; Yoshioka et al., 2002). All of these studies also confirmed the neuroprotective capacity of dopamine agonists. Thus, bromocriptine protects mice against 6-hydroxydopamine and MPTP-induced cell loss (Ogawa et al., 1994b; Muralikrishnan and Mohanakumar, 1998). Cabergoline decreases lipid peroxydation in rat striatum (Finotti et al., 2000) and protects against 6-hydroxydopamine toxicity in mice (Yoshioka et al., 2002). Pramipexole and pergolide scavenge hydroxyl superoxide and nitric oxide radicals (Nishibayashi et al., 1996; Gomez-Vargas et al., 1998) and ropinirole increases the concentration of glutathione, catalase and superoxide dismutase (Iida et al., 1999). The neuroprotective effect of pramipexole has been studied recently in mammals (Schapira, 2002b). Marmosets were given MPTP, but some animals were pre-treated with pramipexole. These animals demonstrated significant neuroprotective effect of pramipexole.

The antioxidant property of dopamine agonists as a group could be explained to some extent by the common hydroxylated benzyl ring, that might be responsible for their free radical scavenger character, but there have been some additional features detected by experiments with distinctive dopamine agonists, which could be also involved in the process of free radical generation and their elimination, like:

- inhibition of monoamine oxidase A and B (MAO-A and MAO-B) by apomorphine (Grunblatt et al., 2001);
- induction of Cu/Zn superoxide dismutase by pergolide (Asanuma et al., 1995);
- activation of glutathione (GSH) system by ropinirole (Tanaka et al., 2001) and cabergoline (Yoshioka et al., 2002);
- activation of catalase and superoxide dismutase by ropinirole (Iida et al., 1999);
- inhibition of the opening of mitochondrial permeability transition pores by pramipexole (Cassarino et al., 1998).

Dopamine agonists seem to have influence also on the regulation of the expression of anti-apoptotic proteins. Bcl-xl and Bcl-2, both potent anti-apoptotic proteins, are increased in mesencephalic cultures and cortical dendritic processes of rats, when cultures or rodents are treated with pramipexole (Carvey and Ling, 2000; Takata et al., 2000). Moreover, transgenic cells overexpressing Bcl-2 can resist dopamine, 6-hydroxydopamine and MPTP induced apoptosis (Offen et al., 1997; Offen et al., 1998). This suggests that some of the neuroprotective effects of dopamine agonists, particularly of pramipexole, might result from the involvement of protooncogenes.

There is still some controversy concerning the role of dopamine receptors in neuroprotection of dopamine agonists. In vitro study has shown that neuroprotective effect of pramipexole does not appear to be dependent on dopamine receptors, since it is not blocked by D2 and D3 anatgonists. On the other hand, pretreatment with D2 antagonist sulpiride prevented the neuroprotective effect of ropinirole in mice (Iida et al., 1999). In this regard it is also interesting to note that not only D2, but also D1 (Noh et Gwag, 1997) and D3 (Iida et al., 1999) agonists and even agonists in their inactive form, like S- apomorphine enantiomer (Grunblatt et al., 2001), may have some neuroprotective effect.

7.2. Clinical Studies

It is hard to draw a firm conclusion from these studies regarding the effect of dopamine agonists in humans. The concentrations of dopamine agonists used in the above experiments were much higher than the levels that might be achieved normally in humans. No wonder that such neuroprotective findings have not been confirmed in human trials yet. However, two studies have been finished recently comparing the progression of the disease in Parkinson's disease patients taking dopamine agonists ropinirole and pramipexole vs. levodopa. Surrogate markers such as 18F–dopa PET and ^{123}Iß-CIT SPECT were used to provide the evidence of neuroprotection. However there still remained some unresolved methodological problems leading to equivocal interpretation of the results.

Ropinirole is a second-generation dopamine agonist. It is a selective D2-type dopamine agonist with a low affinity for D1-type, 5-HT, benzodiazepine and GABA receptors. It has direct and selective stimulant effect on postsynaptic D2 receptors. Most of the orally administered dose is rapidly absorbed from the gastrointestinal tract. The bioavailability is approximately 50%. It shows linear steady-state pharmacokinetics and a good safety profile. T_{max} is about 1,4 hours after dosing and elimination half-life of the compound is 6 hours. The drug is extensively metabolized in the liver by the principal metabolic enzyme cytocrome P450. The major route of excretion of its metabolites is renal.

Ropinirole is efficacious when used as a symptomatic monotherapy in de novo patients with Parkinson's disease (Adler et al., 1997). As adjunct to levodopa in the treatment of Parkinson's disease associated with motor fluctuations, 35% of the patients on ropinirole versus 13% on placebo achieved more than 20% reduction in both levodopa dose and percent time spent "off" (Lieberman et al., 1998). Ropinirole is also efficacious in the prevention of motor complications, reducing the risk of occurrence of dyskinesia, as evidenced by a 5-year of follow-up (Rascol et al., 2000). Dyskinesias developed at a rate 3 times slower in the ropinirole-treated patients compared with the levodopa-only group.

The influence of ropinirole on the progression of Parkinson's disease was measured with 18F-dopa PET in the REAL-PET study (Whone et al., 2002). This was a 2 year, double blind study of 186 patients with de novo Parkinson's disease who were randomized 1:1 to receive ropinirole or levodopa. The primary endpoint was the change in putamen 18F-dopa uptake measured by PET. 73% of the ropinirole and 74% of the levodopa treated patients completed the study. 'Region of interest' analysis of putarmen 18F-dopa uptake showed significantly slower progression in patients on ropinirole. There was a relative difference of 35% between the two treatment arms, the reduction of the uptake beeing smaller for the ropinirole group. Statistical parametric mapping also localized significant reductions in 18F-dopa uptake over two years in the putamen and substantia nigra, but the decrease was again significantly smaller with ropinirole compared to levodopa. After initial improvement in on-treatment UPDRS motor scores for both treatment groups, scores for the ropinirole returned to baseline over two years whereas scores for the levodopa remained improved, indicating that motor function while taking medication was superior in the levodopa than the ropinirole group. Three of the 87 patients receiving ropinirole and 20 of the 75 patients receiving levodopa developed dyskinesia. Investigators concluded that there was a significant difference in the rates of loss of dopamine-terminal function between early Parkinson's disease patients receiving dopamine agonist ropinirole compared with the conventional levodopa therapy. There are however some serious concerns with this study (Ahlskog, 2003). Despite better 18F-dopa PET imaging values, the ropinirole group's parkinsonian motor scores were significantly poorer. Approximately a quarter of the patients in the ropinirole study group dropped out, although equally proportioned between groups and this could bias the outcome depending upon the reason for discontinuation. Finnaly, the imaging outcome could have resulted from drug effects on the regulation of proteins involved in the metabolism or transport of 18F-dopa.

Pramipexole is a nonergot dopamine agonist. It is used for the treatment of Parkinson's disease as a monotherapy and as an adjunct to levodopa. It binds to presynaptic and postynaptic D2 and D3 receptors, but does not have affinity for the dopamine D1 receptor site. It reduces extracellular concentrations of dopamine by inhibiting dopamine synthesis and release. The motor benefits are likely due to dopamine D2 stimulation, whereas its effect on mood may be related to its D3 agonist properties. It has also a relatively high affinity for $\alpha 2$ adrenoreceptors but has little effect on other neurotrasmitter systems.

Pramipexole has linear pharmacokinetics over the usual therapeutic dose range. The plasma elimination half-live is approximately 7 to 9 hours. It is excreted by the renal organic transport system and renal clearance accounts for about 80% of the total clearance of an oral dose.

Pramipexole is efficacious as a monotherapy in de novo patients in controlling motor symptoms over the first two years of treatment of Parkinson's disease (Hubble et al., 1995; Shannon, 1997; Parkinson Study Group, 2000). Longer controlled follow up studies are not yet available. In advanced levodopa-treated Parkinson's disease patients suffering from motor fluctuations, pramipexole is efficacious as adjunct therapy in patients receiving levodopa and in control of motor complications (Guttman et al., 1997; Lieberman et al., 1997; Pinter et al., 1999). It is efficacious in reducing the risk of motor complications (Parkinson Study Group, 2000).

The neuroprotective effect of pramipexole was addressed by the CALM-PD imaging substudy, called CALM-PD-CIT (Parkinson Study Group, 2002). Eighty-two patients were recruited for this substudy from the original CALM-PD group. It was designed as a double-blind randomized clinical trial. Forty-two patients were randomly assigned to pramipexole and forty patients to levodopa. Open-label levodopa could be added subsequently in case of residual disability. The primary outcome variable was the percentage change from baseline in striatal ^{123}Iß-CIT uptake after 46 months. Patients initially treated with pramipexole had significantly less decline in striatal ^{123}Iß-CIT after 2,3 and 4 years of treatment, than the patient randomized to levodopa. Since the study compared two active medications without placebo group, these data cannot directly distinguish whether the difference in the rate of loss of radioligand uptake in the treatment groups results from decrease due to pramipexole, an increase due to levodopa or both.

However, there are some additional problems with the interpretation of the result of this study. DAT is highly regulated protein dependent on factors influencing dopaminergic neurotransimition. DAT and levodopa have both effect on DAT regulation. DAT imaging could therefore reflect a pharmacologic effect on this regulation. The investigators were aware of the potential confounding effects of pharmacologic influences on DAT expression and regulation. They compared seven pramipexole-treated patients to six treated with levodopa and observed no significant difference in ^{123}Iß-CIT uptake at 10 weeks of treatment. With this small number of patients and relatively short term of this assessment they could have missed the real effect demonstrable in longer term (Ahlskog, 2003).

New longitudinal imaging studies are in progress now, designed to demonstrate slowing of progression of Parkinson's disease by dopamine agonists.

8. REFERENCES

Adler, C. H., Sethi, K. D., Hauser, R. A., Davis, T. L., Hammerstad, J. P., Bertoni, J., Taylor, R. L., Sanchez-Ramos, J., and O'Brien, C. F. for the Ropinirole Study Group, 1997, Ropinirole for the treatment of early Parkinson's disease, *Neurology.* **49**:393–399.

Ahlskog, J. E., 2003, Slowing Parkinson's disease progression: Recent dopamine agonist trials, *Neurology.* **60**:381–389.

Appel, S. H., 1981, A unifying hypothesis for the cause of amyotrophic lateral sclerosis, parkinsonism and Alzheimer's disease, *Ann Neurol.* **10**:499–505.

Asanuma, M., Ogawa, N., Nishibayashi, S., Kawai, M., Kondo, Y., and Iwata, E, 1995, Protective effects of pergolide on dopamine levels in the 6-hydroxydopamine-lesioned mouse brain, *Arch Int Pharmacodyn Ther.* **329**:221–230.

Ben-Shachar, D., Zuk, R., and Glinka, Y., 1995, Dopamine neurotoxicity: inhibition of mitochondrial respiration, *J Neurochem.* **64**:718–723.

Betarbet, R., Sherer, T. B., MacKenzie, G., Garcia-Osuna, M., Panov, A. V., and Greenmyre, J. T., 2000, Chronic systemic pesticide exposure reproduces features of Parkinson's disease, *Nat Neurosci.* **3**:1301–1306.

Blunt, S. B., Jenner, P., and Marsden, C. D., 1993, Suppressive effect of l-dopa on dopamine cells remaining in the ventral tegmental area of rats previously exposed to the neurotoxin 6-hydroxydopamine, *Mov Disord.* **8**:129–133.

Boomsma, F., Meerwaldt, J. D., Man in't Veld, A. J., Hoverstadt, A., and Schalekamp, M. A., 1989, Treatment of idiopathic parkinsonism with L-dopa in the absence and presence of decarboxylase inhibitors: effects on plasma levels of L-dopa, dopa dexarboxylase, catecholamines and 3-O-methyl-dopa, *J Neurol.* **236**:223–230.

Brooks, D. J., 1993, PET studies on the early and differential diagnosis of Parkinson's disease, *Neurology.* **43**:S6–S16.

Carter, A. J., and Mueller, R. E., 1991, Pramipexole, a dopamine D2 receptor agonist, decreases the extracellular concentration of dopamine in vivo, *Eur J Pharmacol.* **200**:65–72.

Carvey, P. M., and Ling, Z., 2000, Pramipexole enhances Bcl-xl expression in mesencephalic cultures, *Mov Dis.* 15 Suppl. **3**:17.

Cassarino, D. S., Fall, C. P., Smith, T. S., and Bennett, J. P. Jr., 1998, Pramipexole reduces reactive oxygen species production in vivo and in vitro and inhibits the mitochondrial permeability transition produced by the parkinsonian neurotoxin methylpyridinium ion, *J Neurochem.* **71**:295–301.

Cheng, N., Maeda, T., Kume, T., Kaneko, S., Kochiyama, H., Akaike, A., Goshima, Y., and Misu, Y., 1996, Differential neurotoxicity induced by L-DOPA and dopamine in cultured striatal neurons, *Brain Res.* **16**;743:278–283.

Cotzias, G. C., Miller, S. T., Tang, L. C., and Papavasiliou, P. S., 1977, Levodopa, fertility, and longevity, *Science.* **196**:549–551.

Desagher, S., Glowinski, J., and Premont, J., 1996, Astrocytes protect neurons from hydrogen peroxide toxicity, *J Neurosci.* **16**:2553–2562.

Dexter, D. T., Wells, F. R., Agid, F., Less, A. J., Jenner, P., and Marsden, C. D., 1987, Increased iron content in post-mortem parkinsonian brain, *Lancet.* **2**:1219–1220.

Dziewczapolski, G., Murer, G., Agid, Y., Gershanik, O., and Raisman-Vozari, R., 1997, Absence of neurotoxicity of chronic L-DOPA in 6-hydroxydopamine-lesioned rats, *Neuroreport.* **8**:975–979.

Ferger, B., Teismann, P., and Mierau, J., 2000, The dopamine agonist pramipexole scavenges hydroxyl free radicals induced by striatal application of 6-hydroxydopamine in rats: an in vivo microdialysis study, *Brain Res.* **883**:216–223.

Fink, D. J., DeLuca, N. A., Yamada, M., Wolfe, D. P., and Glorioso, J. C., 2000, Design and application of HSV vectors for neuroprotection, *Gene Ther.* **7**:115–119.

Finotti, N., Castagna, L., Moretti, A., and Marzatico, F., 2000, Reduction of lipid peroxidation in different rat brain areas after cabergoline treatment, *Pharmacol Res.* **42**:287–291.

Frucht, S., Rogers, J. D., Greene, P. E., Gordon, M. F., and Fahn, S., 1999, Falling asleep at the wheel: motor vehicle mishaps in persons taking pramipexole and ropinirole. *Neurology.* **52**:1908–1910.

Gassen, M., Glinka, Y., Pinchasi, B., and Youdim, M. B., 1996, Apomorphine is a highly potent free radical scavenger in rat brain mitochondrial fraction, *Eur J Pharmacol.* **308**:219–225.

Gassen, M., Gross, A., and Youdim, M. B., 1998, Apomorphine enantiomers protect cultured pheochromocytoma (PC12) cells from oxidative stress induced by H2O2 and 6-hydroxydopamine, *Mov Disord.* **13**:661–667.

Gassen, M., Gross, A., and Youdim, M. B., 1999, Apomorphine, a dopamine receptor agonist with remarkable antioxidant and cytoprotective properties, *Adv Neurol.* **80**:297–302.

Gille, G., Rausch, W. D., Hung, S. T., Moldzio, R., Janetzky, B., Hundemer, H. P., Kolter, T., and Reichmann, H., 2002, Pergolide protects dopaminergic neurons in primary culture under stress conditions, *J Neural Transm.* **109**:633–643.

Golbe, L. I., Di Iorio, G., Bonavita, V., Miller, D. C., and Duvoisin, R. C., 1990, A large kindred with autosomal dominant Parkinson's disease, *Ann Neurol.* **27**:276–228.

Gomez-Vargas, M., Nishibayashi-Asanuma, S., Asanuma, M., Kondo, Y., Iwata, E., and Ogawa, N., 1998, Pergolide scavenges both hydroxyl and nitric oxide free radicals in vitro and inhibits lipid peroxidation in different regions of the rat brain, *Brain Res.* **790**:202–208.

Goshima, Y., Ohno, K., Nakamura, S., Miyamae, T., Misu, Y., and Akaike, A., 1993, L-dopa induces Ca(2+)-dependent and tetrodotoxin-sensitive release of endogenous glutamate from rat striatal slices, *Brain Res.* **16**:167–170.

Graham, D. G., 1978, Oxidative pathways for catecholamines in the genesis of neuromelanin and cytotoxic quinones, *Mol Pharmacol.* **14**:633–643.

Grunblatt, E., Mandel, S., Berkuzki, T., and Youdim, M. B., 1999, Apomorphine protects against MPTP-induced neurotoxicity in mice, *Mov Disord.* **14**:612–618.

Grunblatt, E., Mandel, S., Maor, G., and Youdim, M. B., 2001, Effects of R- and S-apomorphine on MPTP-induced nigro-striatal dopamine neuronal loss, *J Neurochem.* **77**:146–156.

Guttman, M., and the International Pramipexole-Bromocriptine Study Group, 1997, Double-blind comparison of pramipexole and bromocriptine treatment with placebo in advanced Parkinson's disease, *Neurology.* **49**:1060–1065.

Hall E. D., Andrus P. K., Oostveen J. A., Althaus J. S., and VonVoigtlander P. F., 1996, Neuroprotective effects of the dopamine D2/D3 agonist pramipexole against postischemic or methamphetamine-induced degeneration of nigrostriatal neurons, *Brain Res.* **742**:80–88.

Han, S. K., Mytilineou, C., and Cohen, G., 1996, L-DOPA up-regulates glutathione and protects mesencephalic cultures against oxidative stress, *J Neurochem.* **66**:501–510.

Hefti, F., Melamed, E., Bhawan, J., and Wurtman, R., 1981, Long term administration of l-dopa does not damage dopaminergic neurons in the mouse, *Neurology.* **31**, 1194–1195.

Hernan, M. A., Takkouche, B., Caamano-Isorna, F., and Gestal-Otero, J. J., 2002, A meta-analysis of coffee drinking, cigarette smoking, and the risk of Parkinson's disease, *Ann Neurol.* **52**:276–284.

Hirsch, E. C., Hunot, S., Damier, P., and Faucheux, B., 1998, Glial cells and inflammation in Parkinson's disease: a role in neurodegeneration? *Ann Neurol.* **44**:S115–S120.

Homann, C. N., Wenzel, K., Suppan, K., Ivanic, G., Kriechbaum, N., Crevenna, R., and Ott, E., 2002, Sleep attacks in patients taking dopamine agonists: review. *BMJ.* **324**:1483–1487.

Hubble, J. P., Koller, W. C., Cutler, N. R., Sramek, J. J., Friedman, J., Goetz, C., Ranhosky, A., Korts, D., and Elvin, A., 1995, Pramipexole in patients with early Parkinson's disease, *Clin Neuropharmacol.* **18**:338–347.

Iida, M., Miyazaki, I., Tanaka, K., Kabuto, H., Iwata-Ichikawa, E., and Ogawa, N., 1999, Dopamine D2 receptor-mediated antioxidant and neuroprotective effects of ropinirole, a dopamine agonist, *Brain Res.* **838**:51–59.

Ishikawa, A., and Tsuji, S., 1996, Clinical analysis of 17 patients in 12 Japanese families with autosomal-recessive type juvenile parkinsonism, *Neurology.* **47**:160–166.

Iwai, A., Masliah, E., Yoshimoto, M., Ge, N., Flanagan, L., de Silva, H. A., Kittel, A., and Saitoh, T., 1995, The precursor protein of non-A beta component of Alzheimer's disease amyloid is a presynaptic protein of the central nervous system, *Neuron.* **14**:467–475.

Jenner, P., 1991, Oxidative stress as a cause of Parkinson's disease, *Acta Neurol Scand Suppl* **136**:6–15.

Jenner, P., Dexter, D. T., Sian, J., Schapira, A. H. V., and Marsden, C. D., 1992, Oxidative stress as a cause of nigral cell death in Parkinson's disease and incidental Lewy body disease, *Ann Neurol.* **32** (Suppl):S82–S87.

Kebabian, J. W., and Calne, D. G., 1979, Multiple receptors for dopamine, *Nature.* **277**:93-96.

Kimmel, H. L., Yoyce, A. R., Carroll, F. I., and Kuhar, M. J., 2001, Dopamine D1 and D2 receptors influence dopamine transporter synthesis and degradation in the rat, *J Pharmacol Exp Ther.* **298**:129–140.

Kitada, T., Asakawa, S., Hattori, N., Matsumine, H., Yamamura, Y., Minoshima, S., Yokochi, M., Mizuno, Y., and Shimizu, N., 1998, Mutations in the parkin gene cause autosomal recessive juvenile parkinsonism, *Nature.* **392**:605–608.

Kitamura, Y., Shimohama S., Kamoshima, W., Ota, T., Matsuoka, Y., Nomura, Y., Smith, M. A., Perry, G., Whitehouse, P. J., and Taniguchi, T., 1998, Alterations of proteins regulating apoptosis, Bcl-2, Bcl-x, Bax, Bak, Bad, ICH-I and CPP32, in Alzheimer's disease. *Brain Res.* **780**:260–269.

Kordower, J. H., Emborg, M. E., Bloch, J., Bloch, J., Ma, S. Y., Chu, Y., Leventhal, L., McBride, J., Chen, E. Y., Palfi, S., Roitberg, B. Z., Brown, W. D., Holden, J. E., Pyzalski, R., Taylor, M. D., Carvey, P., Ling, Z., Trono, D., Hantraye, P., Deglon, N., and Aebischer, P., 2000, Neurodegeneration prevented by lentiviral vector delivery of GDNF in primate models of Parkinson's disease, *Science.* **290**:767–773.

Korsching, S., 1993, The neurotrophic factor concept: a reexamination, *J Neurosci* **13**:2739–2748.

Kreiss, D. S., Bergstrom, D. A., Gonzalez, A. M., Huang, K. X., Sibley, D. R., and Walters, J. R., 1995, Dopamine receptor agonist potencies for inhibition of cell firing correlate with dopamine D3 receptor binding affinities, *Eur J Pharmacol.* **227**:209–214.

Kumar, R., Agarwal, A. K., and Seth, P. K., 1995, Free radical-generated neurotoxicity of 6-hydroxydopamine, *J Neurochem.* **65**:1906.

Lee, C. S., Samii, and A., Sossi, V., 2000, In vivo positron emission tomographic evidence for compensatory changes in presynaptic dopaminergic nerve terminals in Parkinson's disease, *Ann Neurol.* **47**:493–503.

Ling, Z. D., Robie, H. C., Tong, C. W., and Carvey, P. M., 1999, Both the antioxidant and D3 agonist actions of pramipexole mediate its neuroprotective actions in mesencephalic cultures. *J Pharmacol Exp Ther.* **289**:203–210.

Lieberman, A., Ranhosky, A., and Korts, D., 1997, Clinical evaluation of pramipexole in advanced Parkinson's disease: results of a double-blind, placebo-controlled, parallel-group study, *Neurology.* **49:**1162–1168.

Lieberman, A., Olanow, C. W., Sethi K., Swanson, P., Waters, C. H., Fahn, S., Hurtig, H., Yahr, M., and the Ropinirole Study Group, 1998, A multicenter trial of ropinirole as adjunct treatment for Parkinson's disease, *Neurology.* **51:**1057–1062.

Maier Hoehn, M. M., 1983, Parkinsonism treated with levodopa: progression and mortality, *J Neural Transm Suppl.* **19**:253–264.

Masserano, J. M., Gong, L., Kulaga, H., Baker, I., and Wyatt, R. J., 1996, Dopamine induces apoptotic cell death of a catecholaminergic cell line derived from the central nervous system, *Mol Pharmacol.* **50(5)**:1309–1315.

Matsumine, H., Saito, M., Shimoda-Matsubayashi, S., Tanaka, H., Ishikawa, A., Nakagawa-Hattori, Y., Yokochi, M., Kobayashi, T., Igarashi, S., Takano, H., Sanpei, K., Koike, R., Mori, H., Kondo, T., Mizutani, Y., Schaffer, A. A., Yamamura, Y., Nakamura, S., Kuzuhara, S., Tsuji, S., and Mizuno, Y., 1997, Localization of a gene for an abnormal recessive form of juvenile Parkinsonism to chromosome 6q25. 2-27, *Am J Hum Genet.* **60**:588–596.

McGeer, P. L., Itagaki, S., Boyes, B. E., and McGeer, E. G., 1988, Reactive microglia are positive for HLA-DR in the substantia nigra of Parkinson's and Alzheimer's disease brains, *Neurology.* **38**:1285–1291.

Melamed, E., Offen, D., Shirvan, A., Djaldetti, R., Barzilai, A., and Ziv, I., 1998, Levodopa toxicity and apoptosis, *Ann Neurol.* **44**:S149–154.

Mena, M. A., Davila, V., and Sulzer, D., 1997, Neurotrophic effects of L-DOPA in postnatal midbrain dopamine neuron/cortical astrocyte cocultures, *J Neurochem.* **69**:1398–1408.

Michel, P. P., and Hefti, F., 1990, Toxicity of 6-hydroxydopamine and dopamine for dopaminergic neurons in culture, *J Neurosci Res.* **26**:428–435.

Montastruc, J. L., Rascol, O., Senard, J. M., and Rascol, A., 1994, A randomised controlled study comparing bromocriptine to which levodopa was later added, with levodopa alone in previouly untreated patients with Parkinson's disease: a five year follow-up. *J Neurol Neurosurg Psychiatry.* **57**:1034–1038.

Muralikrishnan, D., and Mohanakumar, K. P., 1998, Neuroprotection by bromocriptine against 1-methyl-4-phenyl-1,2,3,6-tetrahydropyridine-induced neurotoxicity in mice, *FASEB J.* **12**:905–912.

Murer, M. G., Dziewczapolski, G., Menalled, L. B., Garcia, M. C,. Agid, Y., Gershanik, O., and Raisman-Vozari, R., 1998, Chronic levodopa is not toxic for remaining dopamine neurons, but instead promotes their recovery, in rats with moderate nigrostriatal lesions, *Ann Neurol.* **43**:561–575.

Mytilineou, C., Han, S. K., and Cohen, G., 1993, Toxic and protective effects of L-dopa on mesencephalic cell cultures, *J Neurochem.* **61**:1470–1478.

Naoi, M., Dostert, P., Yoshida, M., and Nagatsu, T., N-methylated tetrahydro-isoquinolines as dopaminergic neurotoxins, *Adv Neurol* **60**:212–217.

Nishibayashi, S., Asanuma, M., Kohno, M., Gomez-Vargas, M., and Ogawa, N., 1996, Scavenging effects of dopamine agonists on nitric oxide radicals, *J Neurochem.* **67**:2208–2211.

Noh, J. S. and Gwag, B. J., 1997, Attenuation of oxidative neuronal necrosis by a dopamine D1 agonist in mouse cortical cell cultures, *Exp Neurol.* **146**:604–608.

Offen, D., Ziv, I., Sternin, H., Melamed, E., and Hochman, A., 1996, Prevention of dopamine-induced cell death by thiol antioxidants: possible implications for treatment of Parkinson's disease, *Exp Neurol.* **141**:32–39.

Offen, D., Beart, P. M., Cheung, N. S., Pascoe, C. J., Hochman, A., Gorodin, S., Melamed, E., Bernard, R., and Bernard, O., 1998, Transgenic mice expressing human Bcl-2 in their neurons are resistant to 6-hydroxydopamine and 1-methyl-4-phenyl-1,2,3,6- tetrahydropyridine neurotoxicity, *Proc Natl Acad Sci USA.* **12**:5789–5794.

Offen, D., Ziv, I., Panet, H., Wasserman, L., Stein, R., Melamed, E., and Barzilai, A., 1997, Dopamine-induced apoptosis is inhibited in PC12 cells expressing Bcl-2. *Cell Mol Neurobiol.* **17**:289–304.

Ogawa, N., Asanuma, M., Kondo, Y., Kawada, Y, Yamamoto, M., and Mori, A., 1994a, Differential effects of chronic l-DOPA treatment on lipid peroxidation in the mouse brain with and without pretreatment with 6-hydroxydopamine, *Neurosci Lett* .**171**:55–58.

Ogawa, N., Tanaka, K., Asanuma, M., Kawai, M., Masumizu, T., Kohno, M., and Mori, A., 1994b, Bromocriptine protects mice against 6-hydroxydopamine and scavenges hydroxyl free radicals in vitro, *Brain Res.* **657**:207–213.

Olanow, C. W., Fahn, S., Muenter, M, Klawans, H., Hurtig, H., Stern, M., Shoulson, I., Kurlan, R., Grimes, J. D., Jankovic, J., Hoehn, M., Marham, C. H., Duvoisin, R., Reinmuth, O., Leonard, H. A., Ahlskog, E., Feldman, R., Hershey, L., and Yahr, M. D., 1994, A multi-center, double-blind, placebo-controlled trial of pergolide as an adjunct to sinemet in Parkinson's disease. *Mov Disord.* **9**:40–47.

Olanow, C. V., Hauser R. A., Gauger, L., Malapira, T., Koller, W., Hubble, J., Bushenbark, K., Lilienfeld, D., and Esterlitz, J., 1995, The effect of deprenyl and levodopa on the progression of Parkinson's disease. *Ann Neurol.* **38**:771–777.

Papavasiliou, P. S., Miller, S. T., Thal, L. J., Nerder, L. J., Houlihan, G., Rao, S. N., and Stevens, J. M., 1981, Age-related motor and catecholamine alterations in mice on levodopa supplemented diet, *Life Sci.* **28**:2945–2952.

Pardo, B., Mena, M. A., Fahn, S., and De Yebenes, J. G., 1993, Ascorbic acid protects against levodopa-induced neurotoxicity on a catecholamine-rich human neuroblastoma cell line. *Mov Disord.* **8:**278–284.
Pardo, B., Mena, M. A., Casarejos, M. J., Paino, C. L., and De Yebenes, J. G., 1995a, Toxic effects of L-DOPA on mesencephalic cell cultures: protection with antioxidants. *Brain Res.* **5**:133–143.
Pardo, B., Mena, M. A., and de Yebenes, J. G., 1995b, L-dopa inhibits complex IV of the electron transport chain in catecholamine-rich human neuroblastoma NB69 cells, *J Neurochem.* **64**:576–582.
Parkinson, J., 1819, An Essay on the Shaking Palsy, Sherwood, Neely and Jones, London.
Parkinson Study Group, 2000, Pramipexole vs levodopa as initial treatment for Parkinson's disease: A randomised controlled trial, *JAMA.* **284**:1931–1938.
Parkinson Study Group, 2002, Dopamine transporter brain imaging to assess the effects of pramipexole vs levodopa on Parkinson disease progression, *JAMA.* **287**:1653–1661.
Pelton, E. W. 2nd, Kimelberg, H. K., Shipherd, S. V., and Bourke, R. S., 1981, Dopamine and norepinephrine uptake and metabolism by astroglial cells in culture, *Life Sci.* **28:**1655–1663.
Perry, T. L., Young, V. W., Ito, M., Foulks, J. G., Wall, R. A., Godin, D. V., and Clavier, R. M., 1984, Nigrostriatal dopaminergic neurons remain undamaged in rats given high doses of l-dopa and carbidopa chronically, *J Neurochem.* **43**:990–993.
Piercey, M. F., Camacho-Ochoa, M., and Smith, M. W., 1995, Functional roles for dopamine-receptor subtypes, *Clin Neuropharmacol.* **18**:34–42.
Pinter, M. M., Pogarell, O., and Oertel, W. M., 1999, Efficacy, safety and tolerance of the non-ergoline agonist pramipexole in the treatment of advanced Parkinson's disease: a double-blind, placebo controlled, randomised, multicentre study, *J Neuro Neurosurg Psychiatry.* **66**:436–441.
Polymeropoulos, M. H., Higgins, J. J., Golbe, L. I., Johnson, W. G., Ide, S. E., Di Iorio, G., Sanges, G., Stenroos, E. S., Pho, L. T., Schaffer, A. A., Lazzarini, A. M., Nussbaum, R. L., and Duvoisin, R. C., 1996, Mapping of a gene for Parkinson's disease to chromosome 4q21-q23, *Science.* **274**:1197–1199.
Polymeropoulos, M. H., Lavedan, C., Leroy, E., Ide, S. E., Dehejia, A., Dutra, A., Pike, B., Root, H., Rubenstein, J., Boyer, R., Stenroos, E. S., Chandrasekharappa, S., Athanassiadou, A., Papapetropoulos, T., Johnson, W. G., Lazzarini, A. M., Duvoisin, R. C., Di Iorio, G., Golbe, L. I., and Nussbaum, R. L., 1997, Mutation in the alpha-synuclein gene identified in families with Parkinson's disease, *Science.* **276**:2045–2047.
Pothos, E. N., Przedborski, S., Davila, V., Schmitz, Y., and Sulzer, D., 1998, D2-Like dopamine autoreceptor activation reduces quantal size in PC12 cells, *J Neurosci.* **18**:5575–5585.
Przedborski, S., Jackson-Lewis, V., Muthane, U., Jiang, H., Ferreira, M, Naini, A. B., and Fahn, S., 1993, Chronic levodopa administration alters cerebral mitochondrial respiratory chain activity, *Ann Neurol.* **34**:715–723.
Przedborski, S., and Jackson-Lewis, V., 1998, Mechanisms of MPTP toxicity., *Mov Disord.* **13**:35–38.
Quinn, N., Parkes, J. D., Janota, I., and Marsden, C. D., 1986, Preservation of the substantia nigra and locus coeruleus in a patient receiving levodopa (2 kg) plus decarboxylase inhibitor over a four year period, *Mov Disord.* **1**, 65–68.
Rajput, A. H., Fenton, M., Birdi, S., and Macaulay, R., Is levodopa toxic to human substantia nigra? *Mov Disord.* **12**:634–638.
Rascol, O., Brooks, D. J., Korczyn, A. D., De Deyn, P. P., Clarke, C. E., and Lang, A. E., for the 056 Study Group, 2000, A five-year study of the incidence of dyskinesia in patients with early Parkinson's disease who were treated with ropinirole or levodopa, *N Engl J Med.* **342**:1484–1491.
Rinne, U. K., 1983, Problems associated with long-term levodopa treatment of Parkinson's disease, *Acta Neurol Scand.* **95**:19–26.
Rioux, L., Frohna, P. A., Joyce, J. N., and Schneider, J. S., 1997, The effects of chronic levodopa treatment on pre- and postsynaptic markers of dopaminergic function in striatum of parkinsonian monkeys, *Mov Disord.* **12**:148–158.
Saggu, H., Cooksey, J., Dexter, D., Wells, F. R., Lees, A., Jenner, P., and Marsden, C. D., 1989, A selective increase in particular superoxide dismutase activity in Parkinson's substantia nigra, *J Neurochem.* **53**:692–697.
Saji, L. M., Blau, A. D., and Volpe, B. T., 1996, Prevention of transneuronal degeneration of neurons in the substantia nigra reticulate by ablation of the subthalamic nucleus, *Exp Neurol.* **141**:120–129.
Sampaio-Maia, B., Serrao, M. P., and Soares-Da-Silva, P., 2001, Regulatory path-ways and uptake of l-dopa by capillary cerebral endothelial cells, astrocytes and neuronal cells, *Am J Phisiol Cell Physiol.* **280**:C333–C342.
Sanchez-Ramos, J., Overvuk, E., and Ames, B. N., 1994, A marker of oxyradical-mediated DNA damage (8-hydroxy-2'-deoxyguanosine) is increased in nigro-striatum of Parkinson's disease brain, *Neurodegeneration.* **3:**197–204.

Schapira, A. H. V., Cooper, J. M., Dexter, D., Jenner, P., Clark, J. B., and Marsden, C. D., 1989, Mitochondrial complex 1 deficiency in Parkinson's disease, Lancet. **1**:1269.
Schapira, A. H. V., 2002a, Neuroprotection and dopamine agonists, *Neurology.* **58**:S9–S18.
Schapira, A. H. V., 2002b, Dopamine agonists and neuroprotection in Parkinson's disease, *Eur J Neurol* **9**:1–6.
Schwab, R. S., Amador, L. V., and Levine, J. Y., 1951, Apomorphine in Parkinson's disease, *Trans Am Neurol Assoc.* **76**:273–279.
Schwartz, J. C., Giros, B., Martres, M. P., and Sokoloff, P., 1992. The dopamine receptor family: molecular biology and pharmacology, *Semin Neurosci.* **4**:99–108.
Semchuk, K. M., Love, E. J., and Lee, R. G., 1992, Parkinson's disease and exposure to agriculture work and pesticide chemicals, *Neurology.* **42**:1328–1335.
Sethy, V. H., Wu, H., Oostveen, J. A., and Hall, E. D., 1997, Neuroprotective effects of the dopamine agonists pramipexole and bromocriptine in 3-acetylpyridine-treated rats, *Brain Res.* **754**:181–186.
Shannon, K. M., Bennett, J. P., and Friedman, J. H., for the Pramipexole Study Group 1997, Efficacy of pramipexole, a novel dopamine agonist, as monotherapy in mild to moderate Parkinson's disease, *Neurology.* **49**:724–728.
Sian, J., Dexter, D. T., Lees, A. J., Daniel, S., Agid, Y., Javoy-Agid, F., Jenner, P., and Marsden, C. D., 1994, Alterations in glutathione levels in Parkinson's disease and other neurodegenerative disorders affecting basal ganglia, *Ann Neurol.* **36**:348–355.
Sotnikova, T. D., Gainetdinov, R. R., Grekhova, T. V., and Rayevsky, K. S., 2001, Effects of intrastriatal infusion of D2 and D3 receptor preferring antagonists on dopamine release in rat dorsal striatum (in vivo microdialysis study), *Pharmacol Res.* **43**:283–290.
Spencer, J. P. E., Jenner, A., Aruoma, O. I., Evans, P. J., Kaur, H., Dexter, D. T., Jenner, P., Lees, A. J., Marsden, D. C., and Halliwell, B., 1994, Intense oxidative DNA damage promoted by l-DOPA and its metabolites: implications for neurodegenerative disease, *FEBS Lett.* **353:**246–250.
Spencer-Smith, T., Parker, W. D. Jr, and Bennett, J. P., 1994, L-DOPA increases nigral production of hydroxyl radicals in vivo: potential l-DOPA toxicity?, *NeuroReport.* **5:**1009–1011.
Spina, M. B., and Cohen, G., 1988, Exposure of striatal synaptosomes to levodopa elevates levels of oxidized glutathione, *J Pharmacol Exp Ther.* **247**:502–507.
Stocchi, F., Nordera, G., and Marsden, C. D., 1997, Strategies for treating patients with advanced Parkinson's disease with disastrous fluctuations and dyskinesias, *Clin Neuropharmacol.* **20**:95–115.
Tanaka, M., Sotomatsu, A., Kanai, H., and Hirai, S., 1991, Dopa and dopamine cause cultured neuronal death in the presence of iron. *J Neurosci.* **101**:198–203.
Tanaka, M., Sotomatsu, A., Kanai, H., and Hirai, S., 1992, Combined histochemical and biochemical demonstration of nigral vulnerability to lipid peroxidation induced by DOPA and iron, *Neurosci Lett.* **140**:42–46.
Tanaka, K., Miyazaki, I., Fujita, N., Haque, M. E., Asanuma, M., and Ogawa, N., 2001,Molecular mechanism in activation of glutathione system by ropinirole, a selective dopamine D2 agonist, *Neurochem Res.* **26:**31–36.
Takata, K., Kitamura, Y., Kakimura, J., Kohno, Y., and Taniguchi, T., 2000, Increase of bcl-2 protein in neuronal dendritic processes of cerebral cortex and hippocampus by the antiparkinsonian drugs, talipexole and pramipexole, *Brain Res.* **872**:236–241.
Tipton, K. F., and Singer, T., 1993, Advances in our understanding of the mechanism of the neurotoxicity of MPTP and related compounds, *J Neurochem.* **61**:1191–1206.
Ubeda, A., Montesinos, C., Paya, M., and Alcaraz, M. J., 1993, Iron-reducing and free-radical scavenging properties of apomorphine and some related benzylisoquinolines. *Free Radic Biol Med.* **15**:159–167.
Weill, E., 1884, De l'apomorphine dans certains troubles nerveux. *Lyon Med.* **48**:411–419
Whone, A. L., Remy, P., Davis, M. R., Sabolek, M., Nahmias, C., A. Stoessl, J., Watts, R. L., and Brooks, D. J., 2002, The REAL-PET study: slower progression in early Parkinson's disease treated with ropinirole compared with l-dopa, *Neurology.* 58:A82–A83.
Yee, R. E., Cheng, D. W., Huang, S. C., Namavari, M., Satyamurthy, N., and Barrio, J. R., 2001, Blood-brain barrier neuronal membrane transport of 6-[18F]fluoro-L-DOPA, *Biochem Pharmacol.* **62**:1409–1415.
Yoshioka, M., Tanaka, K., Miyazaki, I., Fujita, N., Higashi, Y., Asanuma, M., and Ogawa, N., 2002, The dopamine agonist cabergoline provides neuroprotection by activation of the glutathione system and scavenging free radicals, *Neurosci Res.* **43**:259–267.
Youdim, M. B., Gassen, M., Gross, A., Mandel, S., and Grunblatt, E., 2000, Iron chelating, antioxidant and cytoprotective properties of dopamine receptor agonist; apomorphine, *J Neural Transm.* 58:83–96.
Zou, L., Jankovic, J., Rowe, D. B., Xie, W., Appel, S. H., and Le, W., 1999, Neuroprotection by pramipexole against dopamine- and levodopa-induced cytotoxicity, *Life Sci.* **64**:1275–1285.

COMT INHIBITION IN THE TREATMENT OF PARKINSON'S DISEASE: NEUROPROTECTION AND FUTURE PERSPECTIVES

Vladimir S. Kostić[1]

1. INTRODUCTION

Parkinson's disease (PD) is a progressive neurodegenerative disorder that affects about 1 to 2% of the elderly population (aged > 65 years). It remains that from the clinical point of view, PD's cardinal features are limited to four - tremor, rigidity, akinesia, and postural instability - and that customarily the unambiguous observation of at least two of the first three suffices to pose the clinical diagnosis of PD. Although the progression of PD is slow compared to other degenerative parkinsonian disorders, the annual rate of motor function decline is most rapid early in the course of the disease (during the first 4 to 9 years). This pattern of progression may be correlated with the underlying nigral pathology.[1]

Namely, among the well defined biochemical and anatomical hallmarks of PD, the most dramatic pathological feature is the profound and pervasive reduction in dopamine (DA) content of the brain. In mammals, the dopaminergic pathways of the central nervous system are organized in ascending (i.e., nigrostriatal, mesolimbic, mesocortical, and hypothalamus) and descending (i.e., to the spinal cord) projections.[2] In PD, all ascending pathways, although to different degrees, are affected, whereas it appears that the descending dopaminergic projections to the spinal cord are spared.[3] Two others hallmarks of the neuropathology of PD is the depigmentation of pars compacta of the substantia nigra (SNpc), due to the dramatic loss of nigral dopaminergic neurons that contain conspicuous amounts of pigment called neuromelanine,[4] and the presence of eosinophilic intraneuronal inclusions, called Lewy bodies, in neurons of the SNpc and of several other brain pigmented nuclei.[5] Among the different ascending dopaminergic pathways, by far the most severely affected is the nigrostriatal pathway. This pathway is composed of melanized-dopaminergic neurons whose perikarya are located in the SNpc and which project to the striatum (i.e., caudate nucleus and putamen), where they release DA. In PD, the levels of DA are dramatically reduced in all parts of the nigrostriatal pathway.[6] In addition to

[1] Vladimir S. Kostić, Institute of Neurology CCS. Dr. Subotića 6 Street, 11000 Belgrade, Serbia and Montenegro, Tel: (381)-11-685-596, Fax: (381)-11-684-577, Email: kostic@imi.bg.ac.yu

DA reduction, all other biochemical markers of the dopaminergic pre-synaptic arm of the nigrostriatal pathway are also decreased. These are the levels of DA main metabolites, homovanillic acid (HVA) and dihydroxyphenylacetic acid (DOPAC), activity of DA synthetic enzymes tyrosine hydroxylase (TH) and DOPA decarboxylase (DDC), and a number of DA reuptake sides.[7-10] Conversely, activity of two of the main enzymes metabolizing DA, monoamine oxidase-B (MAO-B) and catechol-*O*-methyl transferase (COMT), are not altered by the disease[7] which is consistent with the fact that they are both located outside dopaminergic neurons.

The general consensus is that a decrement in striatal DA greater than 70% is required before features of PD emerge.[6, 11] This finding is consistent with idea that only profound striatal DA depletion triggers the kind of alterations in the basal ganglia circuitry[12] incriminated in the development of parkinsonism.[13, 14] As previously proposed,[6] the reason for this striking "threshold" effect may lie in the existence of striatal pre- and post-synaptic compensatory mechanisms[15] that develop in sequential fashion as a function of the severity of the lesion.

The mechanisms that underlie the loss of DA-producing neurons in patients with PD are not completely understood, but two main hypotheses that have prevailed have been the toxic and genetic hypothesis.

2. TREATMENT OF PARKINSON'S DISEASE

The main therapeutic goals in PD are (a) to prevent/stop the neurodegenerative process and (b) to alleviate the symptoms of the disease. Although the idea of neuroprotection is quite appealing, so far, compelling results have only been obtained in the fight of symptoms. The basic concept underlying symptomatic control in PD is the restoration of the DA neurotransmission by either replenishing DA in the brain with the DA precursor levodopa or stimulating post-synaptic DA receptors with direct DA agonists (e.g., bromocriptine) or both. Since its introduction in the 1960s, levodopa has remained the single most effective treatment for PD and the mainstay of therapy. Although levodopa can be used as such, nowadays, it is almost exclusively used in combination with a peripheral DDC inhibitor (e.g., carbidopa, benserazide) which potentiates the effects of levodopa allowing about a fourfold reduction in dosage to obtain the same benefit. Moreover, by preventing the formation of peripheral DA, which can act at the area postrema (vomiting center), DDC inhibitors block the development of nausea and vomiting.

When levodopa therapy is initiated, the benefit from levodopa is usually sustained, with general improvement throughout the day and no dose-timing variations; this is the long-duration benefit. The pharmacokinetics of levodopa show a short initial distribution phase with a half-life of 5 to 10 minutes, a peak plasma concentration in about 30 minutes, and an elimination phase of about 90 minutes.[16] The brain levels follow plasma levels.[16] This long-duration benefit of levodopa is attributed to a combination of prolonged storage of DA from exogenous levodopa in residual nigrostriatal nerve terminals and a prolonged postsynaptic effect on DA receptors.

Levodopa is absorbed in the gastrointenstinal tract at the level of the small bowel, using the large neutral amino acid (LNAA) transport system, and then rapidly distributed to other tissues, mainly muscular, with a half-life of 5 to 10 minutes. It crosses the blood-brain barrier (BBB) via the same, LNAA transport system, competing with other plasma

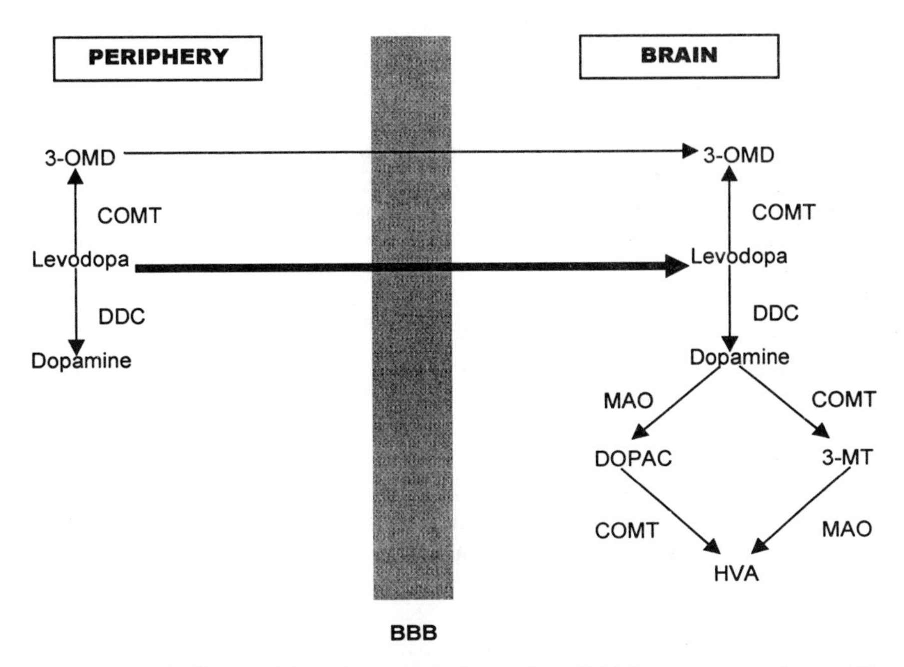

Figure 1. Metabolism of levodopa and dopamine (MAO = monoamine oxidase; COMT=catechol O-methyltransferase; HVA=homovanilic acid; DDC=dopa decarboxylase; 3-OMD=3-O-methyldopa; 3-MT=3-methoxytyramine; DOPAC=dihydroxylphenylacetaldehyde

amino acids. Peripherally, levodopa is rapidly catabolized by DDC and COMT, and is eliminated from plasma with a half-life of approximately 60 to 90 minutes. Since levodopa is co-administered with an inhibitor of DDC (benserazide or carbidopa), alternative degradation by COMT converts the bulk of levodopa to the therapeutically inactive 3-O-methyldopa (3-OMD), thereby limiting the amount of levodopa available to enter the brain. Therefore, even in the presence of a DDC inhibitor, more than 90% of a levodopa dose is metabolized in the periphery.[17] 3-OMD has a significantly longer half life than levodopa (>14 hours compared with 60 to 90 minutes, respectively)[18] and its plasma levels are usually higher than those of levodopa during long-term levodopa therapy. 3-OMD is also transported across the BBB by the LNAA transport system, and, therefore, may compete with levodopa for uptake into the brain. However, in practice, this competion appears to be without significance.[19]

The exact central mechanism of levodopa action is unknown. It is assumed that levodopa is taken up by the residual dopaminergic neurons, decarboxylated to DA by DDC in these surviving cells, and finally synaptically released to stimulate DA receptors on postsynaptic neurons. In the striatum, DA is mainly deaminated by MAO, the main levodopa-catabolizing enzyme in the brain,[20] and methylated by a central COMT (Figure 1). However, the primary way of terminating DA's effects is re-uptake in the synaptic terminals.[17] COMT in the brain converts levodopa to 3-OMD, which cannot be subsequently converted to DA. COMT is an ubiquitous enzyme that catalyses the transfer of the methyl group from the coenzyme S-adenosyl-L-methionine (SAM) to one of the hydroxyl groups of catechols in the presence of Mg^{2+}.[21] The majority of COMT is present in peripheral tissues, particularly the liver, intestinal tract, and kidneys. In the brain, where comparatively low COMT activity is found, COMT resides in astrocytic processes surrounding synapses, capillary walls, and also postsynaptic dendritic spines, while there is no such activity in presynaptic neurons;[18, 22] dopaminergic nigrostriatal neurons are entirely devoid of COMT.[23] The human COMT gene is found at chromosome 22q11.2, and consists of 6 exons.[24] Its expression is regulated by 2 distinct promoters located in exon 3. Two mRNA species with sizes of 1.5 kb and 1.3 kb are synthesized and are translated to give, respectively, membrane-bound (MB-COMT) and soluble (S-COMT) forms of COMT.[25] Most human tissues contain both types of COMT, although in the brain MB-COMT accounts for 70% of the total.

Levels of COMT activity show a trimodal distribution, described as low ($COMT^{L/L}$), intermediate ($COMT^{L/H}$), and high ($COMT^{H/H}$).[26] Therefore, the level of COMT enzyme activity is genetically polymorphic in human tissues. This polymorphism has ethnic differences and may be associated with susceptibility for development of PD in some populations,[27] as well as with influence on clinical response to levodopa in different populations.[28]

COMT may play an important role in the pathophysiology of different human disorders (PD, depression, hypertension, estrogen-induced cancers) since its substrates include catechol estrogens, indolic intermediates in melanin metabolism, xenobiotic catechols, catechol neurotransmitters (e.g., DA and noradrenaline), and drugs (e.g., levodopa).[23, 29]

2.1. Complications Of Long-Term Levodopa Treatment

Although levodopa provides antiparkinsonian benefit throughout the entire course of PD, chronic levodopa treatment is associated with adverse events that limit its utility. The

Table 1. Complications of levodopa therapy

Complications	
Motor fluctuations	
	End of dose ("wearing off")
	Delayed onset of response
	Unpredictable "on-off" phenomenon
	Dose failures
	Freezing
Dyskinesias	
	Peak-dose dyskinesia
	Diphasic dyskinesia
	"Off" period dystonia
Neuropsychiatric problems	
	Hallucinations, delirium, acute confusional state
	Depression (?)
	Behavioral disorders

longer the duration of levodopa therapy and the higher the dose, the greater the likelihood motor complications will occur.[30] After 5 years of levodopa therapy, about 75% of patients with PD have some form of troublesome complication[31] and by far the levodopa most troublesome are response fluctuations, dyskinesias, and behavioral alterations (Table 1).

Disease severity, duration of therapy and the half-life of the dopaminergic agent used to initiate therapy in PD are the variables that correlate with the development of motor fluctuations and dyskinesias best.[32, 33] One of the most parsimonious explanation for at least the wearing-off phenomenon, is the "storage hypothesis".[34] Nigrostriatal DA neurons manifest characteristic, slow (~ 4-5 Hz) single-spike activity, with only occasional interruptions by short-lasting burst of faster (~ 15-20 Hz) spiking in response to salient visual or auditory stimuli.[35] During such activity, intrasynaptic DA levels as a function of neuronal firing rate, remain fairly constant, suggesting that the most physiological approach to DA replacement in PD patients would be to maintain stable normal intrasynaptic concentrations of the transmitter.[36] In the early stages of PD it is possible to fulfill this goal with levodopa therapy, since intrasynaptic DA formed from exogenous levodopa is believed to be largely synthesized and stored in nigrostriatal DA terminals, with a relatively steady rate of release. Such tonic release ensures smooth response to levodopa and explains why the effects initially last for several hours contrasting with its short plasma half-life. With the progression of the neurodegenerative process and further loss of nigrostriatal DA terminals, more levodopa is converted to DA in other DDC-containing cells which lack the capacity to store DA or regulate its release. Now the intrasynaptic concentrations of DA almost reflect the dramatic swings in plasma and brain levodopa availability during standard, periodic dosing regimens. As a consequence, the duration of levodopa's antiparkinsonian effects progressively shortens. Consistent with this view is the demonstration in rats that severity of damage of the nigrostriatal pathway is conversely proportional to the duration of the motor response to levodopa.[37] The "storage hypothesis" suggests that the progressive loss of striatal DA terminals diminishes the normal capacity to buffer fluctuations in striatal DA and as a consequence, fluctuations in plasma levodopa are directly translated into fluctuations in striatal DA, causing DA receptors to be exposed to alternating low and high concentrations of DA (e.g., pulsatile instead of continuous, tonic stimulation).[38] On the other hand, the notion of the "storage hypothesis" as the main culprit has been challenged by the data showing that the motor response to apomorphine is qualitatively and quantitatively identical to that elicited by levodopa.[39] This short-acting DA agonist directly activates DA receptors without depending on the integrity of the pre-synaptic structures. This finding suggests that levodopa-induced motor complications result from changes in post-synaptic structures possibly related to the shift from a tonic to a non-physiological pulsatile stimulation of DA receptors. This idea is supported by the data on improvement of motor fluctuations and dyskinesias in response to continuous infusions of levodopa.[40] Evidence suggests that pulsatile striatal stimulation results in altered regulation of genes and proteins in postsynaptic, "downstream" neurons, leading to changes in neuronal firing patterns, and ultimately to motor complications.[41]

Therefore, continuous dopaminergic stimulation is becoming a central therapeutic concept in the management of PD. Among the most practical ways to achieve such a goal at present is co-administration of levodopa with COMT inhibitor.[42]

2.2. The Role Of COMT Inhibitors In The Treatment Of Parkinson's Disease

The rationale for using a COMT inihibitor is analogous to that for using a DDC inhibitor, i.e., to reduce peripheral degradation of levodopa and thereby allow more levodopa to enter the brain.[33] Since levodopa is administered in combination with DDC inhibitor (carbidopa or benserazide), methylation by COMT is the principal peripheral metabolic pathway. Therefore, COMT inhibition provides a significantly greater proportion of levodopa for entering the brain, where it is decarboxylated to DA. Two reversible COMT inhibitors are currently available, tolcapone and entacapone. Both are active in the periphery, while only tolcapone, being more lipophilic, crosses the BBB. Although previous studies, at the doses used in humans, did not prove a detectable effect on central dopaminergic neurotransmission,[43] results of a recent PET study[44] were compatible with clinical doses of tolcapone having a significant blocking effect on peripheral and central COMT, but not DDC activity in PD. In a microdialysis study in freely moving rats, Napolitano et al.[45] concluded that the blockade of central DA catabolism by tolcapone contributed to the greater increase in striatal DA levels in comparison to entacapone, following levodopa/benserazide administration. However, there are theoretical concerns about administration of a centrally active COMT inhibitor because O-methylation might be physiologically important in the brain's antioxidant defense against catecholamine autooxidation.[46]

Tolcapone and entacapone in clinically relevant doses inhibit erythrocyte COMT activity by 80-90% and 50-75%, respectively.[42] They have antiparkinsonian efficacy only when applied together with levodopa. Both drugs have similar effects on levodopa pharmacokinetics: they increase the plasma levodopa elimination half-life by approximately 50% and the area under curve (AUC; comparing 200 mg doses of each COMT inhibitor, tolcapone increased levodopa AUC by about 80% and entacapone increased it by 40%),[47] without causing a corresponding rise in either the maximal plasma concentration (C_{max}) or the time to reach maximal plasma concentration (T_{max}).[47] Therefore, COMT inhibitors improve bioavailability of levodopa in the brain (e.g., levodopa given in combination with a COMT inhibitor provides increased and smoother interdose plasma levodopa levels and more continuous brain availability compared to levodopa administration alone). In several large clinical studies in PD patients, tolcapone and entacapone have proved effective in prolonging the daily "on" time by 1-2 hours, decrease in duration of "off" time, alleviating the symptoms of motor fluctuations and potentiating the effect of levodopa.[50-52] These effects were achieved in parallel with the reduction of the daily dose of levodopa. In a recent survey of COMT inhibitors,[51] authors identified 17 studies with tolcapone and 7 studies with entacapone that met strict criteria for Level-I studies. This analysis confirmed their efficacy in improving symptomatic control in patients without and with minor motor fluctuations, as well as in the control of fully expressed motor fluctuations.

Side effects associated with the use of COMT inhibitors include gastrointestinal problems, such as diarrhea, and symptoms of hyperdopaminergic activity (nausea, vomiting, orthostatic hypotension, hallucinations, sleep disorders).[33] Abnormal liver function, with fatal outcome in rare cases,[52] has been reported in 1-3% of patients treated with tolcapone, and is rarely observed in entacapone-treated patients.[50] It has been shown that tolcapone, but not entacapone, was a potent uncoupler of oxidative phosphorilation *in vitro* at low micromolar concentrations.[53, 54]

In the review of data on COMT inhibitors in PD that confirm their symptomatic efficacy, the only conclusion regarding the possibility that either tolcapone or entacapone may have a role in prevention of PD progression or prevention of motor complications was "insufficient evidence".[51] Is there anything beyond the symptomatic effects of COMT inhibitors?

3. COMT INHIBITORS FROM THE START OF LEVODOPA THERAPY

Abnormal pulsatile stimulation of denervated DA receptors by dopaminergic drugs with the short half-life (e.g., 60 to 90 minutes for levodopa) contributes to the development of motor complications.[38] Several prospective, double blind, randomized clinical trials of long-acting DA agonists as an initial symptomatic therapy in PD patients were associated with reduced frequency of motor complications in comparison to the treatment of levodopa alone.[55] This effect appears to be a function of their long half-lives and consequent more constant and physiologic stimulation of striatal DA receptors. One of the alternative strategies to extend the short half-life of levodopa, as a critical factor that contributes to an increased risk for motor complications, is to administer levodopa in combination with a COMT inhibitor from the onset of levodopa therapy.[56] This hypothesis, clinically not substantiated, suggests that introduction of levodopa with a COMT inhibitor provides more stable levels of plasma levodopa, thereby diminishing the risk for motor complications. Recently, Jenner et al.[57] reported in the MPTP primate model of PD that this really may the case. In the fully lesioned marmosets, administration of levodopa (qid) resulted in a significant pulsatile antiparkinsonian response and a rapid induction of severe dyskinesias. In contrast, initiation of levodopa with entacapone not only significantly enhanced the antiparkinsonian response, but also significantly reduced the occurrence and severity of dyskinesias. Therefore, it was suggested that introduction of a COMT inhibitor from the start of levodopa therapy in PD may prevent the priming effect(s) underlying involuntary movements and response fluctuations.

4. DEPRESSION IN PARKINSON'S DISEASE

Depression and anxiety are the most common and frequently disabling psychiatric conditions that accompany Parkinson's disease (PD).[58] Although the exact etiology of depression in PD is unclear, available evidence suggests that biochemical changes, psychosocial factors, and situational stressors may all contribute to its development.[59]

Depression may have an impact on several aspects of PD, such as the basic parkinsonian symptomatology, response to drugs, cognition, sleep, fatigability, and appetite. Melamed[60] suggests that "in a patient with a more advanced illness who is optimally managed, a sudden motor deterioration without an obvious cause, ... , may be a result of the development of depression". On multiple regression analyses, depression predicted impaired social, role, and physical functioning in PD patients.[61] Moreover, a longitudinal study by Starkstein et al.[62] showed that those PD patients with initially associated major depression compared to non-depressed patients or those with minor depression had significantly greater cognitive decline, as well as greater deterioration in Activities of Daily Living (ADLs) scores. Therefore, Tom and Cummings[63] suggested that "treatment of depression in patients with PD may significantly slow cognitive decline, deterioration in ADLs and progression to the more advanced stages of the disease". Cognitive functions are also adversely affected by the coexistence of depression. In a prospective cohort study of non-demented patients with PD, the coexistence of depressive features was associated with a significantly greater risk of developing dementia.[64] In recent studies, depression in PD patients was the factor that explained the largest part of the experienced alterations of the quality of life.[65, 66] Interestingly enough, the most commonly "accused" physical disability in PD made only a small contribution to the decrease in the health related quality of life.[65] Therefore, to improve the quality of life in PD patients, it is necessary to make every effort for early recognition and successful treatment of depression.[66]

S-Adenosyl-L-methionine (SAM), molecule synthesized from methionine and ATP, participates as the methyl group donor to a number of metabolic events, including the COMT-dependent metabolism of catecholamines.[67] This molecule has been also associated with depression since (a) states of hypomethylation caused by folate deficiency cause or exacerbate depression, and (b) the methyl group donor property of SAM may improve the depression. In all, almost 40 clinical trials have evaluated the use of SAM in the treatment of depressive symptoms.[68] For instance, one of the more recent meta-analyses[69] concluded that SAM was superior to placebo in treating depressive disorders and approximately as effective as standard tricyclic antidepressants. In the presence of COMT, SAM methylates levodopa and DA, producing 3-OMD and 3-methoxy-tyramine, respectively. SAM levels are decreased in the cerebrospinal fluid of PD patients,[67, 70] while treatment with levodopa further depletes SAM levels in the brain[71, 72] and CSF.[73] One possibility is that coadministration of COMT inhibitors and levodopa may have antidepressant effect through an increase in DA and serotonergic mechanisms.[74] Another interesting possibility is that COMT inhibition by itself may induce increase in SAM, whose decreased levels have been linked to depression.[69] Recently, in an open-label clinical trial, DiRocco et al.[75] reported significant antidepressant efficacy of SAM (800 to 3600 mg per day for a period of 10 weeks, p.o.) in 13 depressed patients with PD, who failed to benefit or had been unable to tolerate other antidepressant drugs. Therefore, it was hypothesized that in PD patients, chronic treatment with levodopa resulted in depletion of SAM in the brain and this state of relative hypomethylation might contribute to the appearance of depressive symptoms.[67] This scenario may be prevented by the introduction of a central COMT inhibitor, suggesting that combination of levodopa with such agent may offer PD patients a broader spectrum of antiparkinsonian efficacy than levodopa alone. In accordance with this suggestion were the data on the potential antidepressant properties of tolcapone in an animal model of depression.[76] Fava et al.[77] in an open study on 21 adult patients reported that tolcapone may be a promising agent in the treatment of major depression.

5. COMT INHIBITORS: ONLY SYMPTOMATIC EFFECTS?

Several lines of evidence suggest that the SNpc in PD is the site of an oxidative stress.[78] Several powerful oxidants are produced in the course of normal metabolism, including hydrogen peroxide, superoxide, peroxyl and hydroxyl, and even nitric oxide (NO). These molecules may cause cellular damage by reacting with nucleic acids, proteins, lipids, and other molecules. Indeed, in the SNpc of parkinsonian patients, there is evidence of increased malondialdehyde and hydroperoxidase that suggests lipid peroxidation, increased carbonyl proteins suggesting oxidized proteins, increased 8-hydroxy-2-deoxyguanesine suggesting DNA degradation, elevation of iron levels, increase in γ-glutamyl transpeptidase activity and diminished reduced glutathione. The possibility that oxidative stress participates in the pathogenesis of PD offers therapeutic strategies based in the use of antioxidant agents in order to provide neuroprotection.

Dopamine metabolism by MAO-B or by autooxidation leads to the formation of hydrogen peroxide, superoxide radicals and several reactive quinones and semiquinones that could contribute to the heightened state of oxidative stress in PD.[78] Neuromelanin within dopaminergic neurons can bind ferric iron and reduce it to its reactive ferrous form.[79] Taken together, these results show that the SNpc, because of its DA and neuromelanin content, is a designated target for oxidative attack. There has been a theoretical concern that levodopa treatment might promote neuronal degeneration in PD because of its potential to generate free radicals after conversion to DA by way of oxidative metabolism. *In vivo* investigations in animals and clinical studies failed to document that levodopa had a detrimental effect on dopaminergic neurons.[80]

Still, since COMT inhibitors increase the delivery of levodopa to the brain, this theoretically might potentiate any oxidative stress induced by levodopa. Gerlach et al.[81] reported that tolcapone, but not entacapone, increased striatal hydroxyl radical production in levodopa/carbidopa treated rats. This increase was explained as a result of an increased rate of MAO-mediated and non-enzymatic (auto-oxidation) DA metabolism following increased central availability caused by reduction in COMT-mediated degradation. However, authors were not able to exclude the possibility of uncoupling mitochondrial oxidative phosphorylation. More recent study of Lyras et al.[82] measured indices of brain protein oxidation, lipid peroxidation and oxidative DNA damage, as well as the integrity of the nigrostriatal projections in normal cynomologus monkeys, and found that chronic high dose levodopa alone or in combination with entacapone did not change them.

High concentrations of levodopa are toxic *in vitro*. Offen et al.[83] showed that addition of purified COMT markedly suppressed toxicity of levodopa to PC12 cells *in vitro*, possibly through the enzymatic metabolism of levodopa to the nontoxic 3-OMD. The protective effect of glial cells against levodopa toxicity may be partly attributable to the fact that glia contain high concentrations of COMT.[84] Storch et al.[85] demonstrate that COMT inhibition attenuates levodopa toxicity toward DA neurons in primary mesencephalic cultures from the rat, but probably not by preventing 3-OMD production or cellular SAM depletion. One of the possible explanations was that COMT inhibition increased the uptake of levodopa or its metabolites from the culture medium into the cytoplasm of glial cells, leading to decrease of extracellular levels of levodopa.[86] Alterations in transmethylation reactions that lead to hyperhomocysteinemia have been suggested in the pathophysiology of neurodegenerative disorders such as PD and Alzheimer's disease (AD).[67, 87] Several groups, including ours (Table 2), reported higher plasma levels of homocysteine on PD patients.[88-94] This is in accordance with the animal data that administration of levodopa increases concentrations of plasma homocysteine and cerebral SAM.[95] Namely, since levodopa is administered with DDC to prevent its peripheral degradation, this increases conversion of levodopa to 3-OMD by COMT. COMT requires SAM as a methyl donor. Demethylation of SAM forms S-adenosylhomocysteine (SAH), which is hydrolyzed to homocysteine. Homocysteine is then metabolized via a transsulfuration pathway, forming cystathionine, or a remethylation cycle, which leads back to methionine (Figure 2).

Concentrations of SAM are decreased in PD, suggesting that transmethylation reactions in levodopa and catecholamine metabolism increase in PD.[92] Hyperhomocysteinemia in PD patients is likely to be secondary to the metabolism of levodopa to produce SAH, rather than to the disease *per se*. Recently, Rogers et al.[94] evaluated 235 patients and concluded that deficiency of folate and vitamin B_{12} levels could not explain the elevated homocysteine levels in these patients. Blandini et al.[91] found a positive correlation between homocysteine and the 3-OMD/levodopa ratio, a variable introduced as an index of the methylated catabolism of levodopa. They hypothesized that the processes of methyl-group transfer involved in the transformation of levodopa into 3-OMD might interfere with the metabolism of homocysteine, causing accumulation of the amino acid. 5, 10-Methylenetetrahydrofolate reductase (MTHFR; Figure 2) is the key rate-limiting enzyme required for the conversion of dietary folate to 5-methyltetrahydrofolate, the methyl group donor required for the remethylation of homocysteine to methionine *in vivo*. A single pair (677 C→T) substitution in the human MTHFR gene increases enzyme thermolability, reduces enzyme activity, and causes hyperhomocysteinemia.[96] Hyperhomocysteinemia might be more severe in PD patients homozygous for the C677T MTHFR (T/T) mutation,[90] but heterozygotes (C/T) also have higher homocysteine levels than those without the mutation.[97] Yasui et al.[90] observed inversed correlation of homocysteine and

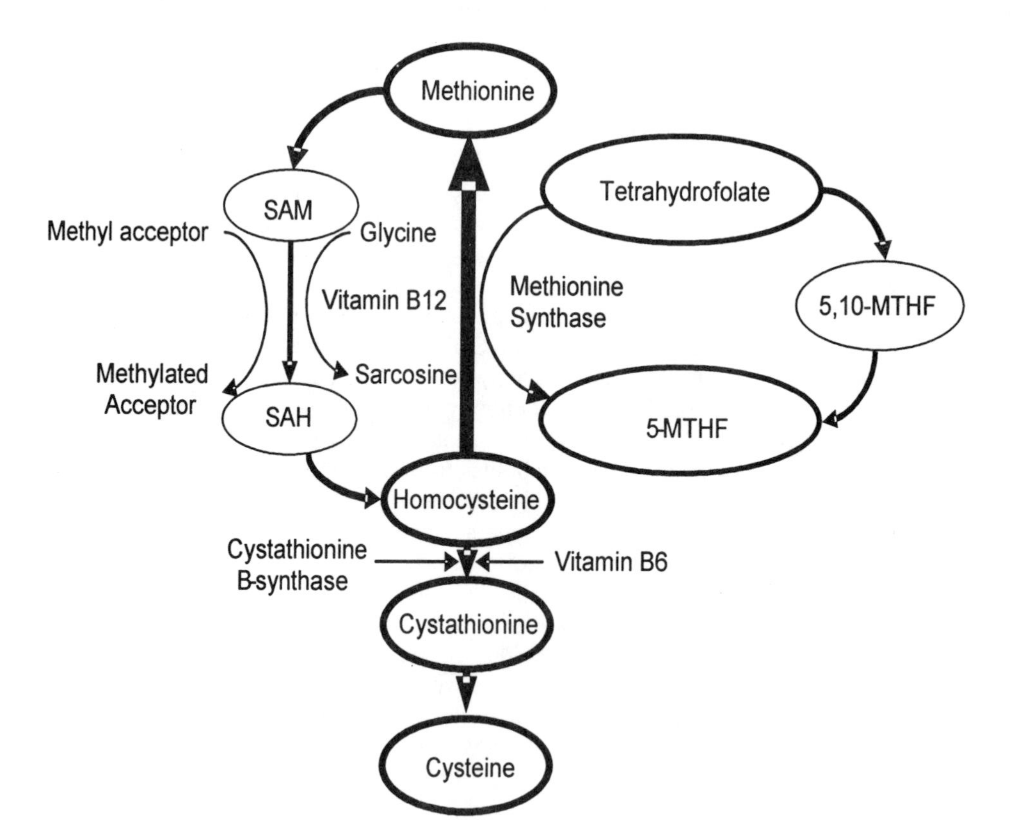

Figure 2. Methionine metabolism pathway (SAM = S-adenosylmethionine; SAH = S-adenosylhomocysteine; MTHF = methylenetetrahydrofolate)

Table 2. Levels (μmol/L) of substrates of the O-methylation cycle in patients with Parkinson's disease (PD)

Substrate	Treated PD patients (n=23)	Untreated PD patients (n=36)	Controls (n=32)
Homocysteine	15.2±4.4*	9.3±2.6	9.1±3.0
S-adenosylmethionine	0.06±0.02*	0.08±0.02	0.09±0.02
S-adenosylhomocysteine	39.8±14.6	47.3±16.4	48.9±22.2

* p<0.01

folate levels in T/T genotype patients. They suggested that increased homocysteine in PD might be related to levodopa, MTHFR genotype, and folate.

Elevated plasma levels of homocysteine increase the risk of atherosclerosis, stroke, and possibly, neurodegenerative disorders, such as PD and AD.[87, 98, 99] Besides proatherogenic and prothrombotic effects,[100] recent studies have shown that homocysteine can be directly toxic to neurons. The mechanisms may involve its prooxidant properties,[101] the activation of NMDA receptors or apoptosis triggered by DNA damage.[102] Homocysteine markedly increases the vulnerability of hippocampal neurons to excitotoxic and oxidative injury in cell culture and *in vivo.*[102] Increased levels of homocysteine in the nucleus of cells may induce DNA strand break by disturbing the DNA methylation cycle or promote the accumulation of DNA damage in neurons by impairing DNA repair.[103] Shi et al.[104] discovered that acidic homocysteine derivatives were potent agonists at several rat metabotropic glutamate receptors. Recent *in vitro* results with homocysteine and its conjugates suggest a certain impact of these compounds on nigral degeneration in PD.[105] When infused directly into the SN or striatum, homocysteine exacerbates MPTP-induced DA depletion, neuronal degeneration and motor dysfunction. Homocysteine exacerbates oxidative stress, mitochondrial dysfunction and apoptosis in human dopaminergic cells exposed to rotenone or the prooxidant Fe^{2+}. These adverse effects of homocysteine on dopaminergic cells were ameliorated by administration of the antioxidant uric acid and by an inhibitor of poly (ADP-ribose) polymerase.[106]

Rogers et al.[94] were the first who reported that levodopa-related hyperhomocysteinemia was associated with the increased risk for coronary artery disease in PD patients. However, regardless of whether increased levels of homocysteine in levodopa-treated PD patients are deleterious due to a direct toxic effects on neurons or through vascular mechanisms, one may hypothesize that decreasing of homocysteine levels may be a rational goal that can be achieved by application of COMT-inhibitors as adjunct to levodopa/DDC therapy.

6. REFERENCES

1. M. Guttman, J. Burkholder J, S.J. Kish, D. Hussey, A. Wilson, J. DaSilva, and S. Houle, [^{11}C]RTI-32 PET studies of the dopamine transporter in early dopa-naive Parkinson's disease: implications for the symptomatic threshold, *Neurology* **48**, 1578-1583 (1997).
2. A. Björklund, and O. Lindvall, Dopamine-containing systems in the CNS, in: *Handbook of chemical neuroanatomy. Classical transmitters in the CNS. Part,* edited by A. Björklund, and T. Hökfelt (Elsevier, Amsterdam, 1984), pp. 55-122.
3. B. Scatton, T. Dennis, R. L'Heureux, J.C. Monfort, C. Duyckaerts, and F. Javoy-Agid, Degeneration of noradrenergic and serotonergic but not dopaminergic neurones in the lumbar spinal cord of parkinsonian patients, *Brain Res.* 380, 181-185 (1986).

4. C.D. Marsden, Neuromelanin and Parkinson's disease. *J. Neural. Transm.* **19**, 121-141 (1983).
5. L.S. Forno, Pathology of Parkinson's disease: the importance of the substantia nigra and Lewy bodies, in: *Parkinson's disease,* edited by G..M. Stern (The Johns Hopkins University Press, Baltimore, 1990), pp. 185-238.
6. O. Hornykiewicz, and S.J. Kish, Biochemical pathophysiology of Parkinson's disease, in: *Parkinson's disease,* edited by M. Yahr, and K.J. Bergmann Raven Press, New York, 1987), pp. 19-34.
7. K.G. Lloyd, L. Davidson, and O. Hornykiewicz, The neurochemistry of Parkinson's disease: effect of Levodopa therapy, *J. Pharmacol. Exp. Ther.* **195**, 453-464 (1975).
8. T. Nagatsu, Changes of tyrosine hydroxylase in parkinsonian brains and in the brains of MPTP-treated mice, *Adv. Neurol.* **53**, 207-214 (1990).
9. X.-H. Zhong, J.W. Haycock, K. Shannak, Y. Robitaille, J. Fratkin, A.H. Koeppen, O. Hornykiewicz, and S.J. Kish, Striatal dihydroxyphenylalanine decarboxylase and tyrosine hydroxylase protein in idiopathic Parkinson's disease and dominantly inherited olivopontocerebellar atrophy, *Mov. Disord.* **10**, 10-17 (1995).
10. J.M. Wilson, A.I. Levey, A. Rajput, L. Ang, M. Guttman, K. Shannak, H.B. Niznik, O. Hornykiewicz, C. Pifl, and S.J. Kish, Differential changes in neurochemical markers of striatal dopamine nerve terminals in idiopathic Parkinson's disease, *Neurology* **47**, 718-726 (1996).
11. Y. Agid, F. Javoy-Agid, and M. Ruberg, Biochemistry of neurotransmitters in Parkinson's disease, in: *Movement Disorders 2,* edited by C.D. Marsden, and S. Fahn (Butterworths, London, 1987), pp. 166-230.
12. C.R. Gerfen, and C.J. Wilson, The basal ganglia, in: *Handbook of chemical neuroanatomy. Integrated systems of the CNS, Part III,* edited by L.W. Swanson, A. Björklund, and T. Hökfelt (Elsevier, New York, 1996), pp. 371-468.
13. M.R. DeLong, Primate models of movement disorders of basal ganglia origin, *Trends Neurosci.* **13**, 281-285 (1990).
14. R.L. Albin, A.B. Young, and J.B. Penney, The functional anatomy of disorders of the basal ganglia, *Trends Neurosci.* **18**, 63-64 (1995).
15. M.J. Zigmond, E.D. Abercrombie, T.W. Berger, A.A. Grace, and E.M. Stricker, Compensations after lesions of the central dopaminergic neurons: some clinical and basic implications, *Trends Neurosci.* **13**, 290-296 (1990).
16. G.F. Wooten, Pharmacokinetics of levodopa, in: *Movement disorders 2,* edited by C.D. Marsden, and S. Fahn (Butterworths, London, 1987), pp. 231-248.
17. U. Trendelenburg, The interaction of transport mechanisms and intracellular enzymes in metabolizing systems, *J. Neural. Transm.* **32**, 3-18 (1990).
18. P.T. Männistö, and S. Kaakkola, Catechol-O-methyltransferase (COMT): biochemistry, molecular biology, pharmacology, and clinical efficacy of the new selective COMT inhibitors, *Pharmacol. Rev.* **51**, 593-628 (1999).
19. J.G. Nutt, Effect of COMT inhibition on the pharmacokinetics and pharmacodynamics of levodopa in parkinsonian patients, *Neurology* **55(suppl 4)**, S33-S37 (2000).
20. V.Glover, Sandler M, Owen F, Dopamine is a monoamine oxidase B substrate in man. Nature 1977;265:80-81.
21. H.C. Guldberg, and C.A. Marsden, Catechol-O-methyl transferase: pharmacological aspects and physiological role, *Pharmacol. Rev.* **27**, 135-206 (1975).
22. T. Karhunen, C. Tilgmann, I. Ulmanen, and P. Panula, Catechol-O-methyltransferase (COMT) in rat brain: immunoelectron microscopic study with an antiserum against rat recombinant COMT protein, *Neurosci. Lett.* **187**, 57-60 (1995).
23. A. Kastner, P. Anglade, C. Bounaix, P. Damier, F. Javoy-Agid, N. Bromet, and Y. Agid, Immunohistochemical study of catechol-O-methyltransferase in the human mesostriatal system, *Neuroscience* **62**, 449-457 (1994).
24. M.H. Grossman, B.S. Emanuel, and M.L. Budarf, Chromosomal mapping of the human catechol-O-methyltransferase gene to 22q11.1-q11.2, *Genomics* **12**, 822-825 (1992).
25. J. Tenhunen, and I. Ulmanen, Production of rat soluble and membrane-bound catechol-O-methyltransferase forms from bifunctional mRNAs, *Biochem. J.* **296**, 595-600 (1993).
26. B. Boudikova, C. Szumlanski, B. Maidak, and R. Weinshilboum, Human liver catechol-O-methyltransferase pharmacogenetics, *Clin. Pharmacol. Ther.* **48**, 381-389 (1990).

27. H. Kunugi, S. Nanko, A. Ueki, E. Otsuka, M. Hattori, F. Hoda, H.P. Vallada, M.J. Arranz, and D.A.Collier, High and low activity alleles of catechol-O-methyltransferase gene: ethnic difference and possible association with Parkinson's disease, *Neurosci. Lett.* **221**, 202-204 (1997).
28. A. Yoritaka, N. Hattori, H. Yoshino, and Y. Mizuno, Catechol-O-methyltransferase genotype and susceptibility to Parkinson's disease in Japan, *J. Neural. Transm.* **104**, 1313-1317 (1997).
29. E.L. Cavalieri, D.E. Stack, P.D. Devanesan, R. Todorovic, I. Dwivedy, S. Higginbotham, S.L. Johansson, K.D. Patil, M.L. Gross, J.K. Gooden, R. Ramanathan, R.L. Cerny, and E.G. Rogaen, Molecular origin of cancer: catechol estrogen-3,4-quinones as endogenous tumor initiators, *Proc. Natl. Acad. Sci. USA* **94**, 10937-10942 (1997).
30. S. Fahn, Adverse effects of levodopa in Parkinson's disease, in: *Handbook of experimental pharmacology, vol. 8*, edited by D.B. Calne (Springer-Verlag, Berlin, 1989), pp. 386-409.
31. V.S. Kostić, S. Przedborski, E. Flaster, and N. Šternić, Early development of levodopa-induced dyskinesias and response fluctuations in young-onset Parkinson's disease, *Neurology* **41**, 202-205 (1991).
32. V.S. Kostić, J. Marinković, M. Svetel, E. Stefanova, and S. Przedborski, The effect of stage of Parkinson's disease at the onset of levodopa therapy on development of motor complications, *Eur. J. Neurol.* **9**, 9-14 (2002).
33. J.G. Nutt, J.H. Carter, E.S. Lea, and G.J.Sexton, Evolution of the response to levodopa during the first 4 years of therapy, *Ann. Neurology* **51**, 686-693 (2000).
34. M.M. Mouradian, J.L. Juncos, G. Fabbrini, and T.N. Chase, Motor fluctuations in Parkinson's disease: pathogenetic and therapeutic studies, *Ann. Neurol.* **22**, 475-479 (1987).
35. W. Schultz, Behavior-related activity of primate dopamine neurons, *Rev. Neurol. (Paris)* **150**, 634-639 (1994).
36. T.N. Chase, and J.D. Oh, Striatal mechanisms and pathogenesis of parkinsonian signs and motor complications, *Ann. Neurol.* **47**, S122-S129 (2000).
37. S.M. Papa, T.M. Engber, A.M. Kask, and T.N. Chase, Motor fluctuations in levodopa treated parkinsonian rats: Relation to lesion extent and treatment duration, *Brain Res.* **662**, 69-74 (1994).
38. W.C. Koller, Levodopa in the treatment of Parkinson's disease, *Neurology* **55(suppl 4)**, S2-S7 (2000).
39. C. Colosimo, M. Merello, A.J. Hughes, K. Sieradzan, and A.J. Lees, Motor response to acute dopaminergic challenge with apomorphine and levodopa in Parkinson's disease: implications for the pathogenesis of the on-off phenomenon, *J. Neurol. Neurosurg. Psychiatry* **60**,634-637 (1996).
40. A.E. Lang, and A.M. Lozano, Parkinson's disease - Second of two parts, *N. Engl. J. Med.* **339**, 1130-1143 (1998).
41. T.N. Chase, The significance of continuous dopaminergic stimulation in the treatment of Parkinson's disease, *Drugs* **55(suppl 1)**, 1-9 (1998).
42. J. Dingemanse, Issues important for rational COMT inhibition, *Neurology* **55(suppl 4)**, S24-S27 (2000).
43. M. Huotari, R. Gainetdinov, and P.T. Männistö, Microdialysis studies on the action of tolcapone on pharmacologically-elevated extracellular dopamine levels in conscious rats, *Pharmacol. Toxicol.* **85**, 233-238 (1999).
44. R. Ceravolo, P. Piccini, D.L. Bailey, K.M. Jorga, H. Bryson, and D.J. Brooks, ^{18}F-Dopa PET evidence that tolcapone acts as a central COMT inhibitor in Parkinson's disease, *Synapse* **43**, 201-207 (2002).
45. A. Napolitano, G. Bellini, E. Borroni, G. Zurcher, and U. Bonuccelli, Effects of peripheral and central catechol-O-methyltransferase inhibition on striatal ectracellular levels of dopamine: a microdialysis study in freely moveing rats, *Parkinsonism. Relat. Dis.* **9**, 145-150 (2003).
46. W. Kuhn, D. Woitalla, M. Gerlach, H. Russ, and T. Muller, Tolcapone and neurotoxicity in Parkinson's disease, *Lancet* **352**, 1313-1314 (1998).
47. K.M. Jorga, COMT inhibitors: pharmacokinetic and pharmacodynamic comparisons, *Clin. Neuropharmacol.* **21(suppl 1)**, S9-S16 (1998).
48. M.C. Kurth, C.H. Adler, M.S. Hilaire, C. Singer, C. Waters, P. LeWitt, D.A. Chernik, E.E. Dorflinger, and K. Yoo, Tolcapone improves motor function and reduces levodopa requirement in patients with Parkinson's disease experiencing motor fluctuations: a multicenter, double-blind, randomized, placebo-controlled trial, *Neurology* **48**, 81-87 (1997).

49. A.H. Rajput, W. Martin, M.G. Saint-Hilaire, E. Dorflinger, and S. Pedder, Tolcapone improves motor function in parkinsonian patients with the "wearing-off" phenomenon: a double-blind, placebo-controlled, multicenter trial, *Neurology* **49**, 1066-1071 (1997).
50. U.K. Rinne, J.P. Larsen, A. Siden, and J. Worm-Petersen, Entacapone enhances the response to levodopa in parkinsonian patients with motor fluctuations, *Neurology* **51**, 1309-1314 (1998).
51. Group of authors, COMT inhibitors, *Mov. Disord.* **17(suppl 4)**, S45-S51 (2002).
52. F. Assal, L. Spahr, A. Hadengue, L. Rubbici-Brandt, P.R. Burkhardt, Tolcapone and fulminant hepatitis, *Lancet* **352**, 958 (1998).
53. E. Nissinen, P. Kaheinen, K. Pentillä, J. Kaivola, I.B. Linden, Entacapone, a novel catechol-O-methyl transferase inhibitor of Parkinson's disease, does not impair mitochondrial energy production, *Eur. J. Pharmacol.* **340**, 287-294 (1997).
54. K. Haasio, K. Lounatmaa, and A. Sukura, Entcapone does nor induce conformational changes in liver mitochondria or skeletal muscle *in vivo, Exp. Toxic. Pathol.* **54**, 9-14 (2002).
55. C.W. Olanow, The role of dopamine agonists in the treatment of early Parkinson's disease, *Neurology* **58(suppl 1)**, S33-S41 (2002).
56. W. Olanow, and J.A. Obeso, Pulsatile stimulation of dopamine receptors and levodopa-induced motor complications in Parkinson's disease: implications for the early use of COMT inhibitors, *Neurology* **55(Suppl 4)**, S72-S77 (2000).
57. P. Jenner, G. Al-Bargouthy, L. Smith, M. Kuoppamaki, M. Jackson, S. Rose, and W. Olanow, Initiation of entacapone with L-dopa further improves antiparkinsonian activity and avoids dyskinesia in the MPTP primate model of Parkinson's disease, *Neurology* **58(Suppl 3)**, A374 (2002).
58. V.S. Kostic, S.R. Filipovic, D. Lecic, D. Momcilovic, D. Sokic, and N. Sternic, Effect of age at onset on frequency of depression in Parkinson's disease, *J. Neurol. Neurosurg. Psychiatry* **57**, 1265-1267 (1994).
59. V.S. Kostić, B.M. Djuričić, N. Šternić, Lj. Bumbaširević, M. Nikolić, and B.B. Mršulja, Depression and Parkinson's disease: possible role of serotonergic mechanisms, *J. Neurol.* **234**, 94-96 (1987).
60. E. Melamed, Neurobehavioral abnormalities in Parkinson's disease, in: *Movement disorders: neurologic principles and practice*, edited by R.L. Watts, and W.C. Koller (McGraw-Hill, New York, 1997) pp. 257-262.
61. S.A. Cole, J.L. Woodard, J.L. Juncos, J.L. Kogos, E.A. Youngstrom, and R.L. Watts, Depression and disability in Parkinson's disease, *J. Neuropsychiatry Clin. Neurosci.* **8**, 20-25 (1996).
62. S.E. Starkstein, H.S. Mayberg, R. Leiguarda, T.J. Preziosi, and R.G. Robinson, A prospective longitudinal study of depression, cognitive decline, and physical impairments in patients with Parkinson's disease, *J. Neurol. Neurosurg. Psychiatry.* **55**, 377-382 (1992).
63. T. Tom, and J.L. Cummings, Depression in Parkinson's disease: pharmacological characteristics and treatment, *Drugs & Aging* **12**, 55-74 (1998).
64. K. Marder, M.X. Tang, L. Cote, Y. Stern, and R. Mayeux, The frequency and associated risk factors for dementia in patients with Parkinson's disease, *Arch. Neurol.* **52**, 695-701 (1995).
65. K.H. Karlsen, J.P. Larsen, E. Tandberg, and J.G. Maeland, Influence of clinical and demographic variables on quality of life in patients with Parkinson's disease, *J. Neurol. Neurosurg. Psychiatry.* **66**, 431-435 (1999).
66. A.-M. Kuopio, R.J. Martilla, H. Helenius, M. Toivonen, and U.K. Rinne, The quality of life in Parkinson's disease, *Mov. Disord.* **15**, 216-223 (2000).
67. T. Bottiglieri, and K. Hyland, S-adenosyl-methionine levels in psychiatric and neurologic disorders, *Acta Neurol Scand* **154(suppl)**, 19-26 (1994).
68. C.W. Fetrow, and J.R. Avila, Efficacy of the dietary supplement S-adenosyl-L-methionine, *Ann. Pharmacother.* **35**, 1414-1425 (2001).
69. G.M. Bressa, S-adenosyl-L-methionine as antidepressant: a meta analysis of clinical studies, *Acta Neurol. Scand.* **154(suppl)**, 7-14 (1994).
70. M. Da Prada, J. Borgulya, A. Napolitano, and G. Zucher, Improved therapy for Parkinson's disease with tolcapone: a central and peripheral COMT inhibitor with an S-adenosylmethionine sparing effect, *Clin. Neuropharmacol.* **17(suppl3)**, 26-27 (1994).
71. R.J. Wurtman, S. Rose, S. Matthyse, J. Stephenson, and R. Baldessarini, L-dihydroxyphenilalanine: effect on S-adenosyl-methionine in the brain, *Science* **169**, 395-397 (1970).
72. R. Surtees, K. Hyland, L-dihydroxyphenilalanine (levodopa) lowers central nervous system S-adenosylmethionine concentrations in humans, *J. Neurol. Neurosurg. Psychiatry* **53**, 569-572 (1990).

73. A. Stock, S. Clarke, C. Clarke, and J. Stock, N-terminal methylation of proteins: structure, function and specificity, *FEBS Lett.* **220,** 8-14 (1987).
74. I. Bellido, A. Gomez-Luque, A. Plaza, F. Ruiz, P. Ortiz, and F. Sanchez de la Cuesta, S-Adenosyl-L-methionine prevents 5-HT(1°) receptors up-regulation induced by acute imipramine in the frontal cortex of the rat, *Neurosci. Lett.* **321,** 110-114 (2002).
75. A. Di Rocco, J.D. Rogers, R. Brown, P. Werner, and T. Bottiglieri, S-Adenosyl-methionine improves depression in patients with Parkinson's disease in an open-label clinical trial, *Mov. Disord.* **15,** 1225-1229 (2000).
76. J.L. Moreau, J. Borgulya, F. Jenck, and J.R. Martin, Tolcapone: a potential new antidepressant detected in a novel animal model of depression, *Behav. Pharmacol.* **5,** 344-350 (1994).
77. M. Fava, J.F. Rosenbaum, A.R. Kolsky, J.E. Alpert, A.A. Nierenberg, M. Spillmann, P. Rensshaw, T. Bottiglieri, G. Moroz, and G. Magni, Open study of the catechol-O-methyltransferase inhibitor tolacapone in major depressive disorder, *J. Clin. Psychopharmacol.* **19,** 329-335 (1999).
78. S. Przedborski, and V. Jackson-Lewis, Experimental developments in movement disorders: update on proposed free radical mechanisms, *Curr. Opin. Neurol.* **11,** 335-339 (1998).
79. H.M. Swartz, T. Sarna, and L. Zecca, Modulation by neuromelanin of the availability and reactivity of metal ions, *Ann. Neurol.* **32 (Suppl.),** S69-S75 (1992).
80. Y. Agid, E. Ahlskog, A. Albanese A, D. Calne, T. Chase, J. De Yebenes, S. Factor, S. Fahn, O. Gershanik, C. Goetz, W. Koller, M. Kurth, A. Lang, A. Lees, C.D. Marsden, E. Melamed, P.P. Michel, Y. Mizuno, J. Obeso, W. Oertel, W. Olanow, W. Poewe, Pollak P, and E. Tolosa, Levodopa in the treatment of Parkinson's disease: a consensus meeting, *Mov. Disord.* **14,** 911-913 (1999).
81. M. Gerlach, A.Y. Xiao, W. Kuhn, R. Lehnfeld, P. Waldmeier, K.H. Sontag, and P. Riederer, The central catechol-O-methyltransferase inhibitor tolcapone increases striatal hydroxyl radical production in L-dopa/carbidopa treated rats, *J. Neural. Transm.* **108,** 189-204 (2001).
82. L. Lyras, B.-Y. Zeng, G. McKenzie, R.K.B. Pearce, B. Halliwell, and P. Jenner, Chronic high dose L-dopa alone or in combination with the COMT inhibitor entacapone does not increase oxidative damage or impair the function of the nigro-striatal pathway in normal cynomologus monkeys, *J. Neural. Transm.* **109,** 53-67 (2002).
83. D. Offen, H. Panet, R. Galili-Mosberg, E. Melamed, Catechol-O-methyltransferase decreases levodopa toxicity *in vitro, Clin. Neuropharmacol.* **24,** 27-30 (2001).
84. E. Hansson, Enzymatic activities of monoamine oxidase, catechol-O-methyltransferase and gamma-aminobutyric acid transaminase in primary astroglial cultures and adult rat brain from different brain regions, *Neurochem. Res.* **9,** 45-57 (1984).
85. A. Storch, H. Blessing, M. Bareiss, S. Jankowski, Z.D. Ling, P. Carvey, and J. Schwarz, Catechol-O-methyltransferase inhibition attenuates levodopa toxicity in mesencephalic dopamine neurons, *Mol. Pharm.* **57,** 589-594 (2002).
86. H. Blessing, M. Bareiss, H. Zettlmeisl, J. Schwarz, and A. Storch, Catechol-O-methyltransferase inhibition protects against 3,4-dihydroxyphenylalanine (DOPA) toxicity in primary mesencephalic cultures: new insights into levodopa toxicity, *Neurochem. Int.* **42,** 139-151 (2003).
87. S. Seshadri, A. Beiser, J, Selhub, P.F. Jacques, I.H. Rosenberg, R.B. D'Agostino, P.W.F. Wilson, and P.A. Wolf, Plasma homocysteine as a risk factor for dementia and Alzheimer's disease, *N. Engl. J. Med.* **345,** 476-483 (2002).
88. P. Allain, A. Le Bouil, E. Cordillet, L, Le Quay, H, Bagheri, and J.L. Montastruc, Sulfate and cysteine levels in the plasma of patients with Parkinson's disease, *Neurotoxicol.* **16,** 527-529 (1995).
89. W. Kuhn, R. Roebroek, H. Blom, D. van Oppenraaij, and T. Muller, Hyperhomocysteinemia in Parkinson's disease, *J. Neurol.* **245,** 811-812 (1998)
90. K. Yasui, H. Kowa, K, Nakaso, T. Takeshima, and K. Nakashima, Plasma homocysteine and MTHFR C677T genotype on levodopa-treated patients with Parkinson's disease, *Neurology* **55,** 437-440 (2000).
91. F. Blandini, R. Fancellu, E. Mortignoni, A. Mangiagalli, C. Pacchetti, A. Samuele, and G. Nappi, Plasma homocysteine and L-dopa metabolism in patients with Parkinson disease, *Clin. Chem.* **47,** 1102-1104 (2001).
92. T. Muller, D. Woitalla, B. Hauptmann, B. Fowler, and W. Kuhn, Decrease in methionine and S-adenosylmethionine and increase of homocysteine in treated patients with Parkinson's disease, *Neurosci. Lett.* **308,** 54-56 (2001).
93. T. Muller, D. Woitalla, B. Fowler, and W. Kuhn, 3-OMD and homocysteine plasma levels in parkinsonian patients. *J. Neural. Transm.* **109,** 175-179 (2002).

94. J.D. Rogers, A. Sanchez-Saffon, A.B. Frol, and R. Diaz-Arastia, Elevated plasma homocysteine levels in patients with levodopa: association with vascular disease, *Arch. Neurol.* **60**, 59-64 (2003).
95. X.X. Liu, K. Wilson, and C.G. Charlton, Effects of L-dopa treatment on methylation in mouse brain: implications for the side effects of L-dopa. *Life Sci.* **66**, 2277-2288 (2000).
96. P. Frosst, H.J. Blom , R. Milos, P. Goyette, C.A. Sheppard, R.G. Matthews, G.J. Boers, M. Den Heujer, L.A. Kluijtmans, and L.P. van den Heuvel, A candidate geneticd risk factor for vascular disease: a common mutation in methylentetrahydrofolate reductase, *Nat. **Genet.*** **10,** 111-113 (1995).
97. W. Kuhn, T. Hummel, D. Woitalla, and T. Muller, Plasma homocysteine and MTHFR C667T genotype in levodopa-treated patients with PD (letter), *Neurology* **56**, 281 (2001).
98. D.S. Wald, M. Law, and J.K. Morris. Homocysteine and cardiovascular disease evidence on causality from a meta-analysis, *Br. Med. J.* **325**, 1202-1206 (2002).
99. S.E. Vermeer, T. Den Heijer, P.J. Koudstaal, M. Oudkerk, A. Hofman, and M.M. Breteler, Incidence and risk factors of silent brain infarcts in the population-based Rotterdam scan study, *Stroke* **34**, 137-146 (2003).
100. T.G. Deloughery, Hyperhomocysteinemia in ischemic stroke, *Sem. Cerebrovasc. Dis. Stroke* **2**, 111-119 (2002).
101. G. Blundell, B.G. Jones, F.A. Rose, and N. Tudball, Homocysteine mediated endothelial cell toxicity and its amelioration, *Atherosclerosis* **122**, 163-172 (1996).
102. I.I. Kruman, C. Culmsee, S.L. Chan, Y. Kruman, Z. Guo, L. Penix, and M.P. Mattson, Homocysteine elicits a DNA damage response in neurons that promotes apoptosis and hypersensitivity to excitotoxicity, *J. Neurosci.* **20**, 6920-6926 (2000).
103. I.I. Kruman, T.S. Kumaravel, A. Lohani, W.A. Pedersen, R.G. Cutler, Y. Kruman, N. Haughey, J. Lee, M. Evans, and M.P. Mattson, Folic acid deficiency and homocysteine impair DNA repair in hippocampal neurons and sensitize them to amyloid toxicity in experimental model of Alzheimer's disease, *J. Neurosci.* **22**, 1752-1762 (2002).
104. Q. Shi, J. Savage, S. Hufesein, L. Rauser, E. Grajkowski, P. Ernsberger, J. Wroblewski, J. Nadeau, and B.L. Roth, L-homocysteine sulfinic acid and other acidic homocysteine derivatives are potent and selective metabotropic glutamate receptor agonists, *J. Pharmacol. Exp. Ther.* **21**, 2344-2348 (2003).
105. T.J. Montine, V. Amarnath, M.J. Picklo, K.R. Sidell, J. Zhang, and D.G. Graham, Dopamine mercapturate can augment dopaminergic neurodegeneration, *Drugs Metabol. Rev.* **32**, 363-376 (2000).
106. W. Duan, B. Ladenheim, R.G. Cutler, I.I. Kruman, J.L. Cadet, and M.P. Mattson, Diatery folate deficiency and elevated homocysteine levels endanger dopaminergic neurons in models of Parkinson's disease, *J. Neurosci.* **80**, 101-110 (2002).

NEUROPROTECTION AND EPILEPSY

Péter Halász and György Rásonyi[1]

1. INTRODUCTION

Epilepsy is a brain disorder affecting 0,5 - 1,0% of the population, characterised by recurrent seizures. Seizures are the result of excessive discharges of neo-or archi-cortical neurons firing in an abnormal synchrony. The symptomes and consequentes of seizures are determined by the function of the brain region from which the abnormal discharge originates, by the degree of spread to other structures of the brain, and by the quantity and ratio of excitatory and inhibitory neurons participating. Seizures are only the tip of the iceberg represented by the disease process itself about which a growing body of knowledge has become available.

During the last years it has become obvious that the current way of treating epilepsy with antiepileptic (AE) drugs is insufficient concerning the modification of the underlying disesease, AEDs provide merely a symptomatic treatment, without a clear influence on the course of the disease (Walker et al., 2002). There is a pressing need to find alternative strategies and to find possibilities to intervene either into the basic processes determining the development of epilepsies or to promote compensatory processes in repairing these dysfunctions.

---

1 Péter Halász and György Rásonyi, National Institute of Psychiatry and Neurology, Epilepsy centre, Budapest, Hűvösvölgyi út 116. H-1021 Hungary.

In neurology the concept of neuroprotection has been developed in the context of cerebrovascular and neurodegenerative diseases (Vajda, 2002) The increasing knowledge about the basic neuronal changes underlying epilepsies allows now to analyse the potential role of neuroprotective agents in epilepsy.

Neuroprotection means protection against cell death, consequently provides remedy against neuronal damage and its consequences. In epilepsy the most frequent constellation is the presence of damage and overexcitation together Increase in excitability may develop after a primary damage as in posttraumatic epilepsy and outburst of epileptic excitability may cause neuronal damage as in cell loss after status epilepticus (De Giorgio et al., 1992) as well as in cases of the so called cytotoxic damages due to extensive glutamatergic involvement (Lipton and Rosenberg, 1994). However, epilepsy may also develop without neuronal damage. The best examples are the primary generalised epilepsies where the epileptic changes in the thalamocortical system develop without any previous damage (Timofeev et al., 1998). An epileptic disorder may develop also in the mesiotemporal structures merely due to an increase of the excitatory input via the repeated entorhinal stimulation of the hippocampus or in the well known kindling model (McNamara et al., 1997). During the course of the epileptic illness repeated epileptic seizures are frequently associated with neuronal damage either by the excytotoxic mechanism (Ben Ari et al., 1980) or by secondary hypoxic and traumatic lesions Therefore although neuroprotection works against only one aspect of a complex cascade of pathological events, it may be a promising option in different steps during the development and course of the epilepsies.

2. CASCADE OF EVENTS LEADING TO CHRONIC PARTIAL EPILEPSY

In he natural course of partial symptomatic epilepsies show certain distinct milestones can be observed in the development of the epileptic disorder. In the history of such epilepsies an initial precipitatory incident (IPI) was frequently found. The most frequent IPIs are complicated febrile convulsions, status epilepticus in TLE and traumatic lesions or hypoxic-iscaemic insults in extratemporal epilepsies. Seizures appear after a long silent period during that a "ripening process" takes place. Thus temporal lobe epilepsy develope in an average of 7.5 years after the initial insult and 16% of patients has a latent period of more than 10 years (French et al., 1993). Similarly the risk of epilepsy exceeds that of the general population for more than 10 years after a serious head injury (Annegers et al., 1998).

An other variation is that after the initial insult early seizures appear within weeks but the late epilepsy develops only after a silent period. The pathophysiology and prognostic significance of the early seizures is not clear enough. In posttraumatic and poststroke epilepsy they obviously signal a transient hyperexcitability forecasting the further changes in the brain associated with the chronic epileptic disorder. However the trials to prevent late consequences with immediate antiepileptic treatment were not successful (Temkin, 2001).

Some workers assume that for the development of chronic epilepsy after an IPI a "second hit" is needed (Walker et al., 2002). It is possible, that even the first seizures after a silent period do refer to a fully developed epileptic process. Probably the seizures themselves pave the way in the build up of the changes underlying the permanently recurring seizures. It is probable that a vitious circle between seizures and the seizure promoting condition is set into motion. According the study of Kwan and Brodie (2000) 60% of those who suffer a first seizure a second one occur, and after treatment with antiepileptic drugs the seizures persist in 15%.

3. POSSIBLE ROLE OF NEUROPROTECTION DURING DIFFERENT STEPS OF THE EPILEPTIC CASCADE

What is the role of neuroprotection in the above delineated evolutional cascade of partial epilepsies? Not all preventive actions are neuroprotective.

The first possibility to prevent the epileptic cascade lies in decreasing of perinatal damages and preventing early status epilepticus as well as complicated febrile seizures. We have no tools up to this moment to prevent the development of synaptic reorganisation leading to chronic epilepsy. The administration of the traditional antiepileptic drugs is not able to prevent epileptogenesis, even when they are effective against the early epileptic excitability expressed in early seizures (Temkin, 2001). The traditional antiepileptic drugs working against seizures could prevent only that part of the epileptogenic cascade where seizure begets seizures. One of the known mechanisms at this point is protection against the excitotoxic process and the consequent apoptosis. The most characteristic feature of the synaptic reorganisation occuring after the epileptogenic insult is proliferation (Sutula et al., 1989) Some of the neural factors promoting proliferation are already known (Repressa, 1993a; b; Niquet et al., 1994; 1995). However we do not know whether the the whole reorganisatory process is really harmful and unanimously promote epileptogenezis or in the contrary there are compensatory events helping to reach control over the overwhelming excitatory processes (Walker et al., 2002).

Nowadays we have several animal models to study the development of different aspects of the epileptic cascade.

Kindling (Goddard, 1967): Repeated low-intensity, rare electrical stimulation (at the beginning responseless) elicits local electrical, later bahavioral epileptic response of an increasing intensity. In this model epileptic hyperexcitation develops without any previous neuronal loss, but neuronal damage may occur as a consequence of overexcitation via excitotoxic mechanism (Tuunanen and Pitkänen, 2000).

Kainic acid model (Ben Ari, 1985; Babb et al., 1995): Local hippocampal administration exerts excitotoxic effect through EAA receptors. Acute phase: status epilepticus like state (10-30 days) Active phase: frequent seizures, latent period (1-3 months) chronic phase with recurrent seizures (3-8 months). Hippocampal cell loss + mossy fiber sprouting.

Pilocarpin model (Cavalheiro et al., 1991; Turski et al., 1987): Acute effect: status epilepticus (24 hours), silent period (4-44 days, in average 2 weeks recurrent seizures (2-15/months) at least for 6 months.

Characteristic cell loss among the hippocampal pyramidal neurons + synaptic reorganisation with mossy fiber sprouting (from the 4th day- max. at the 100th day).

The latter two are chronic mesial temporal lobe epilepsy models closer to the evolution of human epileptic cascades after an early status epilepticus.

Experimental epilepsy developing after local application of FeCl2 (Willmore, 1978) represents a model of the posttarumatic epilepsy with haemorrhagic complication. The early "isolated cortex" type experimental epilepsies may serve as models for subcortical epileptogenic lesions undercutting the cortex from subcortical and horizontal connections. The hypoxic newborn rat model (Jensen, 1992) mimics the human infantile hypoxic–ischaemic epileptogenic insult.

Epilepsy is not always (but in certain forms) a progressive disease. The factors determining the progressive course of epilepsy and those playing role in the possibe prevention of the process is obviously an overlaping field with neuroprotection (Figure 1).

There are four types of events suggesting an escalation of chronic epilepsy. The first is the shortening of intervals between seizures. It is not always the result of some kind of evolution and has complicated ramifications to the field of pharmacoresponsivity. The second is the evolution of the so called "therapy resistence" or "drug refractory" state. The third is a more complex one without animal model. It is the increase of functional deficit signs in EEG and/or neuropsychological and PET studies. The fourth one is the spacial escalation of the epileptic excitability in the form of secondary epileptogenesis and secondary bilateral synchronisation. The phenomenology of these events is worked out mainly in the EEG dimension but certain parallels were found also in clinical symptoms. Recently an evidence of progressive atrophy was found in the ipsilateral hippocampal structure (Fuerst, 2003) or in other brain regions (Liu, 2003) detected by MRI in a group of therapy-resistant patients.

We do not know what is the underlying process in certain progressive mesial temporal lobe epilepsies. Is it determined by the ongoing local reorganisation (or the basic causal disease, if it exists at al?), or is it the consequence of the overt or subclinical seizures? Is it progressing in a "saltatory" way promoted by clusters of seizures? What is the role of genetic factors?

However the progressive course seem to be confined to some epilepsies where coincidence of several unfavorable factors determines the outcome and concerning the majority of epilepsies there are strong arguments in favor of the non-progressive course. Table 1 summarises the arguments for and against progressivity of epileptic disorders.

Table 1. Pro-s and Con-s of epilepsy as a progressive diseases.

Supports progression	Supports stable state
Seizures cause damage in animal models	Frequent seizures may be associated with good prognosis (BCTE)
The duration of epilepsy is correlating with hippocampal damage (Tasch, 1999., Theodore, 1999., Fuerst, 2001., Liu, 2003)	Prognosis is determined more by the etiology than seizure frequency (Semah, 1998., Berg and Shinnar, 1997)
Greater seizure number correlate with worse prognosis in hospital based studies (Reynolds, 1987)	The long lasting untreated epilepsies respond to drug as well as the new ones (Feksi et al., 1991, Placencia et al., 1993,1994)
	Early treatment does not improve the outcome (Musicco et al., 1997)

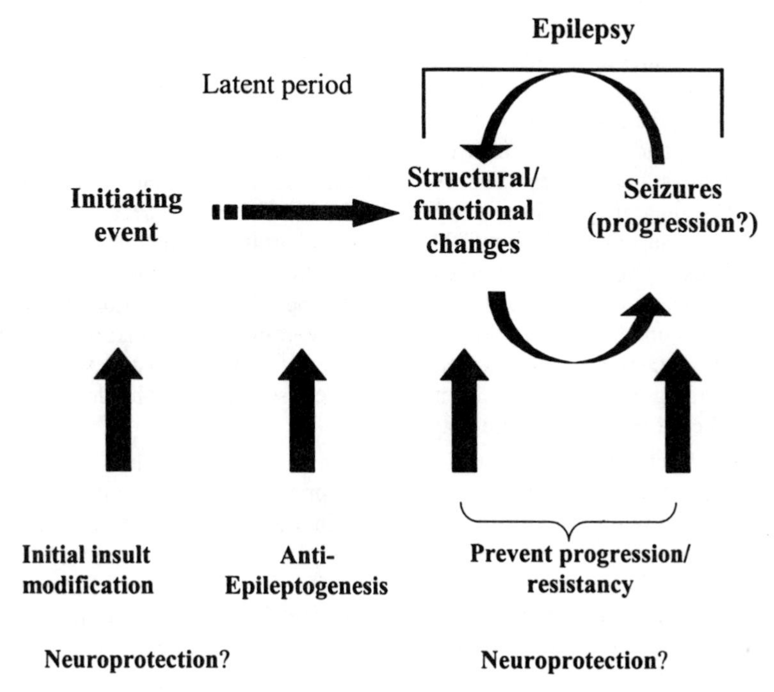

Figure 1. The possible timing of neuroprotection is pointed out here within the developmental scheme of parcial epilepsies beginning with a latency period after an initial event as a result of epileptogenic structural reorganisation and progressing further by a vicious circle caused by the seizures themselves (modified after Walker *et al.* 2002).

4. SEIZURE SUPRESSING VERSUS ANTIEPILEPTOGENIC AND NEUROPROTECTIVE EFFECT – EXPLORING THE AVAILABLE ANTIEPILEPTIC DRUGS

Nowadays we differentiate between seizure supressing (SS) and antiepileptogenic (AEG) and between neuroprotective (NP) and disease modifying (DM) effect. SS is the main effect of the existing drugs against epilepsy. This is the natural consequence of the fact that they were developed against acute seizures. The older laboratory tests measuring „antiepileptic" effect were restricted to acut epileptic seizure models. In the last 15-20 yrs with the development of chronic epilepsy models (kindling, kainate, pilocarpine, SAS) it became possible to look for drugs protecting against the development of the epileptic process after the initial insult. In a stricter sense NP is a defense against cell death but in a broader sense against every irreversible damage causing functional loss. These changes may occur during the initial insult, during the silent period, when epileptogenic network-changes develop, and during the later course of the illness, due to subclinical seizures or the overt seizures themselves. The AEG agents are expected to work against the development of the epileptic disease process and this effect is not necessarily coincident with the seizure supressing effect. DM covers the protection against the progressive course of epilepsy. The difficulties here are that we do not have a unified conception how the epileptic process turns toward a worse outcome. The progression in the disease process, which is underlied by a vitious circle between increasing seizure frequency and excitotoxic consequence of seizures, and the concept of therapy resistancy (Arroyo et al., 2002) are intermingled. Therapy resistancy could be responsible for hindering the seizure supressing effect of the antiepileptic drugs.

All these effects are hardly testable in the human epilepsies. The main difficulty is the that this kind of studies would need long term preventive studies with administration of drugs which is obviously ethically not justified in a population of infants and children. Therefore animal studies were prefered in this field. However here the difficulties lies in the applicability of the results to the human epilepsies. In addition the differences among the different models and among the nature of different IPI-s and among the ontogenical windows where they are situated, makes the issue rather complicated.

Table 2. Antiepileptics with Seizure supressive – Anti-epileptogenic – Neuro-protective effects on different epilepsy models

Effects / Model	Seizure supressive	Anti-epileptogenic (and/or IPI modifying?)	Neuro-protective
	Only ineffective or questionably effective drugs are indicated	Those drugs are not indicated from which data are not available	
Kindling (elicited seizures)	ESX- MK801-	CBZ- (Silver, 1991) PHT- (Schmutz, 1988) VPA+ (Silver, 1991) VGB+ (Shin, 1986) PHB+ (Silver, 1991) TPM+ (Amano, 1998) LEV+ (Löscher, 1998) LTG+ (Stratton, 2003) MK801+ (Applegate, 1997)	Not suitable test for NP
SHS-SE			TPM+(Niebauer, 1999) VGB- (Lothman, 1996)
SAS-SE	CBZ? ESX?	TPM- (Pitkänen, np) VGB- (Halonen, 2001b)	TPM- (Pitkänen, np) VGB- (Halonen, 2001b)
PPS-SE			LTG+ (Halonen, 2001a) FBM+(Mazerati, 2000) CBZ- (Halonen, 2001a) RMC- (Halonen, 1999)
Pilocarpin SE	ESX-	VPA- (Klitgaard, 2001) TPM+ (DeLorenzo, np) VGB- (André, 2001) LEV- (Klitgaard, 2001) KET- (Hort, 1999)	VPA- (Klitgaard 2001) TPM+(DeLorenzo, np) VGB+(André, 2001) LEV+(Klitgaard, 2001) KET+ (Hort, 1999)
Kainic acid SE	 PHB?	VPA+ (Bolanos, 1998) PHB? (Sutula, 1992) PHB- (Bolanos, 1998) GBP? (Cilio, 2001) MK801- (Ebert, 2002)	VPA+(Bolanos, 1998) VGB+(Jolkkonen,1996) VGB- (Pitkänen, 1999) PHB- (Bolanos, 1998) GBP+ (Cilio, 2001) TPM+ MK801+ (Ebert, 2002)

SHS=sustained hippocampal stimulation, SAS= sustained amygdalar stimulation, PPS= perforant path stimulation, SE=status epilepticus, CBZ=carbamazepine, ESX=ethosuccinimide, FBM=felbamate GBP=gabapentin, KET=ketamin (NMDA antagonist), LEV=levetiracetam, LTG=lamotrigine, MK801(NMDA antagonist), PHB=phenobarbiturate, RMC=remacemideTPM=topiramate, VGB=vigabatrine, VPA=valproic acid, '+'=effective, '-'=not effective, '?'=weak or uncertain effect, np=not published data)

In **Table 2** we try to summarise the results achieved by antiepileptic drugs of different protective characteristics on different models. Evidences for disease modifying effect are not yet available. These studies seem to prove that some of the new antiepileptic drugs have neuroprotective effect in different experimental models. The most challenging aspect of these data is that some drugs effective in one of the models are ineffective in an other one. According to the results of such experiments the role of different elements of the epileptogenic process in the different experimental models are rather contradictory:

1. The damage of the principal neurons is not a requisite for epileptogenesis
2. Neither sprouting is not a requisite for epileptogenesis – at least in the pilocarpine model
3. The protection against cell death does not protect against the development of the epileptogenic process.
4. The decrease of the duration and /or severness of the IPI decrease epileptogenesis. Conversely AEG effect could be achieved without the decrease of severeness and or duration of the IPI, intervening either during (kindling and SE models) or after (within a time limit) the IPI.
5. Both the NP and AEG effect is time window dependent
6. Both the NP and AEG effect is dependent – at least in the SE models - of the eliciting form and duration of the IPI

In the majority of the experiments the AEG/NP effect could not be differentiated well enough from the DM effect.

5. MESIAL TEMPOLAR LOBE EPILEPSY (MTLE) - TIME SEQUENCE OF DEVELOPMENT -ANIMAL MODELS OF TLE

In the following we would like to summarise the most accepted current ideas on the development of mesial temporal lobe epilepsy. It is one of the most frequent type of human epilepsies in which the physiopathogenetic aspects, and the possible role of neuroprotection are well developed. This may serve as a model also to the other symptomatic focal epilepsies as the posttraumatic epilepsy or epilepsy after hypoxic-ischaemic insult in infancy.

TLE is perhaps the most common human epileptic condition in adults representing an important part in childhood epilepsies as well. The common scheme in the development of TLE is: a/ early epileptogenic damage (frequently complicated febrile seizures or status epilepticus) affecting mainly the sensitive CA1 and CA3 region of the dentate pyramidal cells, b/ development of hippocampal sclerosis (detectable on MRI), and parallely a synaptic reorganisation; the mossy-fiber sprouting could be observed (Babb et al., 1991) 3/ the behavioral result is the evolvement of recurrent complex partial seizures after some years' latent period. This sequence of events was modeled in the past years by several animal models either by the technique of electrical stimulation (Repressa et al., 1989) (kindling) or chemical interventions kainic acid (Tauck and Nadler, 1985; Repressa et al., 1993a) pilocarpin (Turski et al., 1983) The study of these animal models as well as the resected specimens of human TLE treated by surgery made it possible to understand more about the underlying mechanisms.

6. CELL LOSS, SYNAPTIC REORGANISATION AND HYPEREXCITABILITY IN THE HIPPOCAMPAL NETWORK (THE HEN-EGG PROBLEM)

Let us take a deeper insight in the transformation of the hippocampal network in epileptic samples. The main pathway of excitation in the hippocampal network consists of a trisynaptic circuit in which information from the entorhinal cortex (the main afferent gate of the hippocampal structure) is transmitted sequentially first to dentate granule cells, from there to CA3 pyramidal cells, and lastly from CA3 to CA1 pyramidal cells through Schaffer collaterals. The CA3 cells show strong excitatory arborisation to the neighbouring pyramidal cells and therefore this sector is specially prone to epileptic synchron discharges, while recurrent excitatory synapses between granule cells are not seen in normal conditions.

Cell death, reactive gliosis and mossy fiber sprouting (MFS) are the common features shared by the above mentioned experimental models and human mesial temporal lobe epilepsy (Ben Ari et al., 1980). The causal chain between these features is still not entirely revealed and probably a circular causality is working. Damage of the CA4 pyramidal cells may induce MFS to compensate loss of their main targets, on the other hand cell death in CA3 sector is possible to be caused by paroxysmal activity of the afferent granule cells overreleasing glutamate (Ben Ari et al., 1980) or alternatively the release of Zn^{2+} by mossy fibers synapses could have excitotoxic effect on CA3 postsynaptic cells (Charton, 1985).

The timing of events relevant in development of hippocampal epilepsy seems to be an other important factor. The susceptibility of hippocampal formation for seizing shows characteristic changes during ontogènetic development. One week old rats are much less prone for seizures than those reaching the second and third weeks of life (Mares, 1973., Michelson et al., 1989; Swann and Moshé, 1997). One factor explaining this could be the time difference in maturation of the cell elements in the hippocampus. Granule cells mature up to postnatal day 15, while pyramidal cells are already mature at birth. In a preparation obtaining “granule cells free” hippocampus by applicating postnatal irradiation that selectively destroys dividing cells, epilepsy could not be observed in response to application of epileptogenic agents The early resistence to epilepsy in the hippocampus is also congruent with the phenomenon of “remodeling”of circuits during early life, corresponding to persistence only of those connection which are functioning and the loss of those which are not in function (pruning) according to the principle: “firing together, wiring together”. Therefore the epileptogenicity of hippocampal damage is probably regulated by a “time-window” and that is why too early damage is not so “epileptogenic” compared with those coming later. (Swann and Moshé, 1997) Cell death is present in CA1 and CA4 sectors of the pyramidal neurons and CA2 and partially CA3 is usually preserved. This pattern is similar to the susceptibility to death among condition of status epilepticus even without hypoxia in experimental circumstances.

MFS became in the last years one of the most extensively studied phenomenon in the context of epileptic hippocampal hyperexcitability.It could be detected easily by the Timm stain which reflects the high Zinc content of mossy fiber axons in animal as well as in human specimens (Tauck and Nadler, 1985; Represa et al., 1989; Cavazos et al, 1991). The sprouted mossy fibers of granule cells show new collateral branches grow accross the granule cell layer of dentate gyrus and reach the inner third of the molecular layer and form recurrent synapses with the granule cell dendrites, providing a dense network for syncronised overexcitation. But newly developed synapses onto GABA-ergic neurons have also been found (Ribak and Peterson, 1991) raising the possibility that inhibition should also take part in the concert of the new discharge pattern in which granule cells behave as an epileptic neuron population promoting seizure initiation and propagation through the hippocampal excitatory pathway.

In the last years a large body of evidence supports the statement that temporal lobe epilepsy is very poorly treatable by traditional antiepileptic drugs developed by methods testing seizure propensity in acute seizure models or even by chronic epileptic animal models. These drugs are effective only in protecting against seizures, but they do not prevent the development of epilepsy and they are not able to interfere with deteriorating vicious circles and cascade of events within the framework of synaptic reorganisation Therefore we should have a deeper understanding of the whole machinery of cellular plasticity leading to synaptic reorganisation and new vantage points are needed for therapeutical interventions, probably quite different from the traditional ones. From this point of view the candidates are neuromodulators, the growth factors, neuropeptides and related proteins,structural proteins and second messengers.

Neurons utilize neurotransmitters along with one or more neuropeptides. The release a transmitter or a neuropeptide depends on the frequency of action potentials. Peptides may diffuse for long distances from their site of their release to their receptors. Neuropeptides may facilitate or inhibit the action of neurotransmitters influencing transmitter synthesis, or metabolizing enzymes, or receptors (Kito and Miyoshi, 1991).

Metabotrophic receptors are coupled to a G-protein, which has seven membrane-spanning domains and, by changing shape, activates an enzyme on the inside surface of the membrane, generating a diffusable messenger such as cyclic adenosine monophosphate (AMP), and cyclic guanosine monophosphate (cGMP). This second messenger will, in turn, act on enzymes such as protein kinases to modify the function of target molecules (recepto(rs, ionic channels, or other enzyme or binding proteins) to change cell excitability or gene expression.

Neuromodulators play role in adaptation processes during and after seizure expression. Opioid peptides modulate postictal excitability and behavior both in animal experiments and human epileptic events. In epileptic hippocampi long term changes in opioid peptide receptors and in neuropeptide expressions are commonly seen, and NPY is observed to express in dentate granule cells, where it is not present normally (Rizzi et al 1993) Nitric oxyde (NO) an other important neuromodulator in one hand helps adaptation to greater oxygen demand during seizures by vasodilatation, and works as endogenous anticonvulsant since blockers of NO synthesis enhance seizure activity, but when liberated in excessive amounts can be toxic causing excitotoxic cell death (Wang et al., 1994).

Cyclic AMP as an ubiquitous second messanger and membrane phospholipids induce growth factors and immediate early genes, regulate the function of protein kinases and take part in seizure regulation in many ways (Jope et al., 1992). Calcium also plays important role in transmitter release, long term plasticity, and cell damage, antiepileptics and neuroprotective agents reducing calcium entry into the neurons by blocking the depolarisation of presynaptic terminals.

There are several structural proteins playing also role synaptic reorganisation in development of epilepsy in temporal lobe models. Mossy fiber sprouting is for example probably favored in dentate gyrus by the presence of growth promoting substances as laminin and fibronectin in the granule cell layer (Niquet et al., 1994; 1995). Astrocyte may take place in this procedure as well. Proliferating astrocytes in the hilus express embryonic Neural Cell Adhesion Molecule (NCAM), fibronectin and vitronectin enhanching axonal branching and growth, while reactive astrocytes in CA3 express tenascin antibodies, which is unfavorable substance for mossy fibers.An other major issue is the involvement of growth factors in sprouting of mossy fibers. An increase of Nerve Growth Factor (NGF) expression in dentate gyrus in response to kindling was described (Gall and Isackson, 1989). Seizure dependent changes were observed in levels of fibroblast growth factor (bFGF) and brain derived nerve growth factor (BDNF). Sprouting of mossy fibers was preceded by an important expression of a-tubulin and microtubule associated protein 2 in granule cells (Represa et al., 1993b).

These data show that mossy fiber collateral branch sprouting originate under the influence of throphic factors. Mossy fiber growth and orientation seems to be guided by astrocytes excreting cell adhesion and substrate molecules.

In the light of these results epilepsy should considered as a vicious circle where seizures set in motion a cascade of molecular and genomic changes including early genes, throphic factors, cytoskeletal proteins, and adhesion-substrate molecule beside of transmitter receptors. This kind of changes contribute to development of long-lasting functional and morphological changes summarised as synaptic remodeling, pave the way for evolution of epileptic mechanisms.

7. REORGANISATION OF THE INTERNEURONAL NETWORK IN TLE

Further studies showed that synaptic reorganisation is not restricted to pyramidal cell loss, glial scar, and sprouting of the mossy fibers. A more widespread reorganisation involving the whole interneuronal network develops in the epileptic hippocampus. A large amount of inhibitory interneurons is preserved in the epileptic human dentate gyrus, but their distribution. morphology and synaptic connections differ from controls. In a study of 20 surgically treated temporal lobe epileptic patients the resected hippocampal tissue was examined (Magloczky et al., 2000). Three types of inhibitory interneurons, containing Ca++ binding proteins (calbindin (CB), calretinin (CR), and substance P (SPR) were analysed. The common features of the changes found in the samples are the following: Large hypertrophic CB -containing cells appear in the hilus, their dendritic tree is longer than that in control. The number of CR-positive cells decreases, while the CR-positive supramamillary pathway extends to innervate the entire width of the stratum moleculare. The distribution of SPR-positive cells changes. In the control most are located in the hilus, whereas the epileptic samples show them to be numerous in the stratum moleculare. In addition, they have a bishy and highly varicose dentritic tree. Most of the granule cells lose their CB content and migrate up to the stratum moleculare in the epileptic tissue. The preserved interneurons show signs of high rate metabolic activity. The expansion of the CR-positive supramamillary afferents demonstrates that that in addition to mossy fibre sprouting , there is a second, extrahippocampal source of excess excitation of dentate granule cells. Several morphological features of the changed interneurons seem to be similar to that in the immature state of the hippocampus when neurotrophic factors are active. The increased activity of these cells in hippocampal epilepsy models, could explain the observed morhological changes. These data support the possibility that neurotrophins have a crucial role in the regulation of cell survival during epilepsy should be considered. This would offer new approaches in the development of protective antiepileptic drugs.

8. HISTOLOGICAL CHANGES IN HIPPOCAMPAL SCLEROSIS: IN VIVO VISUAL ASSESSMENT BY MRI-CLINICAL STUDIES CONTRIBUTING TO UNDERSTAND THE ROLE OF HS IN THE PATHOPHYSIOLOGY OF TLE

The histological examination of hippocampal atrophy in epilepsy was first described by Sommer in 1880, who documented the selective vulnerability of the sector later named after him as "Sommer's-sector" The association of hippocampal atrophy and sclerosis with epilepsy and later with TLE was first supported by autopsy findings (Stauder, 1936; Meyer et al., 1954; Margerison and Corsellis, 1966).The most prominent changes occur in the CA1 region of pyramidal cells (Sommer's sector), less severe but important cell loss appears in the CA3 region and in the hilus of the dentate gyrus (CA4 or end folium). The CA2 region and the dentate gyrus are relative resistant. The neuronal loss could be measured nowadays by three-dimensional counting techniques (Braendgaard and Gundersen, 1986; Gundersen et al., 1988; Oorschot, 1994)There is a characteristic fibrous astrogliosis, proportional to the degree of the damage. In the light of the recent research astroglial proliferation seems to be due to the liberation of different plasticity proteins promoting proliferation of glial elements as a part of the reorganisation process, being not simply a scar production. It has been also recognised that in many cases granular cell layer is wider than normal and extend into the molecular layer, or may be duplicated (Houser, 1990).The cause of the granule cell dispersion could be a migration disorder, but the role of damage due to the seizures could not be excluded.The most consistent feature of histological changes in hippocampal sclerosis is mossy fiber sprouting which could be well demonstrated by Timm's staining and/or dynorphin immunohistochemistry. The recurrant sprouted axons of the granule cells contain zinc-labelled vesicles in their excitatory terminals.

Improvement of MR neuroimaging technique allows to detect HS visually, to follow up its morphological changes, and measure it quantitatively. To detect HS scanning should be performed along the axis of the hippocampus in thin (1-3 mm) sections, and application of inversion recovery and T2 -weighted images as well as FLAIR images are necessary. Hippocampal volumetry proved also to be helpful in the hands of other investigators (Jack et al., 1990; Cendes et al., 1993; Adam et al., 1994). Cell loss appears in the form of "atrophy" reflected by decrease of volume and signal in T1 and inversion recovery images, while astrogliosis appears as signal increase in T2 weighted and FLAIR images. MRI-based hippocampal volumes have been shown to correlate with the extent of cell loss. Unilateral HS is found in the 40-60% of patients with intractable TLE, and 60-80% of them have a good outcome after surgery consisting different degree of temporal lobe resections including the resection of the head and certain portion of the hippocampal body as well (Van Paesschen and Révész, 1997)

Bilateral HS has been reported in 50-88% of patients with TLE (Babb et al., 1991). In the acute seizing stage in TLE transient swelling of hippocampal formation was demonstrated by serial MRI investigations (Nohria et al., 1994; Jackson et al, 1995).

Besides classical HS, coexistence of other lesions is frequently demonstrated. This "dual pathology" consist of neoplasms, vascular malformations, post-traumatic scars, cortical dysplasias and microdysgenesis, inflammatory lesions. Tumours are the second largest group of pathology after HS and among them DNT-s, gangliogliomas, oligodendrogiomas and astrocytomas are most frequently found. In these cases febrile convulsions were less frequently found in the patients history. (Raymond et al., 1994). Therefore the combination of HS and developmental anomalies seem to make up an important pathophysiological constellation. Since developmental anomalies are proved to be highly epileptogenic the possibility of secondary evolution of HS by kindling is one of the rational scenarios. The examples of secondary evolution of HS in extratemporal developmental lesions such as localised periventricular heterotopias (Janszky and Halász, 1999) provide further support for this kind of origin.

9. NEUROPROTECTION IN TLE

Pitkänen and Halonen (1998) proposed the post-status epilepsy models to search for antiepileptic and not only anticonvulsant drugs. This model seems to be suitable for this purpose because the latency period between the status and the first occurrence of spontaneous seizures allows to test drugs as prophylactic treatment. Löscher (2002) summarised the hither to performed studies. From a long list of drugs only valproate and topiramate were capable to prevent epilepsy when administered after the status epilepticus (Bolanos et al., 1998) The latter were questioned by the work of Pitkänen using not pilocarpine but sustained electrical stimulation of the amygdala (SAS). Similarly valproate proved to be preventive only in the kainate model and not in the pilocarpine model (Bolanos et al., 1998; Klitgaard et al., 2001). Chronic treatment with levetiracetam or vigabatrin failed to block the development of epilepsy after status epilepticus, but in a study in which the vigabatrin treatment started earlier after pilocarpine administration, a significant neuroprotection was obtained in the hippocampus (André, 2001). In addition to preventing epilepsy valproate and topiramate were also found to reduce the neurodegeneration (degree of hippocampal sclerosis), while phenobarbital and vigabatrin were ineffective in this regard in most of the studies. These findings suggest the existence of a therapeutic window after status epilepticus within which the neuroprotection may be effective. This time window seems to be narrow, ranges in hours rather than days (Ebert et al., 2002). This kind of time course is suggestive for delayed forms of cell death (apoptosis) playing role, beside immediate necrotic processes, in the hippocampal neurodegeneration in TLE.

10. FUTURE PERSPECTIVES

Generations of medical professionels have tried to prevent epileptic seizures for several thousand years. An armady of efficient drugs were developed up to now rendering two thirds of patients seizure free. However the prevention of the epileptogenic process and its progression has been studied only recently. Although promising, the human and experimental data are full of contradictions. We need further animal models showing more similar features to the human epilepsies. They should be suitable to measurably model the consequent functional deficits as well.

The most important risk factors predicting the development and progression of epilepsy should be delineated more precisely. Long term studies are needed with populations showing these features, trying out candidate compounds promising success against these features, based on animal experiments. Finding compounds with antiepileptogenic effect without seizure supressing capacity seems to be rather far. However, the course of epilepsy can be characterised not only by seizure number, and by neuropsychological deficit symptoms, but with fine graded NMRI changes. That way the eventuel disease modifying effect of a drug should be measured in the future, beside or parallel with the seizure supressing effect of it.

It is clear that the prevention of the IPI or diminishing the effect of it has a prophylactic effect Presently we do not have unanimously neuroprotective and/or antiepileptogenic drugs in our hands, but we are not far from discovering them. When it happens will start to change epilepsy from treatable to curable disease.

11. SUMMARY

During the last years it has become obvious that the current way of treating epilepsy with antiepeileptic drugs is insufficient concerning the modification of the underlying disesease and provides merely a symptomatic treatment, without clear influence on the course of the disease. There is a pressing need to find alternative strategies and to find possibilities to intervene either into the basic processes determining the development of epilepsies or to promote compensatory processes in repairing these dysfunctions.The increasing knowledge about the basic neuronal changes underlying epilepsies allows now to analyse the potential role of neuroprotective agents in in epileptogenesis.

In epilepsy the most frequent constellation is the presence of damage and overexcitation together. Increase in excitability may develop after a primary damage as in posttraumatic epilepsy, or outburst of epileptic excitability may cause neuronal damage as in cell loss after status epilepticus or in any case of the so called cytotoxic damage from extensive glutamatergic involvement.

Epilepsy in certain forms is a progressive disease. The factors determining the progressive course and the possibe prevention of it is obviously an overlaping field with neuroprotection. Therefore although neuroprotection works only against certain aspects of a complex cascade of pathological events, might be a promising option in several stadiums during the development and course of epilepsy.

We provide evidences that some of the new antiepileptic drugs have neuroprotective effect on different animal models of chronic partial epilepsies, and how this effect is fitting to the antiepileptogenic, and seizure supressing effect of the same drugs.

REFERENCES

Adam, C., Baulac, M., Saint-Hilaire, J.M., Landau, J., Granat, O., and Laplane, D, 1994, Value of magnetic resonance imaging-based measurements of hippocampal formations in patients with partial epilepsy, *Arch. Neurol.* **51**:130-138.

Amano, K., Hamada, K., Yagi, K., and Seino, M., 1998, Antiepileptic effects of topiramate on amygdaloid kindling in rats, *Epilepsy Res.* **31**:123–128.

André, V., Ferrandon, A., Marescaux, C., and Nehlig, A., 2001, Vigabatrin protects against hippocampal damage but is not antiepileptogenic in the lithium-pilocarpine model of temporal lobe epilepsy, *Epilepsy Res.***47**(1-2):99-117.

Annegers, J. F., Hauser, W. A., Coan, S. P., and Rocca, W. A., 1998, A population-based study of seizures after traumatic brain injuries, *N.Engl.J Med.* **338**:20-24.

Applegate, T. L., Karjalainen. A., and Bygrave, F. L., 1997, Rapid Ca2+ influx induced by the action of dibutylhydroquinone and glucagon in the perfused rat liver, *Biochem J.* **15**:463-467.

Arroyo, S., Brodie, M. J., Avanzini, G., Baumgartner, C., Chiron, C., Dulac, O., French, J. A., and Serratosa, J. M., 2002, Is refractory epilepsy preventable? *Epilepsia* **43**:437-444.

Babb, T. L., Pereira-Leite, J., Mathern, G.W., Pretorius, J. K., 1995, Kainic acid induced hippocampal seizures in rats: comparisons of acute and chronic seizures using intrahippocampal versus systemic injections, *Ital. J. Neurol. Sci.* **16**(1-2):39-44.

Babb, T. L., Kupfer, W. R., Pretorius, J. K., Crandall, P. H., and Levesque, M. F., 1991, Synaptic reorganization by mossy fibers in human epileptic fascia dentata, *Neuroscience* **42**:351-363.

Ben Ari, Y., 1985, Limbic seizure and brain damage produced by kainic acid: mechanisms and relevance to human temporal lobe epilepsy, *Neuroscience* **14**:375-403.

Ben Ari, Y., Tremblay, E., Otterson, O. P., and Maldrum, B. S., 1980, The role of epileptic activity in hippocampus and remote cerebral lesion induced by kainate, *Brain Res.* **191**:79-97.

Berg, A. T. and Shinnar, S., 1997, Do seizures beget seizures? An assessment of the clinical evidence in humans, *J. Clin. Neurophysiol* **14**:102-110.

Bolanos, A. R., Sarkisian, M., Yang, Y., Hori, A., Helmers, S. L., Mikati, M., Tandon, P., Stafstrom, C. E., and Holmes, G. L.,1998, Comparison of valproate and phenobarbital treatment after status epilepticus in rats, **51:**41-48.

Braendgaard, H., Gundersen, H. J.G., 1986, The impact of recent stereological advances on quantitative studies of the nervous system. *J. Neurosci. Met* **18**:39-78.

Cavalheiro, E. A., Leite, J. P., Bortolotto, Z. A., Turski, W. A., Ikonomidou, C., and Turski, L., 1991, Long-term effects of pilocarpine in rats: structural damage of the brain triggers kindling and spontaneous recurrent seizures, *Epilepsia* **32**(6):778-82.

Cavazos, J.E., Golarai, G., and Sutula, T. P., 1991, Mossy fibres reorganization induced by kindling; Time course of development, progression, and permanence, *J. Neurosci.* **11**:2795-2803.

Cendes, F., Andermann, F., Gloor, P., Evans, A., Jones-Gotman, M., Watson, C., Melanson, D., Olivier, A., Peters, T., Lopes-Cendes, I., and Leroux, G., 1993, MRI volumetric measurement of amygdala and hippocampus in temporal lobe epilepsy, *Neurology* **43:**719-725.

Charton, G., Rovira, C., Ben Ari, Y., and Leviel, V., 1985, Spontaneous and evoked release of endogenous Zn^{2+} in the hippocampal MF zone in situ, *Exp. Brain Res.* **58**:202-205.

Cilio, M. R., Bolanos, A. R., Liu, Z., Schmid, R., Yang, Y., Stafstrom, C.E., Mikati, M. A., and Holmes, G. L., 2001, Anti-convulsant action and long-term effects of gabapentin in the immature brain, *Neuropharmacology* **40**:139–147.

DeGiorgio, C. M., Tomiyasu, U., Gott, P. S., and Treiman, D. M.,1992, Hippocampal pyramidal cell loss in human status epilepticus, **33:**23-27.

Devinsky, O., 1999, Patients with refractory seizures, *N Engl J Med* **340**:1565-1570.

Ebert, U., Brandt, C., and Loscher, W., 2002, Delayed sclerosis, neuroprotection, and limbic epileptogenesis after status epilepticus in the rat, *Epilepsia*; **43**(S5):86-95.

Feksi, A. T., Kaamugisha, J., Sander, J. W., Gatiti, S., and Shorvon, S. D., 1991, Comprehensive primary health care antiepileptic drug treatment programme in rural and semi-urban Kenya, ICBERG (International Community-based Epilepsy Research Group) *Lancet* **337**(8738):406-9.

French, J. A., Williamson, P. D., Thadani, V. M., Darcey, T. M., Mattson, R. H., Spencer, S. S., and Spencer, D. D., 1993, Characteristics of medial temporal lobe epilepsy: I. Results of history and physical examination, *Ann.Neurol.* **34**:774-780.

Fuerst, D., Shah, J., Kupsky, W. J., Johnson, R., Shah, A., Hayman-Abello B., Ergh, T., Poore ,Q., Canady, A., and Watson, C., 2001, Volumetric MRI, pathological, and neuropsychological progression in hippocampal sclerosis, *Neurology* **57**(2):184-8.

Fuerst, D., Shah, J., Shah, A., and Watson, C., 2003, Hippocampal sclerosis is a progressive disorder: A longitudinal volumetric MRI study, *Ann. Neurol.***53**(3):413-6.

Gall, C., and Isackson, P. J., 1989, Limbic seizures increase neuronal production of messenger RNA for nerve growth factor, *Science,***245**:758-761.

Goddard, G. V., 1967, Development of epileptic seizures through brain stimulation at low intensity, *Nature* **214**:1020-1021.

Gundersen, H. J. G., Bendtsen, T. F., Korbo, L., Marcussen, N., Moller, A., Nielsen, K., Nyengaard, J. R., Pakkenberg, B., Sorensen, F. B., Vesterby, A., and West, M. J, 1988, Some new, simple and efficient stereological methods and their use in pathological research and diagnosis, *APMIS* **96**:379-394.

Halonen, T., Nissinen, J., Pitkänen, A., 1999, Neuroprotective effect of remacemide hydrochloride in a perforant pathway stimulation model of status epilepticus in the rat, *Epilepsy Res.* **34**:251–269.

Halonen, T., Nissinen, J., and Pitkänen, A., 2001a, Effect of lamotrigine treatment on status epilepticus-induced neu-ronal damage and memory impairment in rat, *Epilepsy Res.* **46**:205–223.

Halonen, T., Nissinen, J., and Pitkänen, A., 2001b, Chronic elevation of brain GABA levels beginning two days after status epilepticus does not prevent epileptogenesis in rats, *Neu-ropharmacology* **40**:536–550.

Hort, J., Brozek, G., Mares, P., Langmeier, M., and Komarek, V., 1999, Cognitive functions after pilocarpine-induced status epilepticus: changes during silent period precede appearance of spontaneous recurrent seizures, *Epilepsia* **40**:1177–1183.

Houser, C. R., 1990, Granule cell dispersion in the dentate gyrus of humans with temporal lobe epilepsy. *Brain Res* **535**:195-204.

Jack, C. R. Jr., Sharbrough, F. W., Twomey, C. K., Cascino, G. D., Hirschorn, K. A., Marsh, W. R., Zinsmeister, A. R., and Scheithauer, B., 1990, Temporal lobe seizures: Lateralization with MR volume measurements of the hippocampal formation, *Radiology* **175**:423-429.

Jackson, G. D., 1995, The diagnosis of hippocampal sclerosis: other techniques. *Magn. Reson. Imaging.* ;**13**(8):1081-93. Review.

Janszky, J., Barsi, P., Halász, P., Erőss, L., and Rásonyi, Gy., 1999, Temporal lobe epilepsy syndrome with peritrigonal nodular heterotopia, *Clin Neurosci.* **52**(1-2):44-50.

Jensen, F. E., Holmes, G. L., Lombroso, C. T., Blume, H. K., and Firkusny, I. R., 1992, Age-dependent changes in long-term seizure susceptibility and behavior after hypoxia in rats, *Epilepsia* **33**:971-980.

Jolkkonen, J., Halonen, T., Jolkkonen, E., Nissinen, J., and Pitkänen, A., 1996, Seizure-induced damage to the hippocampus is prevented by modulation of the GABAergic system, *NeuroReport* **7**:2031–2035.

Jope, R. S., Song, L., and Kolasa, K., 1992, Inositol trisphosphate, cyclic AMP, and cyclic GMP in rat brain regions after lithium and seizures, *Biol Psychiatry* **31**:505-514.

Kito, S., and Miyosi, R., 1991, Effect of neuropeptides on classic types of neurotransmission in the rat central nervous system, in Kito, S., *et al* (eds.), *Neuroreceptors Mechanisms in Brain* Plenum Press, New York, pp. 1-11.

Klitgaard, H., 2001, Levetiracetam: the preclinical profile of a new class of antiepileptic drugs? *Epilepsia* **42**(S4):13-8.

Kwan, P. and Brodie, M. J., 2000, Early identification of refractory epilepsy, *N.Engl.J Med.* **342**:314-319.

Lipton, S. A. and Rosenberg, P. A., 1994, Excitatory amino acids as a final common pathway for neurologic disorders. **330:** 613-622.

Liu, R. S., Lemieux, L., Bell, G. S., Hammers, A., Sisodiya, S. M., Bartlett, P. A., Shorvon, S. D., Sander, J. W., and Duncan, J. S., 2003, Prograssive neocortical damage in epilepsy, *Ann. Neurol.* **53**:312-324.

Loscher, W., 2002, Animal models of epilepsy for the development of antiepileptogenic and disease-modifying drugs. A comparison of the pharmacology of kindling and post-status epilepticus models of temporal lobe epilepsy, *Epilepsy Res.* **50**:105-123.

Loscher, W., Honack, D., and Rundfeldt, C., 1998, Antiepileptogenic effects of the novel anticonvulsant levetiracetam (ucb L059) in the kindling model of temporal lobe epilepsy, *J Pharmacol.Exp.Ther.* **284**:474-479.

Lothman, E.W., Williamson, J.M., Goosens, K.A., Bertram, E.H., 1996. Vigabatrin improves behavioral outcome in a rat model of mesial temporal lobe epilepsy, Abstract in HMR exhibition in connections of *1996 American Epilepsy Society meeting in San Fransisco.*

Maglóczky, ZS., Wittner, L., Borhegyi, ZS., Halász, P., Vajda, J., Czirják, S., and Freund, T., 2000, Changes in the distribution and connectivity of interneurons in the epileptic human dentate gyrus. *Neuroscience* **96**:7-25.

Mares, P., 1973, Ontogenetic development of bioelectrical activity of the epileptogenic focus in rat neocortex *Neuropadiatrie* **4**:434-445.

Margerison, J. H., and Corsellis, J. A. N., 1966, Epilepsy and the temporal lobes. A clinical, electroencephalographic and neuropathological study of the brain in epilepsy with particular reference to the temporal lobes, *Brain* ***89*:499-530.**

Mazarati, A. M., Baldwin, R. A., Sofia, R. D., and Wasterlain, C. W., 2000, Felbamate in experimental model of status epilepticus, *Epilepsia* **41**(2):123–127.

McNamara, J. O., and Wada, J. A., 1997, Kindling model.in Engel J. Jr and Pedly TA eds, *Epilepsy: A Comprehensive Textbook*, Lippincott-Ravan, Philadelphia.,Pp. 419-425.

Meyer, A., Falconer, M. A., and Beck, E., 1954, Pathological findings in temporal lobe epilepsy. *J. Neurol.Neurosurg. Pychia,* **3**:276-285.

Michelson, H. B., Williamson, J. M., Lothman, E. W., 1989, Ontogeny of kindling: the acquisition of kindled responses at different ages with rapidly recurring hippocampal seizures, *Epilepsia,30,672.*

Musicco, M., Beghi, E., Solari, A., and Viani, F., 1997, Treatment of first tonic-clonic seizure does not improve the prognosis of epilepsy, First Seizure Trial Group (FIRST Group, *Neurology* **49**:991-998.

Niebauer, M., Gruenthal, M., 1999, Topiramate reduces neu-ronal injury after experimental status epilepticus, Brain Res. **837**, 263–269.

Niquet, J., Ben Ari, Y., Faissner, A., Represa, A., 1995, Lesion and fibre sprouting in the hippocampus is associed with an increase of tenascin immunoreactivity, an extracellular glycoprotein with repulsive properties. *J. Neurocytol* **24**,611-624.

Niquet, J., Ben Ari, Y., Represa, A., 1994, Glial reaction after seizure induced hippocampal lesion: Immunocytochemical characterization of proliferating glial cell. *J. Neurocytol* **24**,641-656.

Nohria, V., Lee, N., Tien, R. D., Heinz, E. R., Smith JS., DeLong, G, R., Skeen, M. B., Resnick, T. J., Crain, B., and Lewis, D. V., 1994, Magnetic resonance imaging evidence of hippocampal sclerosis in progression: a case report. *Epilepsia* **35**(6):1332-6.

Oorschot, D. E., 1994, Are you using neuronal densities, synaptic densities or neurochemical densities as your definitive data? There is a better way to go, *Prog. Neurobiol* **44**:233-247.

Pitkänen A, Halonen T. Prevention of epilepsy. Trends Pharmacol Sci. 1998 Jul;19(7):253-5.

Pitkänen, A., Nissinen, J., Jolkkonen, E., Tuunanen, J., Halonen, T., 1999, Effects of vigabatrin treatment on status epilepticus-induced neuronal damage and mossy fiber sprouting in the rat hippocampus. *Epilepsy Res.* **33:**67–85.

Placencia, M., Sander, J. W., Roman, M., Madera, A., Crespo, F., Cascante, S., and Shorvon, S. D.,1994, The characteristics of epilepsy in a largely untreated population in rural Ecuador, 57**:** 320-325.

Placencia, M., Sander, J. W., Shorvon, S. D., Roman, M., Alarcon, F., Bimos, C., and Cascante, S.,1993, Antiepileptic drug treatment in a community health care setting in northern Ecuador: a prospective 12-month assessment. **14:** 237-244.

Raymond, A. A., Fish, D. R., Stevens, J. M., Cook, M. J., Sisodiya, S. M., and Shorvon, S. D., 1994, Association of hippocampal sclerosis with cortical dysgenesis in patients with epilepsy, *Neurology,* **44**:,1841-1845.

Represa, A., Le Gall Salle, G., Ben Ari, Y. (1989) Hippocampal plasticity in the kindling model of epilepsy in rats. *Neurosci. Lett* **99**:345-350.

Represa, A., Niquet, J., Charriot Marlangue, C., and Ben Ari, Y., 1993a Reactive astrocytes in the kainic acid damaged hippocampus have the phenotype features of type II astrocytes, *J.Neurocytol,* **22**:299-310.

Represa, A., Pollard, H., Moreau, J., Ghilini, G., Khrestcharisky, M., and Ben Ari, Y., 1993b, Mossy fibres sprouting in epileptic rats is associated with a transient increased expression of tubulin. *Neurosci Lett* **156:**149-152.

Reynolds, E. H., 1987, Early treatment and prognosis of epilepsy, *Epilepsia* **28**(2):97-106.

Ribak, C. B., and Peterson, G. M., 1991, Intragraular mossy fibers in rats and gerbils taken from synapsus with somatic and proximal dendrites of basket cells in the dentate gyrus, *Hippocampus* **1**:355-364.

Rizzi, M., Monno, A., Samanin, R., Sperk, G., and Vezzani, A., 1993, Electrical kindling of the hippocampus is associated with functional activation of neuropeptide Y-containing neurons. *Eur J Neurosci* **5:**1534-1538.

Schmutz, M., Klebs, K., andBaltzer, V., 1988, Inhibition or enhancement kindling evolution by antiepileptics. *J. Neu-ral. Transm.* **72**:245–257.

Semah F, Picot MC, Adam C, Broglin D, Arzimanoglou A, Bazin B, Cavalcanti D, Baulac M., 1998, Is the underlying cause of epilepsy a major prognostic factor for recurrence? *Neurology* **51**(5):1256-62.

Shin, C., Rigsbee, L.C., McNamara, J.O., 1986, Anti-seizure and anti-epileptogenic effect of γ-vinyl γ-aminobutyric acid in amygdaloid kindling, *Brain Res.* **398**:370–374.

Silver, J.M., Shin, C., and McNamara, J.O., 1991, Antiepilepto-genic effects of conventional anticonvulsants in the kin-dling model of epilepsy, *Ann. Neurol.* **29**:356–363.

Stauder, K.H., 1936, Epilepsie und Schlafenlappen. *Arch. Psychiat. Nervenkr* **104**:181-211.

Stratton, S. C., Large, C. H., Cox, B., Davies, G., and Hagan, R. M., 2003, Effects of lamotrigine and levetiracetam on seizure development in a rat amygdala kindling model, *Epilepsy Res.* **53**(1-2):95-106.

Sutula, T., Cascino, G., Cavazos, J., Parada, I., and Ramirez, L., 1989, Mossy fiber synaptic reorganization in the epileptic human temporal lobe, *Ann.Neurol.* **26**:321-330.

Sutula, T., Cavazos, J., Golarai, G., 1992. Alteration of longlasting structural and functional effects of kainic acid in the hippocampus by brief treatment with phenobarbital, *J. Neurosci.* **12**:4173–4187.

Swann, J. and Moshé, L.S., 1997, Developmental issues in animal models, in: J.Engel Jr. and T.A. Pedley (eds.), *Epilepsy: A Comprehensive Textbook,* Lippincott-Raven, Philadelphia, pp. 467-479.

Tasch, E., Cendes, F., Li LM, Dubeau, F., Andermann, F., and Arnold, D.L., 1999, Neuroimaging evidence of progressive neuronal loss and dysfunction in temporal lobe epilepsy, *Ann Neurol.* **45**(5):568-76.

Tauck, D.L., and Nadler, J.V., 1985, Evidence of functional mossy fibre sprouting in hippocampal formation of kainic acid treated rats. *J. Neurosci.* **5**:1016-1022.

Temkin, N. R., 2001, Antiepileptogenesis and seizure prevention trials with antiepileptic drugs: meta-analysis of controlled trials, *Epilepsia* **42**:515-524.

Theodore, W. H., Bhatia, S., Hatta, J., Fazilat, S., DeCarli, C., Bookheimer, S. Y., and Gaillard, W. D., 1999, Hippocampal atrophy, epilepsy duration, and febrile seizures in patients with partial seizures. *Neurology* **52**(1):132-6.

Timofeev, I., Grenier, F., and Steriade, M., 1998, Spike-wave complexes and fast components of cortically generated seizures. IV. Paroxysmal fast runs in cortical and thalamic neurons, *J Neurophysiol.* **80**(3):1495-513.

Turski, W. A., Cavalheiro, E. A., Coimbra, C., da Penha, Berzaghi, M., Ikonomidou-Turski, C., and Turski, L., 1987, Only certain antiepileptic drugs prevent seizures induced by pilocarpine, *Brain Res.* **434**(3):281-305.

Turski, W.A., Cavalheiro, E.A., Schwartz, M., Czuczwar, S.J., Kleinrok, Z., and Turski, L., 1983, Limbic seizures produced by pilocarpine in rats: behavioural electroencephalographic and neuropathological study, *Behav Brain Res* **9:**315-335.

Tuunanen, J. and Pitkanen, A., 2000, Do seizures cause neuronal damage in rat amygdala kindling? **39** : 171-176.

Vajda, F. J. E., 2002, Neuroprotection and neurodegenerative disease, *J. Clin. Neuroscience* **9**(1):4-8.

Van Paesschen, W., Revesz, T., Duncan, J. S., King, M. D., and Connelly, A., 1997, Quantitative neuropathology and quantitative magnetic resonance imaging of the hippocampus in temporal lobe epilepsy, *Ann Neurol.* **42**(5):756-66.

Walker, M. C., White, H. S., and Sander, J. W., 2002, Disease modification in partial epilepsy, *Brain* **125**:1937-1950.

Wang, Q., Theard, A.M., Pelligrino, D.A., Baughman, V., and Hoffmann, W.E., 1994, Nitric oxide (NO) is an endogenous anticonvulsant but not a mediator of the cerebral hyperemia accompanying bicuculline induced seizures in rats, *Brain Res.* **658**:142-148.

Willmore, L. J. and Rubin, J. J., 1984, Effects of antiperoxidants on FeCl2-induced lipid peroxidation and focal edema in rat brain, Exp.Neurol. 83:62-70.

NEUROPROTECTION AND GLATIRAMER ACETATE: THE POSSIBLE ROLE IN THE TREATMENT OF MULTIPLE SCLEROSIS

Tjalf Ziemssen*

1. INTRODUCTION

Multiple Sclerosis (MS) is the most common inflammatory demyelinating disease of the central nervous system (CNS). It is believed to be an immune-mediated disorder in which the myelin sheath or the oligodendrocyte is targeted by the immune system in genetically susceptible people. Oligodendrocytes synthesize and maintain the axonal myelin sheath of up to 40 neighbouring nerve axons in the CNS. Compact myelin consists of a condensed membrane, spiralled around axons to form the insulating segmented sheath needed for saltatory axonal conduction: voltage-gated sodium channels cluster at the unmyelinated nodes of Ranvier, between myelin segments, from where the action potential is propagated and spreads down the myelinated nerve segment to trigger another action potential at the next node.

The pathological hallmark of MS is the demyelinating plaque which consists of infiltrating T lymphocytes and macrophages, damage to the blood-brain barrier and loss of myelin. The composition of the inflammatory infiltrate varies depending on the stage of demyelinating activity. Early symptoms of MS are widely believed to result from this inflammatory axonal demyelination which leads to slowing or blockade of axonal conduction. The regression of symptoms has been attributed to the resolution of inflammatory edema and to partial remyelination.

Although MS is primarily an inflammatory autoimmune disease, it has become evident that axonal loss plays an important role in the pathogenesis of disability for patients with MS[1]. While axonal pathology was elegantly and precisely described in classic MS neuropathology studies more than a century ago[2,3], it has only recently reemerged as a major focus of research[4]. The central question to be addressed is not whether there is axonal loss in MS but when and to what extent does the axonal loss occur. The timing and degree of axonal loss is of importance not only in its relationship to the aetiology of the disease but may well be central to the appearance of clinical symptoms and the progressive

* Tjalf Ziemssen, Max-Planck-Institute of Neurobiology, Department of Neuroimmunology, 82152 Martinsried, Germany; Neurological Clinic, Carl Gustav Carus Universitiy Clinic, 01307 Dresden, Germany

deterioration associated with the disease. The fact that axonal loss is irreversible has important implications for when, and what therapeutic intervention should be used.

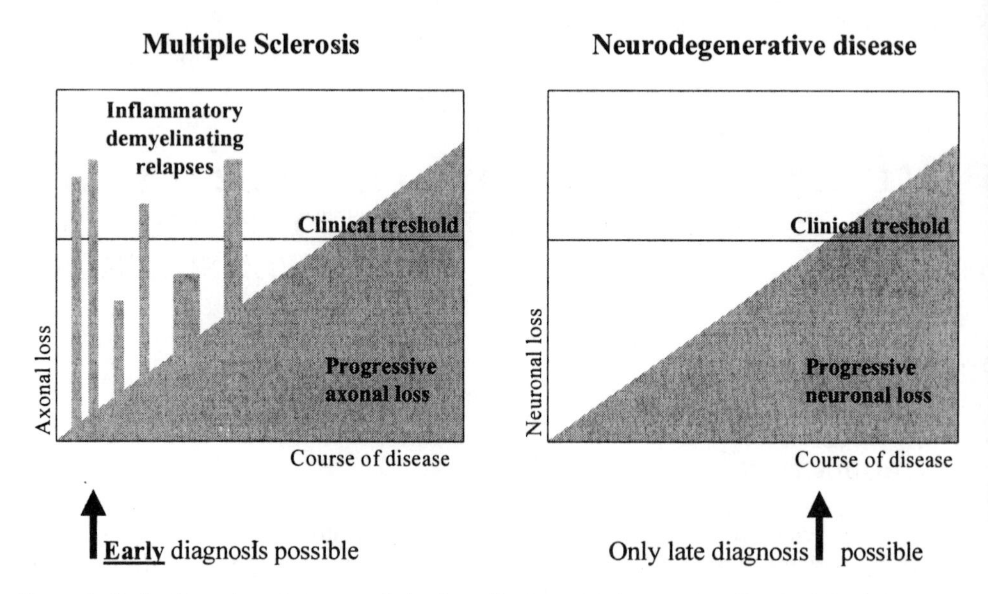

Figure 1: Early diagnosis and therapy of the demyelinating neurodegenerative disorder MS is possible in comparison to other neurodegenerative diseases like M. Parkinson or Alzheimer

It is likely that various mechanisms contribute to axonal damage during different stages of disease. In active lesions, the extent of axonal transsection correlates with inflammatory activity while even there seems to be an inflammation-independent axonal loss[5]. Hence, axonal loss may be caused by inflammatory products of activated immune and glial cells, including proteolytic enzymes, cytokines, oxidative products and free radicals, although the precise molecular mechanisms of axonal damage are poorly understood. In addition the magnitude of axonal loss in chronic MS lesions without pronounced inflammatory infiltrates suggests that mechanisms other than inflammatory demyelination contribute to the degeneration of axons. Several conditions interfere with attemps of axonal regrowth after lesions develop. These include the lack of neurotrophic factors that support growth, the presence of a glial scar (depending on the site of lesion) or the presence of inhibitory molecules that impede axonal growth. Recent evidence shows that axon degeneration following injury has similarities with the cellular mechanisms underlying programmed cell death[5].

The concept of MS as an inflammatory neurodegenerative disease highlights the importance of an active therapeutic approach. But since lesions outnumber clinical relapses by much as 10:1 and, in addition, inflammation-independent axonal degeneration may occur and add to the considerable continuing subclinical pathophysiological process, tissue damage even in the absence of clinical manifestations may take place. Because MS has its characteristic clinical phenotype with clinical relapses and remissions, MS could be diagnosed and treated at an earlier timepoint of the disease in contrast to other neurodegenerative disorders like M. Parkinson or M. Alzheimer (Figure 1). Theoretically, an antiinflammatory and neuroprotective treatment could be started early in MS before most axons and neurons would be lost.

The early and continuous application of disease-modifying therapies offers the possibility that accumulating axonal degeneration and permanent functional disability can be prevented or delayed. The clinical challenge in this respect is therefore the early decision for an individual MS patient on anti-inflammatory and neuroprotective MS therapy.

2. DETECTING AXONAL DAMAGE IN MS

It is well known that there is axonal loss within chronic MS lesions. The use of immunocytochemical methods that stain axonal end-bulbs demonstrates evidence of axonal injury even in acute and early lesions[4]. Axonal injury in MS lesions will lead to both Wallerian degeneration of the axon and also retrograde degeneration of the cell body. The functional consequences of the axon injury will depend on the numbers of injured axons and the topographical organization of the fibres coursing through the lesion[5].

Besides neuropathology, the other technical advance that has drawn attention to axonal loss in MS is the use of magnetic resonance imaging (MRI) and spectroscopy (MRS)[6]. The use of these techniques is helpful in characterizing the underlying pathologic processes in multiple sclerosis. There is consensus that T2-weighted MRI reflects the broad spectrum of pathological changes, including inflammation, edema, demyelination, gliosis and axonal loss. Changes in the number and volume of lesions on T2-weighted MRI (lesion load) are sensitive but nonspecific indicators of disease activity and the response to treatment. There is evidence that – below the detection treshold - the normal-appearing white matter is not normal at all in patients with MS.

Besides inflammatory markers, techniques to quantify images from MRI have revealed significant tissue atrophy in the spinal cord and brain in MS patients. These measures of whole tissue cross-sections or volume do not, however, discriminate between myelin and axonal loss. A powerful technique for the analysis of the biochemical components of tissue in life is magnetic resonance spectroscopy (MRS). The normal proton spectrum in brain tissue is dominated by a signal from N-acetyl aspartate (NAA), an amino acid that appears to be specifically localized to neuronal cell bodies and axons[7]. There are a number of studies demonstrating that the amount of NAA is decreased in MS lesions but importantly also in apparently normal white matter[8,9].

Hypointense lesions on enhanced T1-weighted images ("T1 black holes") have been reported to correspond to areas where chronic severe tissue disruption has occurred[10]. When one considers the evolution of a MS lesion, the T1 black hole usually represents the end stage of the process, when significant demyelination, axonal loss, and reactive gliosis have occurred. The degree of hypointensity appears to correlate with a decrease in the magnetization transfer ratio (MTR) and with axonal loss as quantified by a reduction in the NAA peak on MR spectroscopy or in histopathology[11,12] (Figure 2). This measure is beginning to show a better correlation with disease progression than either T2 disease burden or Gadolinium (Gd)-enhancing lesions in patients with secondary progressive MS. At the time of Gd-enhancement, a significant proportion of lesions will demonstrate some hypointensity on T1-weighted images. Over a period of time, most enhancing lesions become isointense to white matter and their MTR returns to that of normal-appearing white matter. A proportion of Gd-enhancing lesions will remain hypointense and eventually become T1 black holes.

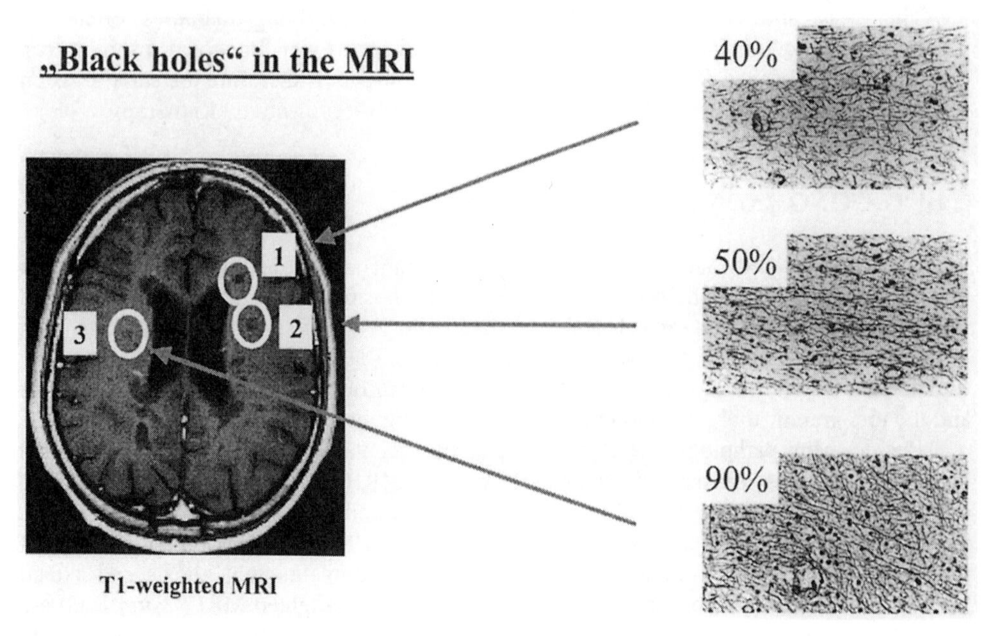

Figure 2: T1 hypointense lesions (black holes) are strongly associated with axonal density, emphasizing their role in monitoring progression in multiple sclerosis.

Roughly 30% to 40% of new lesions will evolve into persistent black holes over short time periods (5 to 12 months) and represent severe and irreversible tissue disruption. Evidence from a large number of postmortem and in vivo MRI studies have substantiated that permanent T1 hypointensive lesions correspond to areas of severe axonal damage and myelin loss. Consistent with these findings, data also indicate that the overall extent of hypointense brain lesions correlates with the degree of MS-related disability[12].

3. NEUROTROPHIC FACTORS AS SPECIAL NEUROTROPHIC AGENTS

Damaged neurons in the CNS attempt to repair themselves although these attemps are usually not succesful. An important component of restoration of function expecially in a disorder like MS in which axons are damaged, must include the potential of axonal repair. One of the more promising approaches to encourage axonal growth is the administration of growth-supporting molecules, particulary specific growth factors.

Historically, nerve growth factor (NGF) was first and for a long time, the only known neurotrophic factor which was primarily best characterized by its anti-apoptotic function on neurons during development. Following the discovery of structurally related proteins with a similar neurotrophic function, the term "neurotrophin" was introduced for this protein family of homodimers with a conserved region containing a cystein bond in the core of the molecule and with duplicate sites for receptor binding[13].

The neurotrophins of the NGF family are not the only proteins with neurotrophic function (Table 1). In recent years, two additional families of protein growth factors have been characterized, which exert strong neurotrophic activity on developing neurons. The first one is the family of the glial-cell-derived neurotrophic factor (GDNF) ligands (GFLs) including GDNF and three related proteins[14]. The second family is formed by the neuropoietic cytokines, which besides other more pleiotropic cytokines includes ciliary neurotrophic factor (CNTF) and leukemia inhibitory factor (LIF)[15]. Although structurally different, these three families are now collectively referred to as neurotrophic factors. In addition, neutroprotective activity has been reported for growth factors not belonging to any of the three neurotrophic factor families, one prominent example beeing insulin-like growth factor (IGF)-1[16].

Table 1: Different protein families with neurotrophic function

NGF-related neurotrophins	Nerve growth factor (NGF) Brain-derived neurotrophic factor (BDNF) Neurotrophin (NT)-3 Neurotrophin (NT)-4/5
GDNF family ligands	Glial-cell-derived neurotrophic factor (GDNF) Neurturin Artemin Persephin
Neuropoetic cytokines	Ciliary neurotrophic factor (CNTF) Leukemia inhibitory factor (LIF)
Miscellaneous factors	Insulin-like growth factor (IGF)-1 Neuregulins (GGF-2)

In a therapeutical context it is important to note that the functions of neurotrophic factors are not restricted to neural development. It is evident that neurotrophins act on mature neurons, most prominently on injured and degenerating nerve cells[17,18]. Neurotrophic factors can protect and rescue neurons in a large number of experimental models. Of particular relevance to MS is the demonstrated ability of BDNF and NT-3 to promote regeneration of long tracts in the spinal cord[19]. In addition, the expression levels of neurotrophins and their receptors are strongly regulated in pathological conditions, thus arguing for a role of these proteins in the response of neurons to traumatic or degenerative processes.

Neurotrophins are also involved in the development and maintenance of glia, including oligodendroglia. NT-3 stimulates the proliferation of oligodendrocyte progenitors in vitro, and both NT-3 and NGF enhance the survival of differentiated oligodendrocytes in cultures[20].

The role of the neurotrophins in demyelinating diseases is presently unclear. Levels of NGF are increased in the CSF and the optic nerve of MS patients and in the brains of animals with the MS model disease, the experimental autoimmune encephalomyelitis (EAE)[21]. There is no information available concerning BDNF or NT-3 in animal and human demyelinating disease. Exogenous NGF can prevent autoimmune demyelination in marmorsets[22]. Whether this is due to inhibition of the immune attack or elicitation of protective responses in oligodendrocytes remains to be determined.

4. THE JANUS FACE OF CNS-DIRECTED AUTOIMMUNE INFLAMMATION

Inflammation is considered to be a key feature in MS pathogenesis. The neurotoxic effects of inflammation are well established and thought to be at least partially responsible for the observed axonal damage. Recently, an increasing body of experimental ecidence supports the view of a dual role of the immune system in CNS-directed autoimmune inflammation. A number of recent studies have proposed that autoimmune inflammation may have neuroprotective effects in the CNS.

On the one hand, MS and its animal models, representing neuroimmunological diseases which arise when immune cells attack the nervous system, provide the paradigm for the deleterious interaction between cells of the immune and nervous system. EAE, the MS animal model, can be induced by active immunization with CNS autoantigens [e.g., myelin basic protein (MBP)], or by the transfer of autoantigen-specific T cells into naive syngeneic recipients[23]. On the other hand, it was recently demonstrated that MBP-specific, encephalitogenic T cells may have seemingly neuroprotective (side-)effects. The neuroprotective and regenerative potential of immune cells was first coined by Michal Schwartz and her colleagues[24].

They demonstrated that autoimmune T cells could protect neurons in an animal model of secondary degeneration after a partial crush injury of the optic nerve[25]. In several experiments, T cells with different specificities (specific for MBP, for the control antigen ovalbumin (OVA), or a heatshock protein (hsp) peptide) were activated by restimulation with their respective antigens in vitro, and then injected into rats immediately after an unilateral optic nerve injury. Seven days after injury, the optic nerves were analyzed by immunohistochemistry for the presence of T cells. Small numbers of T cells could be found in the intact (uninjured) optic nerves of rats injected with anti-MBP T cells which is consistent with previous observations that activated MBP-specific T cells home to intact CNS white matter.

A much more pronounced accumulation of T cells, however, was observed *in the crushed optic nerves* of the rats injected with T cells specific for MBP, hsp peptide, or OVA. The degree of primary and secondary damage to the optic nerve axons and their attached retinal ganglion cells was measured by injecting a neurotracer distal to the site of the optic nerve lesion immediately after the injury, and again after two weeks. The number of labeled retinal ganglion cells as marker for viable axons was significantly greater in the retinas of the rats injected with anti-MBP T cells than in the retinas of rats injected with anti-OVA or anti-hsp peptide T cells. Thus, although all three T-cell lines accumulated at the site of injury, only the MBP-specific, autoimmune T cells had a substantial effect in limiting the extent of secondary degeneration. The neuroprotective effect was confirmed by electrophysiological studies.

The results demonstrate that T-cell autoimmunity can mediate significant neuroprotection after CNS injury. The authors speculate that after injury, 'cryptic' epitopes might become available and might be recognized by endogenous non-encephalitogenic (benign) T cells. After local stimulation, these protective autoreactive T cells could exert their neuroprotective effect. The findings further substantiate the idea that 'natural autoimmunity' can be benign and may even function as a protective mechanism[26]

Macrophages seem to represent another type of immune cell that is capable of mediating neuroprotection and/or stimulating recovery of CNS lesions: The injection of activated macrophages into transected rat spinal cord stimulated tissue repair and partial recovery of motor function[27]. The neuroprotective activity of immune cells is not restricted to the CNS. After experimental axotomy of the facial nerve of immunodeficient SCID mice, the survival of facial motor neurons was severely impaired compared to immunocompetent wild-type mice. Reconstitution of SCID mice with wild-type splenocytes containing T and B cells restored the survival of facial motor neurons in these mice to the level of the wild-type controls[28].

Proinflammatory and neurotoxic factors

TH1 cytokines
TNF
IL-1
Osteopontin
Leukotrienes
Chemokines
MMP
Plasminogen activators
Nitric oxide
Reactive oxygen species
Glutamate
Antibody +/- complement
Cell-mediated cytotoxicity

Destruction

Antiinflammatory and neuroprotective factors

TH2 cytokines
TGF
TNF ?
Soluble TNF receptor
Soluble IL-1 receptor
IL-1 receptor antagonist
Some prostaglandins
Some chemokines
Lipoxins
TIMP
Antithrombin
Neurotrophic factors
like BDNF, NGF

Protection

Figure 3: The Janus face of CNS-directed autoimmune inflammation (adapted from [29])

It is equally evident that a large number of neurotoxic and proinflammatory mediators are produced and released by immune cells (Figure 3)[29]. The neutralization of toxic inflammatory mediators may improve the outcome of MS and experimental models of CNS damage. On the other hand we know that immune cells can supply a number of neuroprotective mediators including neurotrophic factors, anti-inflammatory cytokines and prostaglandins which may reduce tissue damage[30].

Anti-myelin specific antibodies serve as a good example for the Janus face of inflammatory responses in the CNS. On the one hand, they are key players in the demyelinating process[31]. On the other hand, they can promote remyelination and neuroregeneration[32]. There seems to exist a balance between destructive and protective components of inflammation in the CNS.

The idea that inflammatory reactions may not always be harmful under certain conditions even confer neuroprotection and repair has important consequences for the design of immunomodulatory therapies for MS[30]. Undebatably, there is convincing rationale for immunosuppressive treatment when the noxious effects of the inflammatory reaction prevail. Because nonselective immunosuppressive treatments will suppress both destructive and beneficial components of inflammation, therapy is likely to fail when the beneficial effects of CNS inflammation outweigh its negative consequences. It seems for example possible that the lack of beneficial (side) effects of inflammation during the late phase of MS with little inflammation but ongoing axonal loss contributes to the pathogenesis of neurodegeneration. In MS it is unfortunately unclear whether there is a stage of the disease when the inflammatory reaction is more beneficial than harmful.

The concept of the neuroprotective role of inflammation can be extended to neurodegenerative, ischemic and traumatic lesions of the CNS, considering that inflammation is a universal tissue reaction crucial for defense and repair[33].

5. NEUROTROPHIC FACTORS ARE RELEASED BY DIFFERENT IMMUNE CELLS

The precise mechanisms involved in immune-mediated neuroprotection remain to be clarified. A number of recent studies have shown that several types of immune cells and hematogenic progenitor cells express one or more neurotrophic factors. For example, nerve growth factor (NGF) is produced by B cells, which also express the trkA receptor and p75 NGF receptor[34]. Because neutralization of endogenous NGF caused apoptosis of memory B cells, it was concluded that NGF is an autocrine growth factor for memory B cells.

More recently, another neurotrophin, brain-derived neurotrophic factor (BDNF), was found to be expressed in immune cells. BDNF was originally cloned in 1989 as the second member of the neurotrophin family which includes nerve growth factor (NGF) and neurotrophins (NT)-3, -4/5, -6 and -7[35]. Since then, the important role of BDNF in regulating the survival and differentiation of various neuronal populations including sensory neurons, cerebellar neurons and spinal motor neurons has been firmly established[35]. Neurons are the major source of BDNF in the nervous system[35]. BDNF binds to different types of receptors: the tyrosine kinase receptor B (trkB) which exists in two isoforms – the full length receptor (gp145trkB) and the truncated receptor (gp95trkB) lacking the tyrosine kinase domain – and the p75 neurotrophin receptor[36]. It is thought that BDNF and NT4/5 exert their biological function via the full-length form of trkB receptor which expression seems to be restricted to neuronal cell populations.

Immune cells can be a potent source of the neuroprotective factor BDNF in neuroinflammatory disease[30]. Activated human T cells, B cells and monocytes are able to secrete bioactive BDNF after in vitro activation[37]. The BDNF secreted by immune cells is bioactive as it supports neuronal survival in vitro. In histology, BDNF-immunoreactivity was found in T cells and macrophages in active and inactive MS lesions.

Similar observations were recently reported by several other groups of investigators. After experimental injury of the striatum activated macrophages and microglia cells transcribe mRNA for glial cell line-derived neurotrophic factor (GDNF) and BDNF[38]. This could help to explain the sprouting of dopaminergic neurons observed after experimental injury. Transcripts for BDNF and NT-3 and their receptors trkB and trkC were found in subpopulations of human peripheral blood cells[39]. BDNF protein was secreted by cultured T cell clones. Several neurotrophins and their receptors have recently been demonstrated in human bone marrow[40]. More types of immune cells need to be examined for possible functional effects of neurotrophins.

Besides BDNF, there is a robust expression of the full-length BDNF receptor gp145trkB in neurones in the immediate vicinity of multiple sclerosis plaques[41]. Single neurones with clearly pronounced trkB-immunoreactivity close to multiple sclerosis lesions can be observed, suggesting an upregulation of trkB in a proportion of damaged neurones. Additionally, full-length trkB immunoreactivity is present in reactive astrocytes within the lesions. The restriction of trkB expression to neural cell types in multiple sclerosis underscores the possibility that there may be BDNF signalling from infiltrating cells to neurones in neuroinflammatory lesions, as would be necessary for immune cells to support neuronal survival or to provide axonal protection.

Table 2: Cell-type specific expression of BDNF and BDNF full-length receptor TrkB in MS lesions[37,41]

	BDNF expression	TrkB expression
Astrocytes	+	+
Oligodendrocytes	-	-
Neurons	+	+
Inflammatory cells	+	-

The appeal of neurotrophin-mediated neuroprotection in MS lies in the pleiotropy inherent to neurotrophin actions[35]. It is well established that BDNF can prevent neuronal cell death after various pathological insults including experimental transection of axons in the spinal cord[42,43]. Moreover, BDNF can also protect axons against elimination during development, as well as against degeneration after axotomy or in experimental neurodegenerative diseases[44,45]. Furthermore, the preservation of axons provides the basis for neuroregenerative attempts including axonal regeneration and sprouting, which are directly supported by BDNF[43,46]. Apart from influencing survival and regeneration of neuronal elements, BDNF supports remyelination after both peripheral and CNS injury[47]. Finally, BDNF has been shown to downregulate the expression of MHC (major histocompatibility complex) molecules in hippocampal slices, and may thus also act as an immunomodulator[48].

Within the multiple sclerosis lesion immune cells seem to be the major source of BDNF[41]. They are likely to release this substance in the immediate vicinity of nerve cell processes, which - according to the observed trkB expression - are likely to be responsive to the neuroprotective effects of BDNF. This neurotrophin-mediated neuroimmune signalling network could be a major factor that helps to preserve axons in a microenvironment that is clearly capable of exerting significant neurotoxicity. Thus, it should be considered as a beneficial aspect of neuroinflammation that could be worth preserving therapeutically, or even reinforcing using tailored immunomodulatory treatment strategies.

6. NEUROTROPHIC FACTORS AS THERAPY OPTION IN MS

According to current opinion, the treatment of multiple sclerosis has two major objectives, namely (a) suppression of the inflammatory process, and (b) restoration and protection of glial and neuronal function[49]. The potential neuroprotective function of inflammatory cells is relevant to both these treatment goals.

A number of studies have shown that the administration of BDNF protein or the BDNF gene can rescue injured or degenerating neurons and induce axonal outgrowth and regeneration[42,43,47]. Furthermore, BDNF had beneficial effects in several animal models of neurodegenerative diseases[18]. Difficulties in delivering sufficient amounts of BDNF to the site of CNS lesions have so far hampered the successful application of BDNF and other neurotrophic factors for treatment of human diseases[50]. To the best of our knowledge, systemically administered neurotrophic factors do not cross the blood brain barrier. For the treatment of MS, it would seem that they must be administered directly into the CNS.

One promising novel strategy for the delivery of neuroprotective factors relies on the (retroviral) transduction of one or several neurotrophic factors into antigen-specific T cell lines[51]. As the transduced T cells are specific for an autoantigen expressed in the nervous system, they home to the sites where the relevant autoantigen is expressed, recognize their antigen and are then stimulated locally to secrete neurotrophic factor(s)[52]. Unfortunately, these autoimmune T cells could lead to tissue injury like in EAE.

This experimental strategy has a natural counterpart in the secretion of neurotrophic factors by activated immune cells. However, it appears that the neurotrophins secreted by immune cells under natural conditions are often insufficient to prevent injury. It will therefore be worthwhile to further refine the strategies to enhance the production of neurotrophic factors by immune cells and to exploit the homing properties of the immune cells for targeting neuroprotective factors into the nervous system[51,52].

7. DRUG DEVELOPMENT OF GLATIRAMER ACETATE

Glatiramer acetate (Copaxone®, GA), formerly known as copolymer 1, is the acetate salt of a standardized mixture of synthetic polypeptides containing the four amino acids L-alanine, L-glutamic acid, L-lysine and L-tyrosine with a defined molar ratio of 0.14 : 0.34 : 0.43 : 0.09 and an average molecular mass of 4.7 – 11.0 kDa, i.e. an average length of 45-100 amino acids[53,54].

In the 1960s Drs Sela, Arnon and their colleagues at the Weizmann Institute, Israel were involved in studies on the immunological properties of a series of polymers and copolymers which were developed to resemble myelin basic protein (MBP), a myelin protein. MBP in Freund's complete adjuvant induces EAE, the best animal model of MS. They were interested in evaluating whether these polypeptides could simulate the ability of MBP and of fragments and regions of the MBP molecule to induce EAE[55-58]. None of these series was capable of inducing EAE, but several polypeptides were able to suppress EAE in guinea pigs. Copolymer1, later known as GA was shown to be the most effective polymer in preventing or decreasing the severity of EAE. The suppressive effect is a general phenomenon and not restricted to a particular species, disease type or encephalitogen used for EAE induction[59].

Abramsky et al. were the first who treated a group of severe relapsing-remitting MS patients with intramuscular GA 2-3 mg every 2-3 days for 3 weeks, then weekly for 2-5 months[60]. No conclusions could be drawn regarding drug efficacy but there were no significant undesirable side effects. Three clinical trials in the 1980s were performed showing some evidence of efficacy that was adequate for support of the Food and Drug Administration (FDA) approval and a good safety profile[61-63]. However the results of these studies must be interpreted with caution because before 1991 the production of the drug was not standardized[54,64]. Different batches had variable suppressive effects on EAE which could also imply variable effects in MS patients.

In 1991 a phase III multicentre trial with a daily 20 mg dose of a s.c. administered, highly standardized GA preparation was started in the USA. This double-blind placebo controlled study demonstrated that GA significantly reduced the relapse rate without significant side effects[65].

In 1996 GA was approved by the US FDA as a treatment for ambulatory patients with active relapsing-remitting (RR) MS. Since then GA has been licensed for approval in many other countries[59].

8. GLATIRAMER ACETATE: MECHANISM OF ACTION

Until recently the effects of GA on the human immune system and the mechanisms of action of GA were largely unknown. Most of the data have been obtained in animal models so far. Several new papers, however, have shed light on the mechanisms of GA in MS und suggest several major effects of GA on human T cells[66]. In contrast to other immunomodulatory MS therapies which exert its effects in an antigen-nonspecific way, GA appears to preferentially affect immune cells specific for GA, MBP and possibly other myelin autoantigens implicated in the MS disease process.

In contrast to the lack of effect of GA on immune cells isolated from untreated animals, GA induces vigorous polyclonal proliferation of peripheral blood lymphocytes (PBL) from untreated (unprimed) human donors[67-72]. In GA-treated patients, the proliferative response to GA decreases with time[73]. Recent results from our group indicate that this decrease is specific to GA, as it is not observed with recall antigens like tetanus toxoid and tuberculin[71]. Theoretically, the observed decrease in GA-reactive T cells could be due to anergy induction or activation-induced cell death of GA-specific T cells.

GA binds to major histocompatibility complex (MHC) class II and perhaps to MHC class I molecules, thereby competing with the MHC binding of other antigens[75,76]. This effect, which by its nature is antigen-nonspecific, is unlikely to play a role in vivo, since after subcutaneous (s.c.) administration, GA is quickly degraded and thus it is not likely to reach the CNS where it could compete with the relevant auto-antigens for MHC binding[66]. Complexes of GA/MHC can compete with MBP/MHC for binding to the antigen-specific surface receptor of MBP-specific T cells (T cell receptor (TCR) antagonism)[77]. The experimental evidence supporting this effect is controversial[78]. If it occurs, it is unlikely to be relevant in vivo, since GA is unlikely to reach sites where it could compete with MBP.

Table 3: Major immunologic effects of glatiramer acetate (adapted from [66,74])

T cell proliferation in vitro	Suppression of proliferation of MBP-reactive T cells in vitro
Proliferation of PBL during treatment with GA	Decreased proliferation of PBL to GA during treatment
MHC (Major histocompatibility complex) binding	Direct and promiscuous binding of GA to different HLA-DR alleles
T cell migration:	Reduced migration of PBL from GA-treated patients (unknown mechanism); no effect on adhesion molecule expression on human brain microvascular endothelial cells
TH1 -> TH2 shift in PBL	Increased levels of IL-10 in serum and of mRNA for TGF-β and IL-4; reduction of mRNA for TNF-α in PBL
TH1 - > TH2 shift in GA-specific T cells	Shift of GA-reactive T cells from TH1 towards TH2 phenotype during GA treatment
Cross-stimulation of T cells with MBP	Induction by GA of cytokine production in MBP-specific T cells and vice versa
Cross-inhibition of MBP-specific T cell lines	Inhibition of proliferation of T cells specific for MBP and some other antigens
T cell receptor (TCR) antagonism	TCR antagonism with MBP 82–100 (controversial findings)
Altered peptide ligand effect on MBP-specific T cells	Induction of anergy in MBP-specific T cell clones
GA-specific antibodies	Anti-GA-antibodies of IgG2 >> IgG1 isotype without MBP crossreactivity and without neutralizing effect (maximum titer at month 4 after onset of GA-treatment); anti-GA IgG4 antibodies only detectable in GA-treated patients
GA-specific CD8 T cells	Disminished GA-specific CD8+ T cell proliferation in untreated MS patients which is restored by GA treatment
Effects on antigen presenting cells	Inhibition of TNF-α and cathepsin-B production in a monocytic cell line

On the other hand, GA could act in the periphery as an "altered peptide ligand" relative to MBP[79-82]. As a consequence, some of the circulating myelin-specific, potentially pathogenic T cells might become "anergic" or be otherwise changed in their properties, e.g. in their migratory potential. This effect would be relatively antigen-specific and presumably occur in the periphery at the injection sites or in the corresponding draining lymph nodes where MBP-specific T cells might be confronted with GA. Although some in vitro findings support this mechanism, it is not yet known whether the functional properties of MBP-specific T cells are altered in GA-treated patients. It may be of relevance in this context that we were unable to isolate MBP-specific TCL from GA-treated patients[83].

GA treatment induces an in vivo change of the cytokine secretion pattern and the effector function of GA-reactive T helper (TH) cells, a so called TH1-to-TH2-shift[70,71,83-87]. TH cells can be divided into several types based on their characteristics[88-90]. TH1 cells produce proinflammatory cytokines such as IL-2, IL-12, IFN-γ, and tumor necrosis factor (TNF)-α. In contrast , the TH2/TH3 cells produce antiinflammatory cytokines such as IL-4, IL-5, IL-6, IL-10, transforming growth factor (TGF)-β and IL-13. Different lines of evidence suggest that GA treatment changes the properties of the GA-reactive T cells in such a way that they increasingly become TH2-like with time. Using intracellular double-immunofluorescence flow cytometry, we demonstrated that long-term GA-specific T cell lines from untreated MS patients and healthy controls predominantly produce IFN-γ and are to be classified as TH1 cells, whereas GA-reactive T cell lines from GA-treated MS patients predominantly produce IL-4, i.e. behave like TH2 cells[83].

In addition, the study of Farina et al. demonstrated that an automated ELISPOT assay, which is able to detect cytokine production of individual PBL, allows the correct identification of GA-treated and untreated donors in most cases[71]. GA-treated MS patients show (a) a significant reduction of GA-induced proliferation of peripheral blood mononuclear cells, (b) a positive IL-4 ELISPOT response mediated predominantly by CD4+ T cells after in vitro stimulation with a wide range of GA concentrations, and (c) an elevated IFN-γ response partially mediated by CD8+ T cells after stimulation with high GA concentrations. GA-reactive T cells seem to be not physically deleted, but rather they are modified in such a way that they respond to in vitro challenge with GA by secretion of cytokines but not by proliferation. This ELISPOT assay may provide a promising additional tool for monitoring the treatment response in MS patients treated with GA[91].

Chen et al. have recently reported that the described immonological changes (eg. the TH1 -> TH2-shift) are sustained over long treatment periods[92]. While GA treatment up to 9 years, there was a sustained TH2-biased GA response which is in part cross-reactive with MBP. This seems to be consistent with the longterm antiinflammatory effects of the drug and bystander suppression.

Karandikar et al. have recently shown that the proliferative responses to GA are different in distinct subsets of T cells[93]. Whereas GA-induced CD4+ T cell responses were comparable in healthy individuals and MS patients, CD8+ T cell responses were significantly lower in untreated MS patients. Treatment with GA resulted in upregulation of these CD8+ responses with restoration to levels observed in healthy individuals which may suggest a role of these responses in the immunomodulatory effects of the drug.

All GA-treated patients developed GA-reactive antibodies which declined after 6 months and remained low[73]. All antibodies were of the IgG class with IgG1 isotype levels two- to threefold higher than those of IgG2. In this setting, IgG reflected the interaction of B cells with TH1 or TH2 lymphocytes. These data support a GA-induced TH1 to TH2 profile shift. Using a murine model of demyelinating disease, Ure et al. could demonstrate that remyelination of spinal cord axons is promoted by antibodies to GA[94]. These results support the hypothesis that the antibody response in GA-treated patients may be beneficial by facilitating repair of demyelinated lesions.

9. BDNF SECRETION OF GLATIRAMER ACETATE-SPECIFIC T CELLS

As human immune cells can produce brain-derived neurotrophic factor (BDNF)[37] and moreover, the receptor for BDNF, gp145TrkB[41], is expressed in neurons and astrocytes in MS brain lesions, we asked the following questions: (a) Can GA-reactive T lymphocytes produce BDNF, and if so, (b) do TH1- and TH2/TH3-type GA-specific T cells differ in their capacitity to secrete BDNF?

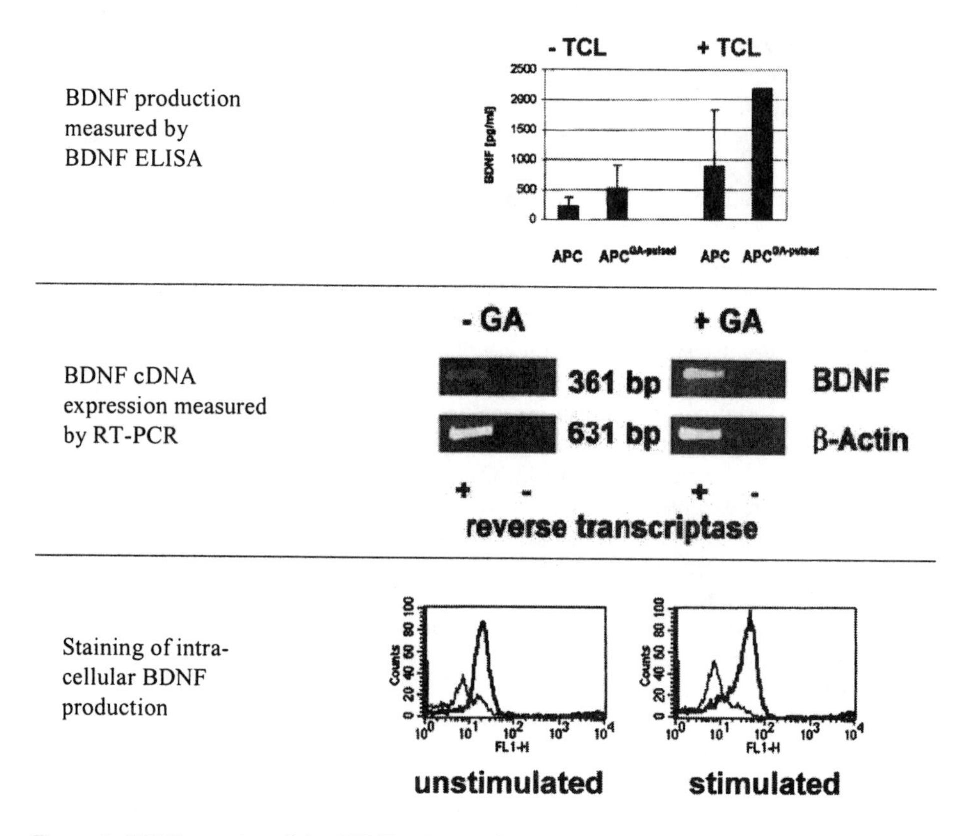

Figure 4: BDNF secretion of the TH2-like GA-specific T cell line BK-M6-COP-7 of a GA-treated patient, measured by BDNF ELISA, by RT-PCR and by intracellular BDNF staining. There is increased BDNF production of GA-specific T cells after activation (taken with permission from Ziemssen et al.[95] (2002) ®Oxford University Press)

To answer these questions, we had to overcome two major technical obstacles: (a) adapting our culture system to prevent added GA from affecting the BDNF enzyme-linked immunosorbent assay (ELISA), and (b) optimizing the intracellular detection of BDNF in individual T-lymphocytes[95]. That ladder enabled us formally to demonstrate that GA-specific TH1-, TH2-, and TH0 cells all have the capacity to produce BDNF. We clearly show that GA-specific, activated TH0, TH1 and TH2 cells produce the neurotrophic factor BDNF – not only at the transcriptional (mRNA) level as detected by RT-PCR, but also at the protein level – using the newly developed assays for BDNF secretion and synthesis of BDNF in individual T cells. The results from all assays consistently showed a low level of basal secretion of BDNF by GA-specific T cells and an increase of BDNF production after stimulation. For our analysis, we selected four well-characterized prototypic GA-specific TCLs. Similar results were obtained with >20 additional GA-specific TCLs[95].

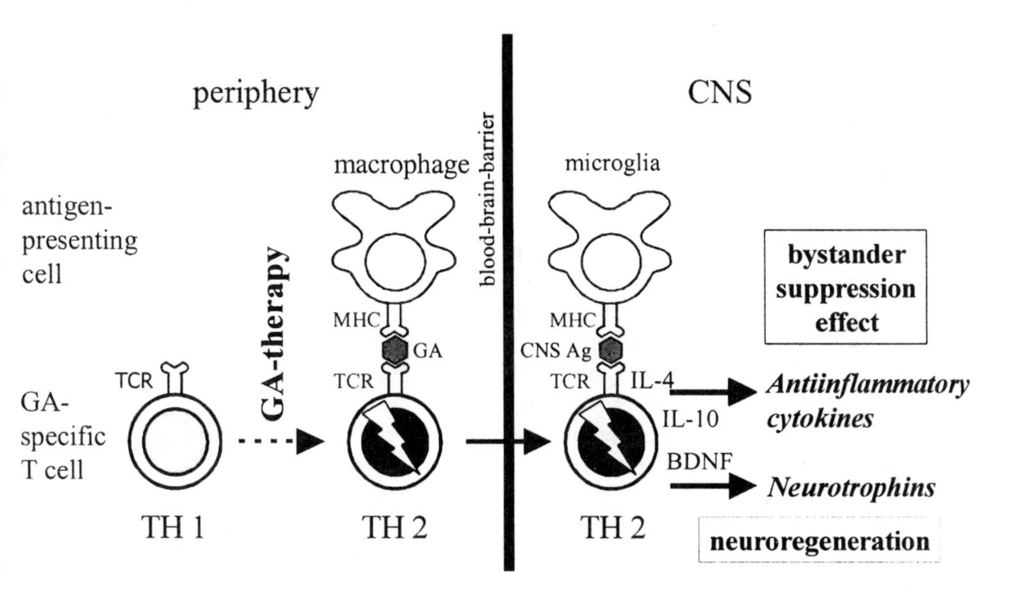

Figure 5: Possible mechanism of action of glatiramer acetate (adapted from [66,74])

Our findings have obvious implications for the presumed mechanism of action of GA. According to current theory, the beneficial effects of GA are basicly mediated by a population of GA-reactive TH2 cells[66,78]. GA treatment induces a gradual shift of GA-reactive T cells from TH1 to TH2. Their constant activation by daily immunization enables them to enter the CNS. Indeed, transferred GA-reactive T cells have been directly demonstrated in the CNS of recipient mice[96]. It is further assumed, that after local recognition of cross-reactive myelin degradation products, the GA-specific TH2 cells are stimulated to secrete antiinflammatory cytokines, which in turn induce bystander suppression in neighbouring encephalitogenic T cells[66,84,85]. Our present findings imply that the locally activated GA-reactive TH2 cells produce not only protective TH2 cytokines, but also BDNF[95].

The following scenario would accommodate for both, the findings in experimental autoimmune encephalomyelitis and the observations in human multiple sclerosis (Figure 5)[66,74].

GA-specific activated T cells can pass through the blood-brain barrier. Inside the CNS, some GA-specific T cells cross-react with products of local myelin turnover presented by local APCs. In our previous studies, 8 ± 15% of GA-reactive TCLs crossreacted with MBP as indicated by cytokine secretion, but not by proliferation[83]. Recognition of crossreactive antigens at the lesion site seems to be necessary for reactivation of the protective T cells and for their neuroprotective effect. In rat models, only MBP- and GA-specific, but not ovalbumin-specific, T cells confer neuroprotection[25,97]. After reactivation in situ, GA-specificTH2/TH3-like regulatory T cells would not only provide antiinflammatory cytokines like IL-4, IL-5, IL-13, and TGF-β, but also BDNF. Furthermore, TH1-like GA-reactive T cells, which are reduced but still present in GA-treated patients[71,83], would also have the capacity to release BDNF in multiple sclerosis lesions[95].

BDNF imported by GA-reactive T cells might help tip the environmental balance in favour of the beneficial, neuroprotective influences. Indeed, in a rat neurodegenerative model, Kipnis and collegues showed that rat immune cells stimulated by immunization with GA secreted neurotrophic factors upon restimulation with GA[97].

The above concept may be extended to other types of immunomodulatory therapy, especially with “altered peptide ligands” (APL)[79]. By definition, these ligands derived from immunogenic peptide antigens by the selective alteration of one or more T-cell receptor contacting residues. In an extrapolation from these recent results to altered peptide ligand (APL) therapies in general, we would expect that APL-specific TH2 and TH1 cells are also capable of producing BDNF as do GA-specific TH2 and TH1 cells. However, for APL therapy to work, it is obviously crucial that neither the TH1 nor the TH2 anti-APL response be encephalitogenic. An ideal APL would induce a nonencephalitogenic TH2 response that mediates local bystander suppression and neuroprotection. If both conditions were met, the TH1 cells could contribute to the therapeutic effect by acting as innocent carriers of neurotrophic factors.

10. NEUROPROTECTIVE EFFECTS OF GLATIRAMER ACETATE IN ANIMAL MODELS

As stated above, it was demonstrated that autoimmune T cells directed against central nervous system-asscociated myelin antigens protect neurons from secondary degeneration[25]. Because it would be desirable to obtain “autoimmune neuroprotection” free of any possible autoimmune disease, the non-encephalitogenic copolymer GA was analyzed in different models of secondary neuronal degeneration. Kipnis et al. demonstrated that the posttraumatic spread of degeneration in the damaged optic nerve can be attenuated by active immunization with GA on the day of injury and by adoptive transfer of GA-specific T cells[97]. In the development of therapies based on autoimmune neuroprotection, the use of

GA seems to involve "safe" antigenic epitopes which will not cause the induction of an autoimmune disease.

GA-reactive T cells were also found to be neuroprotective in other models of CNS injury, where myelin-associated antigens are not active, such as the insult caused by direct exposure of retinal ganglion cells to glutamate toxicity or the death of retinal ganglion cells resulting from increased intraocular pressure in a model of high-tension glaucoma[98]. Following injury, peripheral lymphocytes, regardless of their antigenic specificity, enter the CNS. T cells reactive to myelin antigens are activated at the site of injury or in the cervical lymph nodes which is a prerequisite for neuroprotection. The neuroprotective effects of GA in the neurodegenerative animal models suggest that upon passive transfer or active immunization with GA there is an accumulation and local reactivation of GA-specific T cells because of the crossreactivity with myelin proteins in vivo. Activated GA-reactive T cells can then produce neurotrophic factors besides antiinflammatory factors within the region of injured and damaged neurons and axons.

In addition, Often et al. showed a neuroprotective effect in the MS-model disease, the myelin oligodendrocyte glycoprotein (MOG)-induced EAE[99]. This form of EAE is characterized by axonal loss and neuronal damage in addition to the EAE-typical inflammatory demyelination. As exspected from former EAE studies, the treatment of MOG-EAE with GA lead to a significant lower clinical disease and strongly reduced inflammatory changes. In this study, the investigators were able to demonstrate a strong significant reduction of axonal loss and neuronal damage by GA treatment.

11. CLINICAL MS STUDIES OF GLATIRAMER ACETATE

Four early exploratory open studies were performed in the late 1970s and early 1980s in order to obtain indications of dosing and safety[60-62,100]. In total, 41 patients with relapsing-remitting or secondary progressive MS were enrolled in these studies. The treatment schedules, doses of GA and treatment duration were quite variable from study to study. The maximum dose of 20 mg/day was well tolerated and no severe adverse effects were detected in these studies.

Because there was evidence of some potential benefit, Bornstein et al. started a double-blind, placebo controlled pilot trial[63]. 50 MS patients received either 20 mg GA dissolved in 1 ml saline or just saline for 2 years. There were 62 relapses in the placebo group (average 2.7) and 16 among patients in the GA treated group (average 0.6). In addition, the proportion of relapse-free patients was twice as high in the GA treated group. Only 5 patients of the treatment group showed a confirmed progression of disability, compared to 11 patients in the placebo group (p=0.005). The effects on relapse rate and on disability were more pronounced in patients who had the least disability at entry. This study also reported the occurrence of a post-injection systemic reaction in two patients.

To assess more comprehensively the efficacy of GA in patients with RR MS, a definitive phase III trial was conducted at 11 US medical centres with a total of 251 RR MS patients who received GA at a dosage of 20 mg or placebo by daily s.c. injection for 2 years[65]. As the primary end point, the mean annualized relapse rates were 0.59 for the GA-treated group and 0.84 for the placebo group, a 29% reduction was statistically significant (P=0.007). Trends in the proportion of relapse-free patients and median time to first relapse favoured GA treatment. Patients in both groups with higher disability at entry, measured by the Expanded Disability Status Scale (EDSS)[101], had a higher relapse rate, while the largest reduction in relapse rate between groups occurred in patients with a baseline EDSS of 0-2 (33% versus a reduction of 22% in patients with an EDSS score at entry >2). When the proportion of patients who improved, were unchanged or worsened by 1 EDSS step from baseline to end of study (2 years) was evaluated, significantly more patients on GA improved and more on placebo worsened (p=0.037). The effect of treatment was constant throughout the entire study duration. Patient withdrawals were 19 (15.2%) from the copolymer 1 group and 17 (13.5%) from the placebo group at approximately the same intervals. The treatment was well tolerated. The most common adverse experience was the local injection-site reaction. Rarely, the transient self-limiting immediate post-injection reaction followed the injection in 15.2% of those treated with GA and 3.2% of those treated with placebo.

In an extension of this study up to 35 months with unchanged blinding and study conditions the clinical benefit of GA for both the relapse rate and for neurological disability was sustained[102]. Thus, the reduction of annual relapse rate was 32% in favour of GA (p=0.002). The results of this extension study confirmed the excellent tolrabilitye and safety profile of GA. This study was further on extended as an open-label study with all patients receiving active drug. The published data from approximately 6 years of ongoing trial showed a mean annual relapse rate of the treated patients of 0.42[103]. The rate per year continued to drop and for the sixth year it was 0.23. After 6 years of observation, 152 out of 208 patients (73%) were still participating. 69.3% of the group who stayed on GA without interruption for 5 years or more were neurologically unchanged or improved from baseline by at least one step of the EDSS scale. Recently, the 8 years data of this study were presented demonstrating the sustained efficacy of the GA treatment. Currently, this study, in which 10-year data accrual is planned, is the first to demonstrate the benefits of an immunomodulatory treatment over this clinically highly relevant long-term time period.

12. MAGNETIC RESONANCE IMAGING STUDIES OF GLATIRAMER ACETATE

Several early magnetic resonance imaging (MRI) studies indicated a trend toward benefit with GA, but were limited by the small number of evaluated patients. Wolinsky et al. were able to demonstrate a definite, but modest effect of GA on MRI enhancements of MS patients in the US open-label GA extension MRI trial for relapsing multiple sclerosis[104-106].

A large randomized, double blind, placebo-controlled MRI trial with 239 relapsing-remitting MS patients was conducted to determine the effect, onset and durability of any effect of GA on disease activity monitored with MRI monthly over a 9 months period[107]. Treatment of GA showed as primary outcome measure a significant reduction of 30% in the total number of enhancing lesions on T1-weighted images (p=0.003) and - as secondary outcome measure - a significant reduction of the number of new enhancing lesions (p<0.003), the monthly change in the volume of enhancing lesions (p=0.01) and the change in the volume (p=0.006) and the number (p<0.003) of new lesions seen on T2-weighted images compared to placebo. The relapse rate was also significanty reduced by 33% for GA treated patients (p=0.012).

These data were further substantiated in the open-label phase of the trial that extended the study up to 18 months. Expecially, the MRI lesion burden at the study's end was significantly lower in patients initially randomized to GA compared to placebo (p=0.018).

In addition to its effect on lesion load and volume, GA treatment inhibited lesion evolution into permanent black holes[108]. At each time point, the proportion of new lesions evolving into black holes was lower in the GA-treated group compared to placebo, reaching significance at month 7 (18.9% vs. 26.3%; p=0.04). By month 8, the proportion of lesions evolving into black holes was roughly 50% less in the GA treatment group. In addition, in a small study of 27 patients at the University of Pennsylvania followed for 2 years, the GA-treated group had a threefold lower rate of brain parenchyma loss (0.6% per year vs. 1.8% per year; p=0.0078) than the placebo-treated group[104].

13. CONCLUSION

The requirement for long-term therapy dictates that a successful drug should have sustained safety, tolerability and efficacy. Tolerability is especially important to achieve continued patient compliance. An ideal drug for MS should also be useful for patients with a wide range of disabilities.

The proposed indication for the use of GA is to reduce the relapse rate in patients with relapsing-remitting MS, which has been linked adversely with long-term outcome. Several clinical trials have shown a consistent effect of a reduction of the relapse rate by approximately 30% and of accumulation of disability in relapsing-remnitting MS patients treated with GA. The clinical benefit was supported by positive MRI effects. Clinical experience shows that GA is usually well tolerated and suitable for self administration by patients with multiple sclerosis[109].

The ability of GA to downregulate inflammation and act directly at the lesion site facilitates axonal preservation. This effect may be related to the antiinflammatory effect ("bystander suppression") of GA and the ability of activated GA-reactive T cells to release neurotrophic factors. Because axonal loss is an early and irreversible event in MS, early treatment with drugs that mediate antiinflammatory effects and offer neuroprotection is an important factor in preventing disease progression and disability.

REFERENCES

1. Waxman S.G. Demyelinating diseases – new pathological insights, new therapeutic targets. *N Eng J Med.* 338,323-5 (1998).
2. Charcot J.M. Lectures on the diseases of the nervous system (translated by R.M. May). London: New Sydenham Society (1877).
3. Doinikow B. Über De-Regenerationserscheinungen an Achsenzylindern bei der multiplen Sklerose. *Z Ges Neurol Psychiat* 27,151-78 (1915).
4. Trapp B.D., Peterson J., Ransohoff R.M., Rudick R., Mork S., Bo L. Axonal transsection in the lesions of multiple sclerosis. *N Engl J Med* 338,278-85 (1998).
5. Perry V.H., Anthony D.C. Axon damage and repair in multiple sclerosis. *Phil Trans R Soc London B* 354,1641-7 (1999).
6. Filippi M., Grossman R.I. MRI techniques to monitor MS evolution. The present and the future. *Neurology* 58,1147-53 (2002).
7. Urenjak J., Williams S.R., Gadian D.J., Noble M. Proton nuclear magnetic resonance spectroscopy unambiguously identifies different neural cell types. *J Neurosci* 13,981-9 (1993).
8. Davie C.A., Hawkins C.P., Baker G.J., Brennan A., Tofts P.S., Miller D.H., McDonald W.I. Serial proton magnetic resonance spectroscopy in acute multiple sclerosis lesions. *Brain* 117,49-58 (1994).

9. Fu L., Matthews P.M., De Stephano N., Worsley K.J., Narayanan S., Francis G.S., Antel J.P., Wolfson C., Arnold D.L. Imaging axonal damage in normal appearing white matter. *Brain* 121,103-13 (1998).
10. van Waesberghe J.H., van Walderveen M.A., Castelijns J.A., Scheltens P., Lycklama a Nijeholt G.J., Polman C.H., Barkhof F. Patterns of lesion development in multiple sclerosis: longitudinal observations with T1-weighted spin-echo and magnetization transfer. *Am J Neuroradiol* 19,675-83 (1998).
11. Brück W., Bitsch A., Kolenda H., Brück Y., Stiefel M., Lassmann H. Inflammatory central nervous system emyelination: correlation of magnetic resonance imaging findings with lesion pathology. *Ann Neurol* 42,783-93 (1997).
12. van Waesberghe J.H., Kamphorst W., De Groot C.J., van Walderveen M.A., Castelijns J.A., Ravid R., Lycklama a Nijeholt G.J., van der Valk P., Polman C.H., Thompson A.J., Barkhof F. Axonal loss in multiple sclerosis lesions: magnetic resonance imaging insights into substrates of disability. *Ann Neurol* 46,747-54 (1999).
13. Barde Y.A., Edgar D., Thoenen H. New neurotrophic factors. *Ann Rev Physiol* 45,601-12 (1983).
14. Baloh R.H., Enomoto H., Johnson E.M., Milbrandt J. The GDNF family ligands and receptors – implications for neural development. *Curr Opin Neurobiol* 10, 103-10 (2000).
15. Ip N.Y. The neurotrophins and neuropoietic cytokines: two families of growth factors acting on neural and hematopoietic cells. *Ann N Y Acad Sci* 540,97-106 (1998).
16. Heck S., Lezoualch F., Engert S., Behl C. Insulin-like growth factor-1-mediated neuroprotection against oxidative stress is associated with activation of nuclear factor kappa. *B J Biol Chem* 274,9828-35 (1999).
17. Lindvall O., Kokaia Z., Bengzon J., Elmer E., Kokaia M. Neurotrophins and brain insults. *Trends Neurosci* 17,490-6 (1994).
18. Mitsumoto H., Ikeda K., Klinkosz B., Cedarbaum J.M., Wong V., Lindsay R.M. Arrest of motor neuron disease in wobbler mice cotreated with CNTF and BDNF. *Science* 265,1107-10 (1994).
19. Schnell L., Schneider R., Kolbeck R., Barde Y.A., Schwab M.E. Neurotrophin-3 enhances sprouting of corticospinal tract during development and after adult spinal cord lesion. *Nature* 367,170-3 (1994).
20. Kumar S., Kahn M.A., Dinh L., de Vellis J. NT-3-mediated TrkC receptor activation promotes proliferation and cell survival of rodent progenitor oligodendrocyte cells in vitro and in vivo. *J Neurosci Res* 54,754-65 (1998).
21. Laudiero L.B., Aloe L., Levi-Montalcini R., Buttinelli C., Schilter D., Gillessen S., Otten U. Multiple sclerosis patients express increased levels of beta-nerve growth factor in cerebrospinal fluid. *Neurosci Lett* 147,9-12 (1992).
22. Villoslada P., Hauser S.L., Bartke I., Unger J., Heald N., Rosenberg D., Cheung S.W., Mobley W.C., Fisher S., Genain C.P. Human nerve growth factor protects common marmosets against autoimmune encephalomyelitis by switching the balance of T helper cell type 1 and 2 cytokines within the central nervous system. *J Exp Med* 191,1799-806 (2000).
23. Wekerle H., Kojima K., Lannes-Vieira J., Laßmann H., Linington C. Animal models. *Ann Neurol* 36,S47-53 (1994).
24. Schwartz M., Moalem G., Leibowitz-Amit R., Cohen I.R. Innate and adaptive immune responses can be beneficial for CNS repair. *Trends Neurosci* 22,295-9 (1999).
25. Moalem G., Leibowitz-Amit R., Yoles E., Mor F., Cohen I.Schwartz M. Autoimmune T cells protect neurons from secondary degeneration after central nervous system axotomy. *Nature Med* 5,49-55 (1999).
26. Cohen I.R. The cognitive paradigm and the immunological homunculus. *Immunol Today* 13,490-4 (1992).
27. Rapalino O., Lazarov-Spiegler O., Agranov E., Velan G.J., Yoles E., Fraidakis M., Solomon A., Gepstein R., Katz A., Belkin M., Hadani M., Schwartz M. Implantation of stimulated homologous macrophages results in partial recovery of paraplegic rats. *Nat Med* 4:814-21 (1998).
28. Serpe C.J., Kohm A.P. Huppenbauer C.B., Sanders V.J., Jones K.J. Exacerbation of facial motor-neuron loss after facial nerve transection in severe combined immunodeficient (SCID) mice. *J Neurosci* RC7,1-5 (1999).
29. Kerschensteiner M., Stadelmann C., Dechant G., Wekerle H., Hohlfeld R. Neurotrophic cross-talk between the nervous system and immune system: Implications for inflammatory and degenerative neurological diseases. *Ann Neurol* (in press)
30. Hohlfeld R., Kerschensteiner M., Stadelmann C., Lassmann H., Wekerle H. The neuroprotective effect of inflammation: implications for the therapy of multiple sclerosis. *J Neuroimmunol* 107,161-6 (2000).
31. Steinman L. Multiple Sclerosis: a two-stage disease. *Nat Immunol* 2,762-4 (2001).
32. Warrington A.E., Asakura K., Bieber A.J., Ciric B., Van Keulen V., Kaveri S.V., Kyle R.A., Pease L.R., Rodriguez M. Human monoclonal antibodies reactive to oligodendrocytes promote remyelination in a model of multiple sclerosis. *Proc Natl Acad Sci USA* 97,6820-5 (2000).
33. Wekerle H. Linington C., Lassmann H., Meyermann R. Cellular immune reactivity within the CNS. *Trends Neurosci* 9,271-9 (1986).

34. Torcia M., Bracci-Laudiero L., Lucibello M., Nencioni L., Labardi D., Rubartelli A., Cozzolino F., Aloe L., Garaci E. Nerve growth factor is an autocrine survival factor for memory B lymphocytes. *Cell* 85,345-56 (1996).
35. Lewin G.R., Barde Y.A. Physiology of the neurotrophins. *Annu Rev Neurosci* 19,289-317 (1996).
36. Klein R., Nanduri V., Jing S., Lamballe F., Tapley P., Bryant S., Cordon-Cardo C., Jones K.R., Reichardt L.F., Barbacid M. The trkB tyrosine protein kinase is a receptor for brain-derived neurotrophic factor and neurotrophin-3. : *Cell* 66,395-403 (1991).
37. Kerschensteiner M., Gallmeier E., Behrens L., Leal V.V., Misgeld T., Klinkert W.E., Kolbeck R., Hoppe E., Oropeza-Wekerle R.L., Bartke I., Stadelmann C., Lassmann H., Wekerle H., Hohlfeld R. Activated human T cells, B cells, and monocytes produce brain-derived neurotrophic factor in vitro and in inflammatory brain lesions: a neuroprotective role of inflammation? *J Exp Med* 189,865-70 (1999).
38. Batchelor P.E., Liberatore G.T., Wong J.Y., Porritt M.J., Frerichs F., Donnan G.A., Howells D.W. Activated macrophages and microglia induce dopaminergic sprouting in the injured striatum and express brain-derived neurotrophic factor and glial cell line-derived neurotrophic factor. *J Neurosci* 19,1708-16 (1999).
39. Besser M., Wank R. Cutting edge: clonally restricted production of the neurotrophins brain-derived neurotrophic factor and neurotrophin-3 mRNA by human immune cells and Thl/Th2-polarized expression of their receptors. *J Immunol* 162,6303-6 (1999).
40. Labouyrie E., Dubus P., Groppi A., Mahon F.X., Ferrer J., Parrens M., Reiffers J., de Mascarel A., Merlio J.P. Expression of neurotrophins and their receptors in human bone marrow. *Am J Pathol* 154,405-15 (1999).
41. Stadelmann C., Kerschensteiner M., Misgeld T., Bruck W., Hohlfeld R., Lassmann H. BDNF and gpl45trkB in multiple sclerosis brain lesions: neuroprotective interactions between immune and neuronal cells? *Brain* 125,75-85 (2002).
42. Gravel C., Gotz R., Lorrain A., Sendtner M. Adenoviral gene transfer of ciliary neurotrophic factor and brain-derived neurotrophic factor leads to long-term survival of axotomized motor neurons. *Nat Med* 3,765–70 (1997).
43. Kobayashi N.R., Fan D.P., Giehl K.M., Bedard A.M., Wiegand S.J., Tetzlaff W. BDNF and NT-4/5 prevent atrophy of rat rubrospinal neurons after cervical axotomy, stimulate GAP-43 and Talphal-tubulin mRNA expression, and promote axonal regeneration. *J Neurosci* 17,9583–95 (1997).
44. Sagot Y.,. Rossé T., Vejsada R., Perrelet D., Kato A.C.. Differential effects of neurotrophic factors on motoneuron retrograde labeling in a murine model of motoneuron disease. *J Neurosci* 18,1132–41 (1998).
45. Weibel D., Kreutzberg G.W., Schwab M.E.. Brain-derived neurotrophic factor (BDNF) prevents lesion-induced axonal die-back in young rat optic nerve. *Brain Res* 679,249–54 (1995).
46. Mamounas L.A., Altar C.A., Blue M.E., Kaplan D.R., Tessarollo L., Lyons W.E. BDNF promotes the regenerative sprouting, but not survival, of injured serotonergic axons in the adult rat brain. *J Neurosci* 20,771–82 (2000).
47. McTigue D.M., Horner P.J., Stokes B.T., Gage F.H. Neurotrophin-3 and brain-derived neurotrophic factor induce oligodendrocyte proliferation and myelination of regenerating axons in the contused adult rat spinal cord. *J Neurosci* 18,5354–65 (1998).
48. Neumann H., Misgeld T., Matsumuro K., Wekerle H. Neurotrophins inhibit major histocompatibility class II inducibility of microglia: involvement of the p75 neurotrophin receptor. *Proc Natl Acad Sci USA* 95,5779–84 (1998).
49. Compston A. Future prospects for the management of multiple sclerosis. *Ann Neurol* 36,S146-50 (1994).
50. Sagot Y., Vejsada R., Kato A.C. Clinical and molecular aspects of motor-neuron diseases: Animal models, neurotrophic factors and Bcl-2 oncoprotein. *Trends Pharmacol Sci* 18,330-7 (1997).
51. Kramer R., Zhang Y., Gehrmann J., Gold R., Thoenen H., Wekerle H. Gene transfer through the blood-nerve barrier: NGF-engineered neuritogenic T lymphocytes attenuate experimental autoimmune neuritis. *Nat Med* 1,1162-6 (1995).
52. Flügel A., Matsumuro K., Neumann H., Klinkert W.E., Birnbacher R., Lassmann H., Otten U., Wekerle H. Anti-inflammatory activity of nerve growth factor in experimental autoimmune encephalomyelitis: inhibition of monocyte transendothelial migration. *Eur J Immunol* 31,11-22 (2001).
53. Teitelbaum D., Arnon R., Sela M. Copolymer 1: from basic research to clinical application. *Cell Mol Life Sci* 53,24-28 (1997).
54. Arnon R., Sela M., Teitelbaum D. New insights into the mechanism of action of copolymer 1 in experimental allergic encephalomyelitis and multiple sclerosis. *J Neurol* 243 (Suppl 1),S8-13 (1996).
55. Teitelbaum D., Webb C., Bree M., Meshorer A., Arnon R., Sela M. Suppression of experimental allergic encephalomyelitis in Rhesus monkeys by a synthetic basic copolymer. *Clin Immunol Immunopathol* 3,256-62 (1974).
56. Teitelbaum D., Webb C., Meshorer A., Arnon R., Sela M. Suppression by several synthetic polypeptides of experimental allergic encephalomyelitis induced in guinea pigs and rabbits with bovine and human basic encephalitogen. *Eur J Immunol* 3,273-9 (1973).

57. Teitelbaum D., Webb C., Meshorer A., Arnon R., Sela M. Protection against experimental allergic encephalomyelitis. *Nature* 240,564-6 (1974).
58. Teitelbaum D., Meshorer A., Hirshfeld T., Arnon R., Sela M. Suppression of experimental allergic encephalomyelitis by a synthetic polypeptide. *Eur J Immunol* 1,242-8 (1971).
59. Teitelbaum D., Sela M., Arnon R. Copolymer 1 from the laboratory to FDA. *Isr J Med Sci* 33,280-4 (1997).
60. Abramsky O., Teitelbaum D., Arnon R. Effect of a synthetic polypeptide (COP 1) on patients with multiple sclerosis and with acute disseminated encephalomeylitis. Preliminary report. *J Neurol Sci* 31,433-8 (1977).
61. Bornstein M.B., Miller A.I., Slagle S., Arnon R., Sela M., Teitelbaum D. Clinical trials of copolymer I in multiple sclerosis. *Ann N Y Acad Sci* 436,366-72 (1984).
62. Bornstein M.B., Miller A.I., Teitelbaum D., Arnon R., Sela M. Multiple sclerosis: trial of a synthetic polypeptide. *Ann Neurol* 11,317-9 (1982).
63. Bornstein M.B., Miller A., Slagle S., Weitzman M., Drexler E., Keilson M. et al. A placebo-controlled, double-blind, randomized, two-center, pilot trial of Cop 1 in chronic progressive multiple sclerosis. *Neurology* 41,533-539 (1999).
64. Johnson K.P. A review of the clinical efficacy profile of copolymer 1: new U.S. phase III trial data. *J Neurol* 243 (Suppl 1),S3-S7 (1996).
65. Johnson K.P., Brooks B.R., Cohen J.A., Ford C.C., Goldstein J., Lisak R.P. et al. Copolymer 1 reduces relapse rate and improves disability in relapsing-remitting multiple sclerosis: results of a phase III multicenter, double-blind placebo-controlled trial. The Copolymer 1 Multiple Sclerosis Study Group. *Neurology* 45,1268-76 (1995).
66. Neuhaus O., Farina C., Wekerle H., Hohlfeld R. Mechanisms of action of glatiramer acetate in multiple sclerosis. *Neurology* 56,702-8 (2001).
67. Brosnan C.F., Litwak M., Neighbour P.A., Lyman W.D., Carter T.H., Bornstein M.B. et al. Immunogenic potentials of copolymer I in normal human lymphocytes. *Neurology* 35,1754-9 (1985).
68. Qin Y., Zhang D.Q., Prat A., Pouly S., Antel J. Characterization of T cell lines derived from glatiramer-acetate-treated multiple sclerosis patients. *J Neuroimmunol* 108,201-6 (2000).
69. Burns J., Krasner L.J., Guerrero F. Human cellular immune response to copolymer I and myelin basic protein. *Neurology* 136,92-4 (1986).
70. Duda P.W., Schmied M.C., Cook S.L., Krieger J.I., Hafler D.A. Glatiramer acetate (Copaxone) induces degenerate, Th2-polarized immune responses in patients with multiple sclerosis. *J Clin Invest* 105,967-76 (2000).
71. Farina C., Then Bergh F., Albrecht H., Meinl E., Yassouridis A., Neuhaus O., Hohlfeld R. Treatment of multiple sclerosis with Copaxone (COP): Elispot assay detects COP-induced interleukin-4 and interferon-gamma response in blood cells. *Brain* 124,705-19 (2001).
72. Duda P.W., Krieger J.I., Schmied M.C., Balentine C., Hafler D.A. Human and murine CD4 T cell reactivity to a complex antigen: recognition of the synthetic random polypeptide glatiramer acetate. *J Immunol* 165,7300-7 (2000).
73. Brenner T., Arnon R., Sela M., Abramsky O., Meiner Z., Riven-Krietman R. et al. Humoral and cellular immune responses to Copolymer 1 in multiple sclerosis patients treated with Copaxone. *J Neuroimmunol* 115,152-60 (2001).
74. Ziemssen T., Neuhaus O., Farina C., Hartung H.P., Hohlfeld R. Treatment of multiple sclerosis with glatiramer acetate – new information about ist mechanisms of action, pharmacokinetics, adverse effects and clinical studies [German]. *Nervenarzt* 73,321-31 (2002).
75. Fridkis-Hareli M., Strominger J.L. Promiscuous binding of synthetic copolymer 1 to purified HLA-DR molecules. *J Immunol* 160,4386-97 (1998).
76. Ragheb S, Lisak R.P. The lymphocyte proliferative response to glatiramer acetate in normal humans is dependent on both major histocompatibility complex (MHC). *J Neurol* 247 (Suppl 3): III/119 [abstract] (2000).
77. Aharoni R., Teitelbaum D., Arnon R., Sela M. Copolymer 1 acts against the immunodominant epitope 82-100 of myelin basic protein by T cell receptor antagonism in addition to major histocompatibility complex blocking. *Proc Natl Acad Sci USA* 96,634-9 (1999).
78. Gran B., Tranquill L.R., Chen M., Bielekova B., Zhou W., Dhib-Jalbut S. et al. Mechanisms of immunomodulation by glatiramer acetate. *Neurology* 55,1704-14 (2000).
79. Nishimura Y., Chen Y.Z., Kanai T., Yokomizo H., Matsuoka T., Matsushita S. Modification of human T-cell responses by altered peptide ligands: a new approach to antigen-specific modification. *Intern Med* 37,804-17 (1998).
80. Teitelbaum D., Milo R., Arnon R., Sela M. Synthetic copolymer 1 inhibits human T-cell lines specific for myelin basic protein. *Proc Natl Acad Sci USA* 89,137-41 (1992).

81. Webb C., Teitelbaum D., Arnon R., Sela M. In vivo and in vitro immunological cross-reactions between basic encephalitogen and synthetic basic polypeptides capable of suppressing experimental allergic encephalomyelitis. *Eur J Immunol* 3,279-86 (1973).
82. Teitelbaum D., Aharoni R., Arnon R., Sela M. Specific inhibition of the T-cell response to myelin basic protein by the synthetic copolymer Cop 1. *Proc Natl Acad Sci USA* 85,9724-8 (1988).
83. Neuhaus O., Farina C., Yassouridis A., Wiendl H., Then Bergh F., Dose T. et al. Multiple sclerosis: comparison of copolymer-1- reactive T cell lines from treated and untreated subjects reveals cytokine shift from T helper 1 to T helper 2 cells. *Proc Natl Acad Sci USA* 97,7452-7 (2000).
84. Aharoni R., Teitelbaum D., Sela M., Arnon R. Bystander suppression of experimental autoimmune encephalomyelitis by T cell lines and clones of the Th2 type induced by copolymer 1. *J Neuroimmunol* 91,135-46 (1998).
85. Aharoni R., Teitelbaum D., Sela M., Arnon R. Copolymer 1 induces T cells of the T helper type 2 that crossreact with myelin basic protein and suppress experimental autoimmune encephalomyelitis. *Proc Natl Acad Sci USA* 94,10821-6 (1997).
86. Miller A., Shapiro S., Gershtein R., Kinarty A., Rawashdeh H., Honigman S. et al. Treatment of multiple sclerosis with copolymer-1 (Copaxone): implicating mechanisms of Th1 to Th2/Th3 immune-deviation. J Neuroimmunol 92,113-21 (1998).
87. Dabbert D., Rosner S., Kramer M., Scholl U., Tumani H., Mader M. et al. Glatiramer acetate (copolymer-1)-specific, human T cell lines: cytokine profile and suppression of T cell lines reactive against myelin basic protein. *Neurosci Lett* 289,205-8 (2000).
88. Paul W.E., Seder R.A. Lymphocyte responses and cytokines. *Cell* 76,241-51 (1994).
89. Allen J.E., Maizels R.M. Th1-Th2: reliable paradigm or dangerous dogma? *Immunol Today* 18,387-92 (1997).
90. Mosmann T.R., Sad S. The expanding universe of T-cell subsets: Th1, Th2 and more. *Immunol Today* 17,138-46 (1996).
91. Farina C., Wagenpfeil S., Hohlfeld R. Immunological assay for assessing the efficacy of glatiramer acetate (Copaxone) in MS. A pilot study. *J Neurol* 249,1587-92 (2002).
92. Chen M., Conway K., Johnson K.P., Martin R., Dhib-Jalbut S. Sustained immunological effects of Glatiramer acetate in patients with multiple sclerosis treated for over 6 years. *J Neurol Sci* 201,71-7 (2002).
93. Karandikar N.J., Crawford M.P., Yan X., Ratts R.B., Brenchley J.M., Ambrozak D.R., Lovett-Racke A.E., Frohman E.M., Stastny P., Douek D.C., Koup R.A., Racke M.K. Glatiramer acetate (Copaxone) therapy induces CD8(+) T cell responses in patients with multiple sclerosis. *J Clin Invest* 109,641-9 (2002).
94. Ure D.R., Rodriguez M. Polyreactive antibodies to glatiramer acetate promote myelin repair in murine model of demyelinating disease. *FASEB J* 16,1260-2 (2002).
95. Ziemssen T., Kümpfel T., Klinkert W.E., Neuhaus O., Hohlfeld R. Glatiramer acetate-specific T-helper 1- and 2-type cell lines produce BDNF: implications for multiple sclerosis therapy. Brain-derived neurotrophic factor. *Brain* 125,2381-91 (2002).
96. Aharoni R., Teitelbaum D., Leitner O., Meshorer A., Sela M., Arnon R. Specific Th2 cells accumulate in the central nervous system of mice protected against experimental autoimmune encephalomyelitis by copolymer 1. *Proc Natl Acad Sci USA* 97,11472-7 (2000).
97. Kipnis J., Yoles E., Porat Z., Cohen A., Mor F., Sela M., Schwartz M. T cell immunity to copolymer 1 confers neuroprotection on the damaged optic nerve: possible therapy for optic neuropathies. *Proc Natl Acad Sci USA* 97,7446-51 (2000).
98. Schori H., Kipnis J., Yoles E., WoldeMussie E., Ruiz G., Wheeler L.A., Schwartz M. Vaccination for protection of retinal ganglion cells against death from glutamate cytotoxicity and ocular hypertension: implications for glaucoma. *Proc Natl Acad Sci USA* 98,3398-403 (2001).
99. Offen D. et al. Treatment with glatiramer acetate induces neuroprotective effects in chronic MOG-induced EAE. *J Neurol* 249 Suppl. 1 [abstract] (2002)Bornstein M.B., Miller A., Slagle S., Weitzman M., Crystal H., Drexler E. et al. A pilot trial of Cop 1 in exacerbating-remitting multiple sclerosis. *N Engl J Med* 317,408-14 (1987).
101. Kurtzke J.F. Rating neurologic impairment in multiple sclerosis: an expanded disability status scale (EDSS). *Neurology* 33,1444-52 (1983).
102. Johnson K.P., Brooks B.R., Cohen J.A., Ford C.C., Goldstein J., Lisak R.P. et al. Extended use of glatiramer acetate (Copaxone) is well tolerated and maintains its clinical effect on multiple sclerosis relapse rate and degree of disability. Copolymer 1 Multiple Sclerosis Study Group. *Neurology* 50,701-8 (1998).
103. Johnson K.P., Brooks B.R., Ford C.C., Goodman A., Guarnaccia J., Lisak R.P. et al. Sustained clinical benefits of glatiramer acetate in relapsing multiple sclerosis patients observed for 6 years. Copolymer 1 Multiple Sclerosis Study Group. *Mult Scler* 6,255-66 (2000).

104. Ge Y., Grossman R.I., Udupa J.K., Fulton J., Constantinescu C.S., Gonzales-Scarano F. et al. Glatiramer acetate (Copaxone) treatment in relapsing-remitting MS: quantitative MR assessment. *Neurology* 54,813-7 (2000).
105. Mancardi G.L., Sardanelli F., Parodi R.C., Melani E., Capello E., Inglese M. et al. Effect of copolymer-1 on serial gadolinium-enhanced MRI in relapsing remitting multiple sclerosis. *Neurology* 50,1127-1133 (1998).
106. Wolinsky J.S., Nurayana P.A., Johnson K.P., and the Copolymer 1 Multiple Sclerosis study group and the MRI Analysis Center. United States open-label glatiramer acetate extension trial for relapsing multiple sclerosis. *Mult Scler* 7,33-41 (2001).
107. Comi G., Filippi M., Wolinsky J.S. European/Canadian multicenter, double-blind, randomized, placebo-controlled study of the effects of glatiramer acetate on magnetic resonance imaging--measured disease activity and burden in patients with relapsing multiple sclerosis. European/Canadian Glatiramer Acetate Study Group. *Ann Neurol* 49,290-7 (2001).
108. Filippi M., Rovaris M., Rocca M.A., Sormani M.P., Wolinsky J.S., Comi G. European/Canadian Glatiramer Acetate Study Group. Glatiramer acetate reduces the proportion of new MS lesions evolving into "black holes". *Neurology* 57,731-3 (2001).
109. Ziemssen T., Neuhaus O., Hohlfeld R. Risk-Benefit Assessment of Glatiramer Acetate in Multiple Sclerosis. *Drug Safety* 24,979-90 (2001).

CAUSES AND CONSEQUENCES OF DISTURBANCES OF CEREBRAL GLUCOSE METABOLISM IN SPORADIC ALZHEIMER DISEASE:
Therapeutic Implications

Siegfried Hoyer*

1. INTRODUCTION

Nosologically, Alzheimer disease (AD) is no one single disorder. Evidence is provided that a small proportion of 5% to 10% of all Alzheimer cases is caused by missense mutations in presenilin 1 or 2 genes on chromosomes 14 and 1, or in APP gene on chromosome 21 leading to autosomal dominant familial AD with early onset. This difference of inheritance serves as the basis of the amyloid cascade hypothesis which is limited to the above mutations. The latter hypothesis explains the increased formation of the APP derivative βA4 which aggregates to amyloid (Hardy and Selkoe, 2002). However, in constrast, the great majority of all Alzheimer cases (95% to 90%) was found to be sporadic in origin and of late onset. βA4 has not been proven to be necessary for the generation and the development of this neurodegenerative disorder (Joseph et al., 2001). Thus, the amyloid cascade hypothesis may not be accepted for sporadic Alzheimer disease (SAD). Instead, susceptibility genes may contribute to the onset of the latter type of AD. Best known are allelic abnormalities on the APOE-gene on chromosome 19 responsible for both anticipated onset and increase in severity of both inherited and sporadic AD. Candidate susceptibility genes for SAD are assumed to be on chromosomes 4, 6, 10 and 20 (Bertram et al., 2000a; Pericak-Vance et al., 2000). Other candidate genes did not show any association or linkage with AD (Bertram et al., 2000b).

Susceptibility genes may participate in the origin of disorders becoming evident late in life and showing a chronic and progressive course. Such a genetic predisposition together with adult lifestyle risk factors may then cause the disease (Holness et al., 2000). Age has been found to be a major risk factor for SAD because of multiple age-related changes at the cellular and molecular levels (for review, Hoyer 1995; Hoyer, 2000a).

*Department of Pathochemistry and General Neurochemistry, University of Heidelberg, Im Neuenheimer Feld 220/221, Heidelberg, Germany 69120

Thus, there is good evidence, that SAD originates in a concerted action of early-life events and lifestyle risk factors.

It has been shown that damage to the fetal brain at the third trimenon/stage of pregnancy may induce a genetic predisposition which will become clinically evident when adult lifestyle risk factors come into play. Recently, it has been clearly demonstrated that infants of diabetic mothers are at increased risk for hippocampal damage and corresponding memory impairments (Nelson et al., 2000). Similarly, in experimental rats, prenatal stress in the third out of three stages of pregnancy induced increased glucocorticoid secretion from young adulthood to senescence accompanied by behavioral abnormalities in aged animals (Vallée et al., 1999).

Morphologically, the behavioral abnormalities may be due to losses in synapses of different kinds in the hippocampus (Hayashi et al., 1998; Takahashi, 1998) one of the few brain regions that continues to develop postnatally. Thus, the hippocampal regulation of the HPA-axis (see below) may become disturbed particularly with respect to the responsiveness to stress stimuli in later life (Biagini et al., 1998).

Physiologically, glucose is the major nutrient of the healthy, adult, mammalian brain.There is increasing evidence that neuronal glucose metabolism is antagonistically controlled by insulin and cortisol (see below). During normal aging, the regulation of the neuronal glucose metabolism has been demonstrated to be reduced due to a reduction of the neuronal insulin signal transduction pathway. Moreover, in SAD, it has been shown that the neuronal insulin receptor is desensitized similarly to non-insulin dependent diabetes mellitus. This abnormality along with a reduction in brain insulin concentraion is assumed to induce a cascade-like process of disturbances in cellular glucose metabolism and related metabolism associated with the formation of both amyloidogenic derivatives and hyperphosphorylated tau protein. In this short update, glucose metabolism and its control during aging and in SAD are detailed, and recent data on the therapeutic mechanism and efficacy of Ginkgo biloba extract (EGb 761) on oxidative glucose metabolism and behavior are discussed.

2. GLUCOSE METABOLISM AND ITS REGULATION

2.1. The Healthy Adult Brain

Glucose is the source of functionally important metabolites such as acetyl-CoA and ATP (Garland and Randle, 1964; Erecinska and Silver, 1989). The energy-rich compound acetyl-CoA is used 1. for further oxidation to ATP in the tricarboxylic acid cycle (more than 95% of acetyl-CoA), 2. for the formation of the neurotransmitter acetylcholine (1% to 2% of acetyl-CoA) (Gibson et al, 1975), and 3. for the formation of cholesterol in the 3-hydroxy-3-methylglutarly-CoA cycle (Michikawa and Yanagisawa, 1999). Cholesterol is the main sterol in membranes, and it also serves as the basic compound from which neurosteroids derive (Rupprecht and Holsboer, 1999). Glia-derived cholesterol has been demonstrated to promote synaptogenesis in nervous tissue (Mauch et al., 2001).

Cerebral glucose metabolism is tightly linked to free fatty acids via acetyl-CoA (Singh et al., 1989), and to amino acids via the tricarboxylic acid cycle (Sacks, 1957; 1965).

The main goal of oxidative glucose metabolism is the formation of ATP which represents the driving force for nearly all cellular and molecular work, in particular:

- folding of proteins (Braakman et al., 1992; Gething and Sambrook, 1992),
- sorting of proteins (Rothman and Wieland, 1996),
- transport of proteins (Rothman, 1996),
- neurotransmission, i.e., the synapse is the site of the highest energy utilization (Kadekaro et al., 1985),
- function of the endoplasmic reticulum/Golgi apparatus and trans-Golgi network (Seksek et al., 1995; Verde et al., 1995; Demaurex et al., 1998; Lannert et al., 1998),
- maintenance of transmembranaceous ion fluxes (Erecinska and Silver, 1989).

2.2. Action of Insulin

Several findings clearly demonstrated that neuronal glucose metabolism is antagonistically controlled by insulin and cortisol (for review, Hoyer, 1995; 2002). Substantial evidence has been gathered in support both of the transport of pancreatic insulin to the brain and of its production in the CNS (Wozniak et al., 1993). Insulin mRNA was found to be distributed in a highly specific pattern with highest densities in pyramidal cells. Neither insulin mRNA nor synthesis of the hormone was observed in glial cells (Devaskar et al., 1994). Two different types of insulin receptors have been found in adult mammalian brain: a peripheral type detected on glia cells, and a neuron-specific brain type with high concentrations on neurons (Adamo et al., 1989), particularly in the synaptic neuropile (Baskin et al., 1994). Binding of insulin to the extracellular α-subunit of its receptor induces autophosphorylation of the intracellular β-subunit by phosphorylation of the receptor's intrinsic tyrosine residues for receptor activation. The receptor's activity is known to be additionally regulated by the action of both phosphotyrosine phosphatases (dephosphorylation) (Goldstein, 1993), and serine kinases (Häring, 1991). Both, glucocorticoids and catecholamines cause insulin receptor desensitization either by inhibition of phosphorylation of the tyrosine residues or by phosphorylation of serine residues (Häring et al., 1986; Giorgino et al., 1993). Beside these two mechanisms, the cytokine tumor necrosis factor-α (TNF-α) has been demonstrated to decrease both the insulin-stimulated insulin receptor autophosphorylation (Hotamisligil et al., 1994) and the phosphorylation of the insulin receptor substrates 1 and 2 (Hotamisligil et al., 1994; Peraldi et al., 1996; Valverde et al., 1998). Otherwise, TNF-α caused the phosphorylation of serine and threonine residues of insulin receptor substrate-1 which, in turn, induces insulin resistance by means of an inhibition of insulin receptor tyrosine kinase activity (Hotamisligil et al., 1996). Thus, several different mechanisms help to regulate the activity of the insulin receptor.

Recent findings give rise to the assumption that the trafficking of the amyloid precursor protein (APP) is under the control of insulin and the insulin receptor tyrosine kinase (Solano et al., 2000; Gasparini et al., 2001). Insulin increased dose-dependently the extracellular levels of both the APP derivatives Aβ40 and Aβ42, and reduced the intracellular concentrations of both peptides. The insulin mediated reduction of intracellular Aβ was stimulated by its increasing egress from the Golgi apparatus and the trans Golgi-network (for details see below). Interestingly, the insulin receptor tyrosine kinase activity was found to be essential for the effect of insulin on Aβ trafficking i.e. the reduction of intracellular Aβ40 and Aβ42. Insulin has also been shown to regulate the phosphorylation state of tau protein by influencing the activity of phosphorylating enzymes. A deficit in insulin concentration increased the activity of glycogen synthase kinase-3 (Hong and Lee, 1997) which was found to cause tau-hyperphosphorylation (Mandelkow et al., 1992).

2.3. Action of Glucocorticoids

Glucocorticoids have been demonstrated to act adversely to insulin on neuronal glucose utilization (Horner et al., 1990; Virgin et al., 1991; Sapolsky, 1994). Long-term application of corticosterone reduced the activities of glycolytic key enzymes in both rat parietotemporal cerebral cortex and hippocampus, and diminished energy-rich phosphates more severely in hippocampus than in cerebral cortex (Lannert, 1998). In agreement with these findings is the effect of long-term cortisol treatment which damaged nerve cells preferentially in the hippocampus. Hippocampal neurons have been demonstrated to functionally inhibit the HPA-axis-mediated release of cortisol from the adrenal gland. Neuronal damage in the hippocampus then leads to a long-lasting hypercortisolism (Sapolsky et al., 1991; Sapolsky, 1994). It, thus, becomes obvious that insulin and cortisol act as functional antagonists in the brain.

2.4. The Aging Brain

Epidemiologically, the prevalence of SAD increases from 0.5% at the age of 60 years to nearly 50% at 85 years and older (Ott et al., 1995). In centenarians, the rate of individuals with moderate to severe cognitive deficits was found to be about 60%, i.e., increase of the prevalence rate flattened in the 10th decade of life (Kliegel et al., 2001). Thus, age is the main risk factor for SAD, and multiple inherent changes in fundamental metabolic principles at the cellular, molecular and genetic levels in cerebral glucose/energy metabolism, its control and related pathways are set into motion with aging.

As has beend demonstrated recently (Frölich et al, 1998), the neuronal insulin signal transduction system has been found to undergo reduction. The concentration of insulin, density of insulin receptors and the activity of the tyrosine kinase have been shown to fall beyond the age of 60 years. Upon glucose stimulation, the insulin concentration in dialysate of the hypothalamus of aged rats was half the level of young animals (Gerozissis et al., 2001). Clear evidence exists that a higher basal cortisol concentraion becomes effective with aging (Lupien et al, 1994). The increased cortisol concentration in plasma has been found to be mirrored in the cerebrospinal fluid of healthy elderly people who showed enhanced CSF cortisol levels (Swaab et al., 1994). In parallel, the number of corticosteroid receptors type I has been shown to be selectively reduced with aging in the hippocampus (Reul et al., 1988). As a consequence, functional/structural changes may results in the hippocampus and may lead to a reduction of its major inhibitory control of cortisol secretion (Sapolsky et a., 1983; 1986; 1991). The decrease in the inhibitory function may then disinhibit the HPA-axis (Cizza et al., 1994) resulting in an increase of the basal tone of the HPA-axis leading to hypercortisolemia which finally may compromise the function of the neuronal insulin receptor (Giorgino et al., 1993). Thus, a marked imbalance between insulin function and cortisol effects exists in the aging brain (Figure 1).

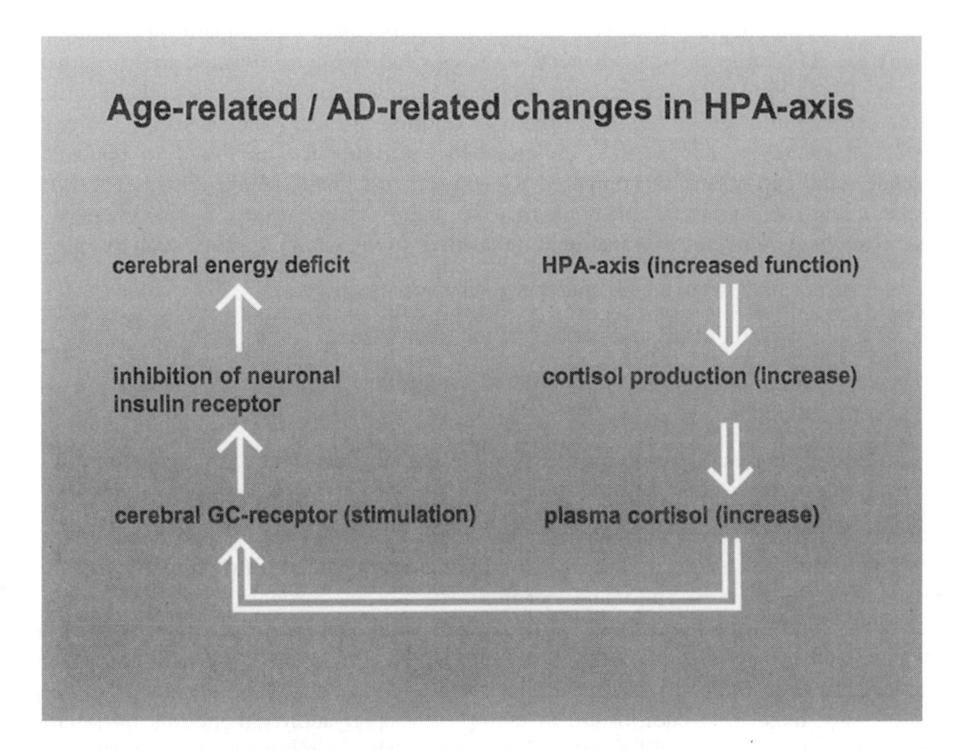

Figure 1. With aging, and more pronounced in SAD, the HPA-axis exerts an increased function resulting in an enhanced cortisol production and in hypocortisolemia. Either the neurotoxic effect of cortisol itself on neurons, or the stimulation of glucocorticoid (type II) receptors, or both, may inhibit the function of the neuronal insulin receptor resulting in a disturbance of the cerebral oxidative glucose metabolism, and in an energy deficit.

The age-related variations in the regulation systems of neuronal glucose metabolism may cause a reduction of the glycolytic glucose breakdown from young adulthood to senescence (Hoyer, 1995). As a concequence pyruvate production decreased whereas lactate formation was found to be increased causing changes in the cytoplasmic redox potential and in intracellular pH in terms of an acidic shift (Hoyer and Krier, 1986; Roberts and Sick, 1996).

Another consequence is the reduced formation of acetyl-CoA and, thus, the diminished synthesis of acetylcholine (Gibson et al., 1981; Bowen, 1984) associated with a decreased function of the cholinergic system (Bigl et al., 1987). This may lead to a shift in the balance between noradrenergic and acetylcholinergic innervation of the microvasculature in terms of an increase in the sympathetic tone based on an increase of the noradrenaline concentration in cerebral cortex with aging and a prolongation of the release of noradrenaline after stress (Ida et al., 1982; Harik and McCracken, 1986; Perego et al., 1993). The age-related increase in cerebral noradrenaline may be another factor, beside cortisol, to compromise the function of the neuronal insulin receptor (Häring et al., 1986).

As far as the synthesis of the energy-rich phosphates is concerned, decreases of ATP and CrP between 5% to 10% were observed in senescence as compared to adulthood. Otherwise, ATP turnover was enhanced with aging indicating an elevated energy demand (Hoyer, 1985; 1992). Recent findings provided evidence for an age-associated decrease in the expression of genes related to proteins working in energy metabolism (Cho et al., 2002). Interestingly, exposure to an enriched environment was found to oppositely regulate gene expression as compared to aging at rest (Jiang et al., 2001). This short update shows that aging of the brain may be mainly characterized, beside changes of several single parameters, by functional imbalances of regulatory systems such as:

- energy production (reduced) and energy turnover (increased),
- insulin action (reduced) and cortisol action (increased),
- acetylcholine action (reduced) and noradrenaline action (increased), indicating a sympathetic tone,
- shift in the gene expression profile from the anabolic site (reduced) to catabolic site (increased)

to name some out of many other changes.

These changes/shifts may indicate an uncoupling of synchronization which has been demonstrated to exist in biological systems (Mirollo and Strogatz; 1990). This model may correspond to the increase in entropy which is an elemental , inherent principle of chemical and biological processes (Hess, 1983; 1990; Prigogine, 1989). In the physical sciences, the term criticality is used respectively to describe a self-organized metalabile steady state (metalabile equilibrium in entropy). Smaller additonal internal or external events, even one that is ineffective in itself, may change biological and/or biophysical properties of the aging brain. Such events may shift a system from supercriticality to criticality to subcriticality/catastrophic reaction (growing entropy) (Bak et al., 1988; Held et al., 1990), i.e. a disease in medical terms. In this respect, age is the main risk factor for neurodegenerative brain disorders.

3. GLUCOSE METABOLISM AND ITS REGULATION IN SPORADIC ALZHEIMER DISEASE (SAD)

3.1. Glucose Consumption and its Control

Early and severe abnormalities were found in cerebral glucose metabolism which worsened in parallel with the dementia symptoms. Otherwise, cerebal oxygen consumption was less severely diminished as compared to glucose consumption (for review, Hoyer, 1996; Hoyer et al., 1991). This disproportion between oxygen and glucose may indicate that substrates other than glucose are oxidized to form energy (see below). In contrast to early-onset AD, the late-onset type was associated with less prominent local abnormalities in glucose utilization but was accentuated in cortical areas such as temporo-parietal and frontal association cortices (Mielke et al., 1992; Herholz et al., 2002), especially in severe dementia (Forster et al., 1984; Duara et al., 1986). This hypometabolism in cerebral cortex was found to be particularly accentuated in brain structures with both high glucose need and insulin sensitivity (for review, Henneberg and Hoyer, 1995).

In this context, a disturbance of the insulin signal transduction pathway in SAD brain was demonstrated in that both insulin concentration and the activity of the insulin receptor tyrosine kinase were decreased whereas insulin receptor density was found to be upregulated (Frölich et al., 1998). The glucose transport protein 2 (GLUT 2), coupled with glucokinase, participates in the glucose-sensing mechanism of insulin secretion. GLUT2 mRNA has been found in some brain areas (Leloup et al., 1994), and specific inhibition of GLUT2 has been shown to suppress insulin secretion (Leloup et al., 1998). It is tempting to assume that glucose hypometabolism due to a primary insulin receptor desensitization in the brain is followed by a reduced insulin secretion/concentration in SAD brain (see above). The experimental inhibition of the neuronal insulin receptor by means of intracerebroventricular application of the diabetogenic substance streptozotocin induced decreases in local glucose consumption in brain structures with a high glucose need (Hawkins et al., 1983) and insulin sensitivity (Henneberg and Hoyer, 1995). This decrease in brain cortical and hippocampal areas may correspond to the abnormalities in cerebral glucose metabolism in SAD (see above) (Duelli et al., 1994).

It is not yet clear which mechanisms may cause the desensitization of the neuronal insulin receptor in SAD brain. As in non-nervous tissue, cortisol and catecholamines, in particular noradrenaline, may be candidates (Häring, 1991; Giorgino et al., 1993). Indeed, drastically enhanced levels of cortisol were found in the cerebrospinal fluid of sporadic Alzheimer patients as compared to normal aging and middle-aged adulthood (Swaab et al., 1994). As pointet out above, the disinhibition of the HPA-axis with aging may result in hypercortisolemia due to an increase of its basal tone (Cizza et al., 1994). Also, noradrenaline was found to be higher in cerebrospinal fluid of Alzheimer patients correlating with severity of dementia, compared to controls (Peskind et al., 1998) which obviously upregulates the cAMP-second messenger system (Martinez et al., 1999). The latter data may point to a further increase in the sympathetic tone as compared to normal aging.

3.2. Disturbances in Glycolysis and Tricarboxylic Acid Cycle

The diminished cerebral glucose utilization in SAD may be due to reduced capacities of key enzymes working in glycolytic glucose breakdown (for review, Hoyer, 1988) although the results are somewhat inconsistent and controversial (Bigl et al., 1996; 1999). Obviously, there is a fall in enzyme activity in neurons but an increase in astrocytes (Bigl et al., 1999. The metabolic capacity to oxidize glucose may be assumed to be diminished: pyruvate dehydrogenase (Perry et al., 1980; Sorbi et al., 1983), and α-ketoglutarate dehydrogenase (Mastrogiacomo et al., 1993) were drastically reduced. In experimental animals, the activities of glycolytic key enzymes, of pyruvate dehydrogenase and α-ketoglutarate dehydrogenase were found to be decreased after inhibition of the neuronal insulin receptor by intracerebroventricular application of streptozotocin (Plaschke and Hoyer, 1993; Lannert, 1998). Therefore, it may be assumed that, in SAD brain, the reduction in the activities of key enzymes working in glucose breakdown is also caused by the damage of the neuronal insulin signaling pathway (see above).

In the beginning of SAD, the metabolic pattern of cerebral oxygen and glucose consumption was found to be characterized by a disproportion between these two parameters (Hoyer et al., 1991; Fukuyama et al., 1994). However, the deficit in neuronal glucose availability may be partially and transiently balanced by the utilization of endogenous brain substrates to meet the energy demand of the brain: glucoplastic amino acids such as glutamate and fatty acids (Hoyer and Nitsch, 1989; Nitsch et al., 1992; Pettegrew et al., 1995). As a side effect of glutamate utilization, neurotoxic ammonia is formed in SAD brain (Hoyer et al., 1990), as also was found when glucose is lacking in the brain (Benzi et al., 1984). These conditions are different from hyperammonemia in which ammonia is taken up by the brain accompanied by an increase of both cerebral glucose utilization and glutamine output (Papenberg et al., 1975). As a consequence of increased ammonia in the brain, mitochondrial dehydrogenases such as α-ketoglutarate dehydrogenase (at lower ammonia concentration) and isocitrate dehydrogenase and malate dehydrogenase (both at higher ammonia concentration) were found to be inhibited (Lai and Cooper, 1991).

3.3. Glucose-Related Metabolism

The diminished activity of the pyruvate dehydrogenase complex yields reduced levels of acetyl-CoA (Perry et al., 1980). As a consequence of the latter and due to the reduced activity of acetylcholine transferase, the synthesis of acetylcholine in the presynaptic neuron is markedly diminished (Sims et al., 1983a). It is noteworthy that the degeneration of the cholinergic system correlates with the progress of disturbed mental capacities in Alzheimer patients (Baskin et al., 1999). The reduced cellular availability of acetyl-CoA may be assumed to be also a source of the decreased formation of both intracellular cholesterol and of neurosteroids. Recent studies clearly showed a reduced level of cholesterol in brain tissue membranes and postmortem CSF of Alzheimer patients (Svennerholm and Gottfries, 1994; Mulder et al., 1998; Eckert et al., 2000). Cholesterol is the main sterol in membranes which contribute strongly to normal cell functions and synaptogenesis (Mauch et al., 2001). Any major damage of membrane integrity through a loss of membrane constituents, such as e.g. cholesterol, is incompatible with normal cellular function. Severe membrane damage generally leads to cell death (Michikawa and Yanagisiwa, 1999; for review Klein, 2000).

Another pathophysiologic consequence of the markedly perturbed glucose metabolism is the fall of ATP production from glucose by around 50% in the beginning of SAD. The oxidative utilization of substrates other than glucose improved ATP formation to 80% of normal, but there was a decline thereafter throughout the course of the disease (Hoyer, 1992). A fall in ATP formation in AD brain was also found in other investigations (Sims et al., 1983b; Brown et al., 1989). In contrast to SAD, no significant drop in ATP formation could be found in early-onset familiar AD (Hoyer, 1992) which may mean that ATP-dependent processes such as protein synthesis (Buttgereit and Brand, 1995) and other cellular and molecular processes (see above) may not be involved in the pathogenesis of this pathological condition. In experimental animals, the inhibition of the neuronal insulin receptor function caused a long-term fall in ATP concentration in cerebral cortex accompanied by behavioral abnormalities (Lannert and Hoyer, 1998).

3.4. Endoplasmic Reticulum, Golgi Apparatus, trans Golgi Network

Functions such as folding and sorting of proteins are performed in these intracellular compartments (Braakman et al., 1992; Gething and Sambrook, 1992; Rothman and Wieland, 1996). This work is highly ATP-dependent and is ensured by a pH of around 6 (Seksek et al., 1995; Verde et al., 1995; Demaurex et al., 1998; Lannert et al., 1998). Also, binding of chaperones to the unfolded state of proteins, its promotion to correct folding and assembly of secretory proteins, and the dissociation of protein aggregates in the lumen of the endoplasmis reticulum are ATP-dependent processes. A depletion of cellular ATP has been found to prevent dissociation of chaperone/proteine complexes and, thus, to block secretion of these proteins (Dorner et al., 1990). Misfolded or malfolded proteins accumulate in the endoplasmic reticulum (Kaufman, 1999), indicating "endoplasmic reticulum stress". Normally, the response to endoplasmic reticulum stress is the increasing transcription of its resident chaperones. However, mutated (misfolded) proteins downregulate the "unfolded protein response" what has been assumed to occur in genetic Alzheimer disease (Thomas et al., 1995; Taubes, 1996; for review, Imaizumi et al., 2001).

With respect to sporadic Alzheimer disease, the size of the Golgi apparatus has been demonstrated to be smaller than normal indicating a hypometabolic state in SAD involving the neuronal activity in general and protein synthetic activity in particular (Salehi et al., 1994; 1995; Salehi and Swaab, 1999). Cerebral hypometabolism in SAD has been characterized above in that an ATP deficit exists caused by a disturbance in the insulin signal transduction cascade. The diminished availability of ATP reduced protein synthesis (Buttgereit and Brand, 1995; see also above), i.e. the processing of the amyloid precursor protein (APP) is involved in that its secreted form is reduced and the potentially amyloidogenic C-terminal fragment is enhanced (for review, Hoyer 2002b). Both, neuronal insulin deficit and the blockade of the insulin signal transduction at the insulin receptor were found to reduce the release of secreted APP (APPs) and βA4 whereby insulin did not affect the APP trafficking between endplasmic reticulum and Golgi apparatus (Solano et al., 2000; Gasparini et al., 2001). Both APP derivatives, APPs and βA4 , are retained in the cell what has been found in SAD neurons (Gouras et al., 2000). βA4 has been demonstrated to accumulate for longer in axons even after unspecific damage without upregulation of the APP gene (Iwata et al., 2002). Neurons accumulating βA4 undergo lysis to form amyloid plaques (D'Andrea et al., 2001). The latter process may be stimulated by advanced glycation end products deriving from a disturbed glucose metabolism (Münch et al., 1998).

Provided, the (non-mutated), APP is nevertheless misfolded due to the energy deficit, it is not only retained in the endoplasmic reticulum alone but can move to the intermediate compartment and to the Golgi apparatus to be recycled back to the endoplasmic reticulum to associate with the chaperone glucose related protein (GRP 78) (Hammond and Helenius, 1994). The proteolysis of such chaperone/proteine complexes depends of the availability of ATP (Okada et al., 1991). The diminished ATP in SAD (see above) may inhibit proteolysis, and, thus, may result in a disease (Perlmutter, 1999). Interestingly, GRP 89 has been found to be increased in sporadic Alzheimer brain (Hamos et al., 1991).

Outside of the endoplasmic reticulum/Golig apparatus/trans Golgi network, both ATP and insulin have been shown to regulate the phosphorylation state of tau protein by influencing the activities of phosphorylating enzymes. A reduction of ATP activates both protein kinases erk36 and erk40 (Röder and Ingram, 1991) which in turn causes tau-hyperphosphorylation (Bush et al., 1995). A deficit in insulin concentration increases the activity of glycogen synthase kinase-3 (Hong and Lee, 1997) which also causes tau-hyperphosphorylation (Mandelkow et al., 1992).

These data clearly indicate that the abnormalities in the energy state and the insulin signal transduction cascade are of central significance in SAD brain.

4. PHARMACOTHERAPEUTIC APPROACHES

Several strategies have been proposed for the treatment of Alzheimer disease. Only some out of many others shall be discussed here.

Vaccination against the deposition of βA4 in the brain is based on results from transgenic mice overexpressing mutated APP, thus, representing the genetic type of AD. In some instances, vaccination reduced βA4 deposition, sometimes accompanied by an improvement of reduced mental capacities. However, data from patients suffering from sporadic AD are lacking.

Acetylcholinesterase inhibitors. This therapy is based on the assumption that the cholinergic presynapse is particularly damaged in SAD resulting in a reduced production of acetylcholine (Sims et al., 1983a; Wurtman, 1992). The inhibition of acetylcholinesterase in the synaptic cleft may then increase the presynaptic acetylcholine concentration. However, it has been demonstrated that this therapy led to a stimulation of postsynaptic receptors what increased the expression of the acetylcholinesterase gene (Sapolsky 1998; Kaufer et al., 1998; von der Kammer et al., 1998) resulting in an increase of acetylcholinesterase in the CSF of sporadic AD patients (Davidsson et al., 2001). Thus, the proposed effect of acetylcholinesterase inhibitors cannot be validated.

Statins. In vitro-studies provided evidence that cholesterol depletion by statins inhibited the generation of βA4 (Simons et al., 1998). In unselected patients with the diagnosis "probable AD", the prevalence of dementia was reduced by 60% to 70% when treated with lovostatin and pravastatin, but not with simvastatin (Wolozin et al., 2000). It was emphasized that statins reduce the risk of dementia in general, but not of SAD in particular (Jick et al., 2000). However, in a recent study performed on moderately demented SAD patients, treatment with simvastatin showed a sligth improvement of dementia symptoms but only in combination with acetylcholinesterase inhibitors (Simons et al., 2002).

Ginkgo biloba extract (EGb 761). Several in vitro- and in vivo- studies have provided clear evidence that EGb 761 improves both the damaged energy state of the brain and the function of neuronal membranes (for review, Hoyer et al., 1999). Furthermore, EGb 761 has been demonstrated to modulate insulin receptor binding (Löffler et al., 2001) and some RNAs of transcription factors and growth factors (Watanabe et al., 2001).

With respect to the pharmacologic efficacy of EGb 761, an interesting aspect has been discussed in that EGb 761 affects the gene expression of the peripheral benzodiazepine receptor of the adrenocortical mitochondria associated with a decrease in circulating cortisol concentration (Amri et al., 1996). Thus, it is tempting to assume that EGb 761 possesses an anti-stress effect by reducing hypercortisolemia and its detrimental effects on brain function particularly in SAD (see above; Figure 2). In SAD patients, EGb 761 alleviated dementia symptoms (LeBars et al., 1997; 2000).

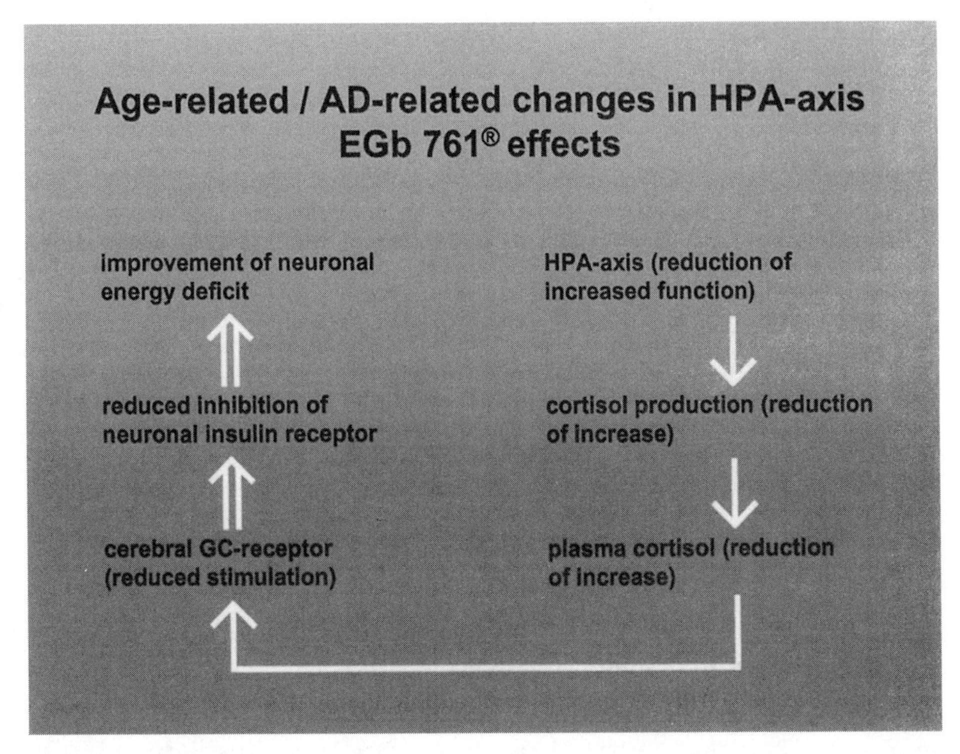

Figure 2. EGb 761 is assumed to possess an anti-stress effect by reducing hypercorticosolemia. As a consequence, the detrimental effect of cortisol on the function of the neuronal insulin receptor, causing disturbances in oxidative glucose metabolism and energy production may be minimized.

5. SUMMARY

Alzheimer disease is not a single disorder. Etiologically, two different types or even diseases exist: inheritance in 5% to 10% of all Alzheimer cases versus 90% to 95% AD cases whith sporadic origin (SAD). Different susceptibility genes along with adult lifestyle risk-factors- in the case of SAD the risk factor aging- may be assumed to cause the latter disorder. There is evidence that a disturbance in the insulin signal transduction pathway may be a central and early pathophysiologic event in SAD. Both, hypercortisolemia and increased adrenergic activity, in both old age and SAD may render the function of the neuronal insulin receptor vulnerable resulting in a diminished production of ATP. The reduced availability of ATP may damage the function of the endoplasmic reticulum/Golgi apparatus/trans Golgi network generating misfolded and malfolded proteins retained in the cell. In SAD, amyloid precursor protein is found to accumulate intracellularly thus not representing the cause but a driving force in the pathogenesis of SAD. Additionally, both disturbed insulin signaling and reduced ATP forward the hyperphosphorylation of tau protein. Thus, abnormalities in oxidative brain metabolism lead to the formation of two main morphologic hallmarks of SAD: senile plaques and neurofibrillary tangles. Therefore, the therapeutic goal in SAD should be the improvement of the neuronal energy state. Findings from both basic and clinical studies showed that Ginkgo biloba extract (EGb 761) may be appropiate to approach that goal.

REFERENCES

Adamo, M., Raizada, M.K., and LeRoith, D., 1989, Insulin and insulin-like growth factor receptors in the nervous system, Mol. Neurobiol. 3: 71-100.

Amri, H., Ogwuegbu, S.O., Boujard, N., Drieu, K., and Papadopoulos, V., 1996, In vivo regulation of the peripheral-type benzodiazepine receptor and glucocorticoid synthesis by the Ginkgo biloba extract EGb 761 and isolated ginkgolides, Endocrinology 137: 5707-5718.

Bak, P., Tang, C. and Wiesenfeld, K., 1988, Self-organized criticality, Phys. Rev. A38: 365-374.

Baskin, D.S. Browning, J.L., Pirozzolo, F.J., Korporaal, S., Baskin, J.A., and Appel, S.H., 1999, Brain choline acetyltransferase and mental function in Alzheimer disease, Arch. Neurol. 56: 1221-1223

Baskin, D.G., Schwartz, M.W., Sipols, A.J., D'Alessio, D.A., Goldstein, B.J., and White, M.F., 1994, Insulin receptor substrate-1 (IRS-1) expression in rat brain, Endocrinology 134: 1952-1955

Benzi, G., Pastoris, I., Villa, R.F., and Giuffrida-Stella, A.M., 1984, Effect of aging on cerebral cortex energy metabolism in hypoglycemia and posthypoglycemic recovery, Neurobiol. Aging 5: 205-212.

Bertram, L., Blacker, D., Mullin, K., Keeny, D., Jones, J., Basu, S., Yhu, S. McInnis, M.G., Go, R.C.P., Vekrellis, K. et al., 2000a, Evidence for genetic linkage of Alzheimer's disease to chromosome 10q, Science 290: 2302-2303.

Bertram, L., Blacker, D., Crystal, A., Mullin, K., Kenney, D., Jones, J., Basu, S., Yhu, S., Guenette, S., McInnis, M., et al., 2000b, Candidate genes showing no evidence for association or linkage with Alzheimer's disease using family-based methodologies, Exp. Gerontol. 35: 1353-1361.

Biagini, G., Pich, E.M., Carani, C., Marrama, P., and Agnati, L., 1998, Postnatal maternal separation during the stress hyporesponsive period enhances the adrenocortical response to novelty in adult rats by affecting feedback regulation in the CA1 hippocampal field, Int. J. Devl. Neurosci. 16: 187-197.

Bigl, V., Arendt, T., Fischer, S., Werner, M., and Arendt, A., 1987, The cholinergic system in aging, Gerontology 33: 172-180.

Bowen, D.M., 1984, Cellular aging: selective vulnerability of cholinergic neurons in human brain, Monogr. Dev. Biol. 17: 42-59.

Braakman, J., Helenius, J., and Helenius, A., 1992, Role of ATP and disulfide bonds during protein folding in the endoplasmic reticulum, Nature 356: 260-262.

Brown, G.G. Levine, S.R., Gorell, J.M., Pettegrew, J.W., Gdowksi,, J.E., Bueri, J.A. Helpern, J.A., and Welch, K.M.A., 1989, In vivo 31P-NMR profiles of Alzheimer disease and multiple subcortical infarct dementia, Neurology 39: 1423-1427.

Bush, M.L., Niyashiro, J.S., and Ingram, V.M., 1995, Activation of a neurofilament kinase, a tau kinase and tau phosphatase by decreased ATP levels in nerve growth factor-differentiated PC12 cells, Proc. Natl. Acad. Sci. USA 92: 1962-1965.

Buttgereit, F., and Brand, M.D., 1995, A hierarchy of ATP- consuming processes in mammalian cells, Biochem. J. 312: 163-167.

Cho, K.S., Choi, J., Ha., C.M., Son, Y. J., Choi, W.S., and Lee, B.K., 2002, Comparison of gene expression in old versus young rat hippocampus by cDNA array, Neuro Report 13: 285-289.

Cizza, G., Calogero, A.E., Brady, L.S., Bagdy, G., Bergamini, E., Blackman, M.R., Chrousos, G. P., and Gold, P.W., 1994, Male Fischer 344/N rats show a progressive central impairment of the hypothalamic-pituitary-adrenal axis with advancing age, Endocrinology 134: 1611-1620.

D'Andrea, M.R., Nagele, R.G., Wang, H.-Y., Peterson, P.A., and Lee, D.H.S., 2001, Evidence that neurones accumulating amyloid can undergo lysis to form amyloid plaques in Alzheimer's disease, Histopathology 38: 120-134.

Davidsson, P., Blennow, K., Andreasen, N., Eriksson, B., Minthon, L., and Hesse, C., 2001, Differential increase in cerebrospinal fluid-acetylcholinesterase after treatment with acetylcholinesterase inhibitors in patients with Alzheimer's disease, Neurosci. Lett. 300: 157-160.

Demaurex, N., Furuya, W., D'Souza, S., Bonifacino, J.S., and Grinstein, S., 1998, Mechanism of acidification of the trans-Golgi network (TGN). In situ measurements of pH using retrieval of TGN 38 and furin from the cell surface, J. Biol. Chem. 273: 2044-2051.

Devaskar, S.U., Giddings, S.J., Rajakumar, P.A., Carnaghi, L.R., Menon, R.K., and Zahn, D.S., 1994, Insulin gene expression and insulin synthesis in mammalian neuronal cells, J. Biol. Chem. 269: 8445-8454.

Dorner, A.J., Wasley, L.C., and Kaufman, R.J., 1990, Protein dissociation from GRP 78 and secretion are blocked by depletion of cellular ATP levels, Proc. Natl. Acad. Sci. USA 87: 7429-7432.

Duara, R., Grady, C., Haxby, J., Sundaram, S., Cutler, N.R., Heston, L., Moore, A., Schlageter, N., Larson, S., and Rapoport, S.I., 1986; Positron emission tomography in Alzheimer's disease, Neurology 36: 879-887.

Duelli,R.,Schröck,H., Kuschinsky,W., and Hoyer,S., 1994, Intracerebroventricular injection of streptozotocin induces discrete local changes in cerebral glucose utilization in rats, Int. J. Devl. Neurosci. 12: 737-743.

Eckert, G.P., Cairns, N.J., Maras, A., Gattaz, W.F., and Müller, W.E., 2000, Cholesterol modulates the membrane-disordering effects of beta-amyloid peptides in the hippocampus: specific changes in Alzheimer's disease, Dementia Geriatr. Cogn. Disord. 11: 181-186.

Erecinska, M., and Silver, I.A., 1989, ATP and brain function, J. Cereb. Blood Flow Metab. 9: 2-19.

Foster, N.L., Chase, T.N., Mansi, L., Brooks, R., Fedio, P., Patronas, N.J. and DiChiro, G., 1984, Cortical abnormalities in Alzheimer's disease, Ann. Neurol. 16: 649-654.

Frölich, L., Blum-Degen, D., Bernstein, H.G., Engelsberger, S., Humrich, J., Laufer, S., Muschner, D., Thalheimer, A., Türk, A., Hoyer, S., et al., 1998, Insulin and insulin receptors in the brain in aging and sporadic Alzheimer's disease, J. Neural Transm. 105: 423-438.

Fukuyama, H., Ogawa, M., Yamauchi, H., Yamaguchi, S., Kimura, J., Yonekura, Y., and Konishi, J., 1994, Altered cerebral energy metabolism in Alzheimer's disease: a PET study, J. Nucl. Med. 35: 1-6.

Garland, P.B., and Randle, P.J., 1965, Control of pyruvate dehydrogenase in the perfused rat heart by the intracellular concentration of acetyl-coenzyme A, Biochem. J. 91: 76C-77C

Gasparini, L., Gouras, K.G., Wang, R., Gross, R.S., Beal, M.F., Greengard, P., and Yu, H., 2001, Stimulation of β-amyloid precursor protein trafficking by insulin reduces intraneuronal β-amyloid and requires mitogen-activated protein kinase signaling, J. Neurosci. 21: 2561-2570.

Gerozissis,, R., Rouch, C., Lemierre, S., Nicolaidis, S., and Orosco, M., 2001, A potential role of central insulin in learning and memory related to feeding, Cell. Mol. Neurobiol. 21: 389-401.

Gething, M.-J., and Sambrook, J., 1992, Protein folding in the cell, Nature 355: 33-45.

Gething, M.-J., McCammon, K., and Sambrook, J., 1986, Expression of wild-type and mutant forms of influenza hemagglutinin: the role of folding in intracellular transport, Cell 46: 939-950.

Gibson, G.E., Jope, R., and Blass, J.P., 1975, Decreased synthesis of acetylcholine accompanying impaired oxidation of pyruvic acid in rat brain minces, Biochem. J. 148: 17-23.

Gibson, G.E., Petersen, C., and Jenden, D.J., 1981, Brain acetylcholine synthesis declines with senescence, Science 213: 674-676.

Giorgino, F., Almahfouz, A., Goodyear, L.J., and Smith, R.J., 1993, Glucocorticoid regulation of insulin receptor and substrate IRS-1 tyrosine phosphorylation in rat skeletal muscle in vivo, J. Clin. Invest. 91: 2020-2030.

Goldstein, B.J., 1993, Regulation of insulin receptor signaling by protein-tyrosine dephosphorylation, Receptor 3: 1-15.

Gouras, G.K., Tsai, J., Naslund, J., Vincent, B., Edgard, M., Greenfield, J.P., Haroutunian, V., Buxbaum, J.S., Xu, H., Greengard, P., and Relkin, N.R., 2000, Intraneuronal βA42 accumulation in human brain, Am. J. Pathol. 156: 15-20.

Hammond, C., and Helenius, A., 1994, Quality control in secretory pathway: retention of a misfolded viral membrane glycoprotein involves cycling between the ER, intermediate compartment, and Golgi apparatus, J. Cell Biol. 126: 41-52.

Hamos, J.E. Oblas, B., Pulaski Salo, D., Welch, W. J., Bole, D.G., and Drachman, D.A. 1991, Expression of heat shock proteins in Alzheimer's disease, Neurology 41: 345-350.

Häring, H.U., 1991, The insulin receptor: signalling mechanism and contribution to the pathogenesis of insulin restistance, Diabetologica 34: 848-861.

Hardy, J., and Selkoe, D.J., 2002, The amyloid hypothesis of Alzheimer's disease: Progress and problems on the road to therapeutics, Science 297: 353-356.

Harik, S.I., and McCracken, K.A., 1986, Age-related increase in presynaptic noradrenergic markers of the rat cerebral cortex, Brain Res. 381: 125-130.

Hawkins, R.A., Mans, A.M., Davis, D.W., Hibbard, L.S., and Lu, D.M., 1983, Glucose availability to individual cerebral structures is correlated to glucose metabolism, J. Neurochem. 40: 1013-1018.

Hayashi, A., Nagaoka, M., Yamada, K., Ichitani, Y., Miake, Y., and Okado, N., 1998, Maternal stress induces synaptic loss and developmental disabilities of offspring, Int. J. Devl. Neurosci. 16: 209-216.

Held, G.A., Solina, D.H., Keane, D.T., Haag, W.J., Horn, P.M., and Grinstein, G., 1990, Experimental study of critical-mass fluctuations in an evolving sandpile, Physic. Rev. Lett. 69: 1120-1123.

Henneberg, N., and Hoyer, S., 1995, Desensitization of the neuronal insulin receptor: a new approach in the etiopathogenesis of late-onset sporadic dementia of the Alzheimer type (SDAT)?, Arch. Gerontol. Geriatr. 21: 63-74.

Herholz, K., Salmon, E., Perani, D., Baron, J.-C., Holthoff, V., Frölich, L., Schönknecht, P., Ito, K., Mielke, R., Kalbe, E., et al., 2002, Discrimination between Alzheimer dementia and controls by automated analysis of multicenter FDG PET, NeuroImage 17: 302-316.

Hess, B., 1983, Non-equilibrium dynamics of biochemical processes, Hoppe-Seylers Z. Physiol. Chem. 364: 1-20.

Hess, B., 1990, Order and chaos in chemistry and biology, Fresenius J. Anal. Chem. 337: 459-468.

Holness, M.J., Langdown, M.L., and Sugden, M.C., 2000, Early-life programming of susceptibility to dysregulation of glucose metabolism and the development of typ 2 diabetes mellitus, Biochem. J. 349: 657-665.

Hong, M., and Lee, V.M.Y., 1997, Insulin and insulin-like growth factor-1 regulate tau phosphorylation in cultured human neurons, J. Biol. Chem. 272: 19547-19553.

Horner, H.C., Packan, D.R., and Sapolsky, R.M., 1990, Glucocorticoids inhibit glucose transport in cultured hippocampal neurons and glia, Neuroendocrinology 52: 57-64.

Hotamisligil, G.A., Murray, D.L., Choy, L.N., and Spiegelman, B.M., 1994, Tumor necrosis factor α inhibits signaling from the insulin receptor, Proc. Natl. Acad. Sci. USA 91: 4854-4858.

Hotamisligil, G.S., Peraldi, P., Budavari, A., Ellis, R., White, M.F., and Spiegelman, B.M., 1996, IRS-1-mediated inhibition of insulin receptor tyrosine kinase activity in TNF-α- and obesity-induced insulin resistance, Science 271: 665-668.

Hoyer, S., 1985, The effect of age on glucose and energy metabolism in brain cortex of rats, Arch. Gerontol. Geriatr. 4: 193-203.

Hoyer, S., 1988, Glucose and related brain metabolism in dementia of Alzheimer type and its morphological significane, Age 11: 158-166.

Hoyer, S., 1992, Oxidative metabolism in Alzheimer brain. Studies in early-onset and late-onset cases, Mol. Chem. Neuropathol. 16: 207-224.

Hoyer, S., 1995, Age-related changes in cerebral oxidative metabolism. Implications for drug therapy, Drugs Aging 6: 210-218.

Hoyer, S., 1996, Oxidative metabolism deficiencies in brain of patients with Alzheimer's disease, Acta Neurol. Scand. Suppl. 165: 18-24.

Hoyer, S., 1998, Is sporadic Alzheimer disease the brain type of non-insulin dependent diabetes mellitus? A challenging hypothesis, J. Neural Transm. 105: 415-422.

Hoyer, S., 2002a, The aging brain. Changes in the neuronal insulin/insulin receptor signal transduction cascade trigger late-onset sporadic Alzheimer disease (SAD). A mini-review, J. Neural. Transm. 109: 991-1002.

Hoyer, S., 2002b, The brain insulin signal transduction system in sporadic (type II) Alzheimer disease: an update, J. Neural Transm. 109: 341-360.

Hoyer, S., and Krier, C., 1986, Ischemia and the aging brain. Studies on glucose and energy metabolism in rat cerebral cortex, Neurobiol. Aging 7: 23-29.

Hoyer, S., and Nitsch, R., 1989, Cerebral excess release of neurotransmitter amino acids subsequent to reduced cerebral glucose metabolism in early-onset dementia of Alzheimer type, J Neural. Transm. (Gen. Sect.) 75: 227-232.

Hoyer, S., Nitsch, R., and Oesterreich, K., 1990, Ammonia is endogenously generated in the brain in the presence of presumed and verified dementia of Alzheimer type, Neurosci. Lett. 117: 358-362.

Hoyer, S., Nitsch, R., and Oesterreich, K., 1991, Predominant abnormality in cerebral glucose utilization in late-onset dementia of the Alzheimer type: a cross-sectional comparison against advanced late-onset dementia and incipient early-onset cases, J. Neural. Transm. (PD-Sect.) 3: 1-14.

Hoyer, S., Lannert, H., Nöldner, M., and Chatterjee, S.S., 1999, Damaged neuronal energy metabolism and behavior are improved by Ginkgo biloba extract (EGb 761), J. Neural. Transm. 106: 1171-1188.

Ida, Y., Tanaka, M., Kohno, Y., Nakagawa, R., Iimori, K., Tsuda, A., Hoaki, Y., and Nagasaki, N., 1982, Effects on age and stress on regional noradrenaline metabolism in the rat brain, Neurobiol. Aging 3: 233-236.

Imaizumi, K., Miyoshi, K., Katayama, T., Yoneda, T., Taniguchi, M., Kudo, T., and Tohyama, M., 2001, The unfolded protein response and Alzheimer's disease, Biochim. Biophys. Acta 1536: 85-96.

Iwata, A., Chen, X.-H., McIntosh, T.K., Browne, K.D., and Smith, D.H., 2002, Long-term accumulation of amyloid-β in axons following brain trauma without persistent upregulation of amyloid precursor protein genes, J. Neuropathol. Exp. Neurol. 61: 1056-1068.

Jiang, C.H., Tsien, J.Z., Schultz, P.G., and Hu, K., 2001, The effects of aging on gene expression in the hypothalamus and cortex of mice, Proc. Natl. Acad. Sci. USA 98: 1930-1934.

Jick, H., Zornberg, G.L., Jick, S.S., Seshadri, S., and Drachman, D.A., 2000, Statins and the risk of dementia, Lancet 356: 1627-1631.

Joseph, J., Shukitt-Hale, B., Denisova, N.A., Martin, A., Perry, G., and Smith, M.A., 2001, Copernicus revisited: amyloid beta in Alzheimer's disease, Neurobiol. Aging 22: 131-146.

Kadekaro, M., Crane, A.M., and Sokoloff, L., 1985, Differential effects of electrical stimulation of sciatic nerve on metabolic activity in spinal cord and dorsal root ganglion in the rat, Proc. Natl. Acad. Sci. USA 82: 6010-6013.

Kaufer, D., Friedman, A., Seidman, S., and Soreq, H., 1998, Acute stress facilitates long-lasting changes in cholinergic gene expression, Nature 393: 373-377.

Kaufman, R.J., 1999, Stress signaling from the lumen of the endoplasmic reticulum: coordination of gene transcriptional and translational controls, Genes Dev. 13: 1211-1233.

Klein, J., 2000, Membrane breakdown in acute and chronic neurodegeneration: focus on choline-containing phospholipids, J. Neural Transm. 107: 1027-1063.

Kliegel, M., Rott, C., d'Heureuse, V., Becker, G. and Schönemann, P., 2001, Dementia in the very old is not a necessity: results from the Heidelberg Centenarian Study, Z. Gerontopsychol-psychiatrie 14: 169-180.

Lai, J.C.K., and Cooper, A.J.L., 1991, Neurotoxicity of ammonia and fatty acids: differential inhibition of mitochondrial dehydrogenases by ammonia and fatty acyl coenzyme A derivatives, Neurochem. Res. 16: 795-803.

Lannert, H., 1998, Effekte von Streptozotozin, Streptozotozin/Estradiol und Kortikosteron auf Lernverhalten, Gedächtnisfunktion und Energiestoffwechsel im zerebralen Kortex und Hippokampus der adulten männlichen Ratte, Thesis, University of Heidelberg.

Lannert, H., and Hoyer, S., 1998, Intracerebroventricular administration of streptozotocin causes long-term diminutions in learning and memory abilities and in cerebral energy metabolism in adult rats, Behav. Neurosci. 112: 1199-1208.

Lannert, H., Gorgas, K., Meißner, I., Wieland, F.T., and Jeckel, D., 1998, Functional organization of the Golgi-apparatus in glycosphingolipid biosynthesis: Lactosylceramide and subsequent glycosplingolipids are formed in the lumen of the late Golgi, J. Biol. Chem. 273: 2939-2946.

Le Bars, P.L., Katz, M.M., Berman, N., Itil, T.M., Freedman, A.M., and Schatzberg, A.F., 1997, A placebo-controlled, double-blind, randomized trial of an extract of Ginkgo biloba for dementia. North American EGb Study Group, J. Am. Med. Ass. 278: 1327-1332.

Le Bars, P.L., Kieser, M., and Itil, K.Z., 2000, A26-week analysis of a double-blind, placebo-controlled trial of the ginkgo biloba extract EGb 761 in dementia, Dement. Geriatr. Cogn. Disord. 11: 230-237.

Leloup, C, Arluison, M., Lepetit, N., Cartier, N., Marfaing-Jallat, P., Ferre, P., and Penicaud, L., 1994, Glucose transporter 2 (GLUT2): expression in specific brain nuclei, Brain Res. 638: 221-226.

Leloup, C. Orosco, M., Serradas, P., Nicolaidis, S., and Penicaud, L. 1998, Specific inhibition of GLUT2 in arcuate nucleus by antisense oligonucleotides suppress nervous control of insulin secretion, Mol. Brain Res. 57: 275-280.

Löffler, T., Lee, S.K., Nöldner, M., Chatterjee, S.S., Hoyer, S. and Schliebs, R., 2001, Effect of Ginkgo biloba extract (EGb 761) on glucose metabolism-related markers in streptozotocin-damaged rat brain, J. Neural Transm. 108: 1457-1474.

Lupien, S., Lecors, A., Lussier, I., Schwartz, G., Nair, N., and Meany, M., 1994, Basal cortisol levels and cognitive deficits in human aging, J. Neurosci. 14: 2893-2903.

Mandelkow, E.M., Drewes, G., Biernat, J., Gustke, N., van Lint, J., Vandenheede, J.R., and Mandelkow, E., 1992, Glycogen synthase kinase-3 and the Alzheimer-like state of microtubule-associated protein tau, FEBS Lett. 314: 315-321.

Martinez, M., Fernandez, E., Frank, A., Guaza, C., de la Fuente, M., and Hernandez, A., 1999, Increased cerebral spinal fluid cAMP levels in Alzheimer's disease, Brain Res. 846: 265-267.

Mastrogiacomo, F., Bergeron, C., and Kish, S.J., 1993, Brain α-ketoglutarate dehydrogenase complex activity in Alzheimer's disease, J. Neurochem. 61: 2007-2014.

Mauch, D.H., Nägler, K., Schumacher, S., Göritz, E.C., Otto, A., Pfrieger, F.W., 2001, CNS synaptogenesis promoted by glia-derived cholesterol, Science 294: 1354-1357.

Michikawa, M., and Yanagisawa, K., 1999, Inhibition of cholesterol production but not of nonsterol isoprenoid products induces neuronal cell death, J. Neurochem. 72: 2278-2285.

Mielke, R., Herholz, K., Grond, M., Kessler, J., and Heiss, W.D., 1992, Differences of regional cerebral glucose metabolism between presenile and senile dementia of Alzheimer type, Neurobiol. Aging 13: 93-98.

Mirollo, R.E., and Strogatz, S.H., 1990, Synchronization of pulse-coupled biological oscillators, SIAM J. Appl. Math. 50: 1645-1649.

Münch, G., Schinzel, R., Loske, C., Wong, A., Durany, N., Li, J.J., Vlassara, H., Smith, M.A., Perry, G., and Riederer P., 1998, Alzheimer's disease-synergistic effects of glucose deficit, oxidative stress and advanced glycation endproducts, J. Neural Transm. 105: 439-462.

Mulder, M., Ravid, R., Swaab, D.,F., de Kloet, E.R., Haasdijk, E.D., Julk, J., van der Boom, J. and Havekes, L.M., 1998, Reduced levels of cholesterol, phospholipids, and fatty acids in cerebrospinal fluid of Alzheimer disease patients are not related to apolipoprotein E4, Alzheimer Dis. Assoc. Disord. 12: 198-203.

Nelson, C.A., Wewerka, S., Thomas, K.M., Tribby-Walbridge, S., de Regnier, R., and Georgiff, M., 2000, Neurocognitive sequelae of infants of diabetic mothers, Behav. Neurosci. 114: 950-956.

Nitsch, R.M., Blusztajn, J.K., Pittas, A.G., Slack, B.E., Growdon, J.A., and Wurtman, R.J., 1992, Evidence for a membrane defect in Alzheimer disease brain, Proc. Natl. Acad. Sci. USA 89: 1671-1675.

Okada, M., Ishikawa, M., and Mizushima, Y., 1991, Identification of a ubiquitin-and ATP-dependent protein degradation in rat cerebral cortex, Biochim. Biophys. Acta 1073: 514-520.

Ott, A., Breteler, M.M.B., van Harskamp, F., Claus, J.J., van der Cammen, T.J.M., Grobbee, D.E., and Hofman, A., 1995, Prevalence of Alzheimer's disease and vascular dementia: association with education. The Rotterdam study, Br. J. Med. 310: 970-973.

Papenberg, J., Lanzinger, G., Kommerell, B., and Hoyer, S., 1975, Comparative studies of the electroencephalogram and the cerebral oxidative metabolism in patients with liver cirrhosis, Klin. Wschr. 53: 1107-1113.

Peraldi, P., Hotamisligil, G.S., Buurman, W.A., White, M.F., and Spiegelman, B.M., 1996, Tumor necrosis factor (TNF)-α inhibits insulin signaling through stimulation of the p55TNF receptor and activation of sphingomyelinase, J. Biol. Chem. 271: 13018-13022.

Perego, C., Vetrugno, C.C., De Simoni, M.G., and Algeri, S., 1993, Aging prolongs the stress-induced release of noradrenaline in rat hypothalamus, Neurosci. Lett. 157: 127-130.

Pericak-Vance, M.A., Grubber, J., Bailey, L.R., Hedges, D., West, S., Sentoro, L., Kemmerer, B., Hall, J.L., Saunders, A.M., Roses, A.D., et al. 2000, Identification of novel genes in late-onset Alzheimer's disease, Exp. Gerontol. 35: 1343-1352.

Perlmutter, D.H., 1999, Misfolded proteins in the endoplasmine reticulum, Lab. Invest. 79: 623-638.

Perry, E.K., Perry, R.H., Tomlinson, B.E., Blessed, G., and Gibson, P.H., 1980, Coenzyme A-acetylating enzymes in Alzheimer's disease: possible cholinergic „compartment" of pyruvate dehydrogenase, Neurosci. Lett. 18: 105-110.

Peskind, E.R., Elrod, R., Dobie, D.J., Pascualy, M., Petrie, E., Jensen, C., Brodkin, K., Murray, S., Veith, R.C., and Raskind, M.A., 1998, Cerebrospinal fluid epinephrine in Alzheimer's disease and normal aging, Neuropsychopharmacology 19: 465-471.

Pettegrew, J.W., Klunk, W.E., Kanal, E., Panchalingam, K., and McClure, R.J., 1995, Changes in brain membrane phospholipid and high-energy phosphate metabolism precede dementia, Neurobiol. Aging 16: 973-975.

Plaschke, K., and Hoyer, S., 1993, Action of the diabetogenic drug streptozotocin on glycolytic and glycogenolytic metabolism in adult rat brain cortex and hippocampus, Int. J. Devl. Neurosci. 11: 477-483.

Prigogine, I., 1989, What is entropy?, Naturwissenschaften 76: 1-8.

Reul, J.M.H.M., Rothuizen, J., and Dekloet, E.R., 1991, Age-related changes in the dog hypothalamic-pituitary-adrenocortical system: neuroendocrine activity and corticosteroid receptors, J. Steroid. Biochem. Molec. Biol. 40: 63-69.

Roberts, E.L. jr., and Sick, T.J., 1996, Aging impairs regulation of intracellular pH in rat hippocampal slices, Brain Res. 735: 339-342.

Röder, H.M., and Ingram, V.M., 1991, Two novel kinases phosphorylate tau and the KSP site of heavy neurofilament subunits in high stoichiometric ratios, J. Neurosci. 11: 3325-3342.

Rothman, J.E., 1996, The protein machinery of vesicle budding and fusion, Protein Science 5: 185-194.

Rothman, J.E., and Wieland, F.T., 1996, Protein sorting by transport vesicles, Science 272: 227-234.

Rupprecht, R., and Holsboer, F., 1999, Neuroactive steroids: mechanism of action and neuropsychopharmacological perspectives, Trends Neurosci. 22: 410-416.

Sacks, W., 1957, Cerebral metabolism of isotopic glucose in normal human subjects, J. Appl. Physiol. 10: 37-44.

Sacks, W., 1965, Cerebral metabolism of doubly labelled glucose in human in vivo, J. Appl. Physiol. 20: 117-130.

Salehi, A., and Swaab, D.F., 1999, Diminished neuronal metabolic activity in Alzheimer's disease, J. Neural Transm. 106: 955-986.

Salehi, A., Heyn, S., Gonatas, N.K., and Swaab, D.F., 1995, Decreased protein synthetic activity of the hypothalamic tubero mamillary nucleus in Alzheimer's disease as suggested by smaller Golgi apparatus, Neurosci. Lett. 193: 29-32.

Salehi, A., Lucassen, P.J., Pool, C.W., Gonatas, N.K., Ravid, R., and Swaab, D.F., 1994, Decreased neuronal activity in the nucleus basalis of Meynert in Alzheimer's disease as suggested by the size of the Golgi apparatus, Neuroscience 59: 871-880.

Sapolsky, R.M., 1994, Glucocorticoids, stress and exacerbation of excitotoxic neuron death, Sem. Neurosci. 6: 323-331.

Sapolsky, R.M., 1998, The stress of Gulf War syndrome, Nature 393: 308-309.

Sapolsky, R.M., Krey, L.C., and McEwen, B.S., 1983, The adrenocortical stress-response in the aged malge rat: impairment of recovery from stress, Exp. Gerontol. 18: 55-64.

Sapolsky, R.M., Krey, L.C., and McEwen, B.S., 1986, The neuroendocrinology of stress and aging. The glucocorticoid cascade hypothesis, Endocr. Res. 7: 284-301.

Sapolsky, R.M., Zola-Morgan, S., and Squire, L.R., 1991, Inhibition of glucocorticoid secretion by the hippocampal formation in the primate, J. Neurosci. 11: 3695-3704.

Seksek, O., Biwersi, J., and Verkman, A.S., 1995, Direct measurement of trans-Golgi pH in living cells and regulation by second messengers, J. Biol. Chem. 270: 4967-4970.

Simons, M., Keller, P., de Strooper, B., Beyreuther, K., Dotti, C., and Simons, K., 1998, Cholesterol depletion inhibits the generation of β-amyloid in hippocampal neurons, Proc. Natl. Acad. Sci. USA 95: 6460-6464.

Simons, M., Schwärzler, I., Lütjohann, D., von Bergmann, K., Beyreuther, K., Dichgans, J., Wormstall, H., Hartmann, T., and Schulz, J.B., 2002, Treatment with simvastatin in normocholesterolemic patients with Alzheimer's disease: A 26-week randomized, placebo-controlled, double-blind trial, Ann. Neurol. 52: 346-350.

Sims, N.R., Bowen, D.M., Allen, S.J., Smith, C.C.T., Neary, D., Thomas, D.J., and Davison, A.N., 1983a, Presynaptic cholinergic dysfunction in patients with dementia, J. Neurochem. 40: 503-509.

Sims, N.R., Bowen, D.M., Neary, D., and Davison, A.N., 1983b, Metabolic processes in Alzheimer's disease: adenine nucleotide content and production of $^{14}CO_2$ from (U^{14} C) glucose in vitro in human neocortex, J. Neurochem. 41: 1329-1334.

Singh, H., Usher, S., and Poulos, A., 1989, Mitochondrial and peroxisomal beta-oxidation of stearic and lignoceric acids by rat brain, J. Neurochem. 53: 1711-1718.

Solano, D.C., Sironi, M., Bonfini, C., Solerte, S.B., Govoni, S., and Racchi, M., 2000, Insulin regulates soluble amyloid precursor protein release via phosphatidyl inositol 3 kinase-dependent pathway, FASEB J. 14: 1015-1022.

Sorbi, S., Bird, E.D., and Blass, J.P., 1983, Decreased pyruvate dehydrogenase complex activity in Huntington and Alzheimer brain, Ann. Neurol. 13: 72-78.

Svennerholm, L., and Gottfries, C.G., 1994, Membrane lipids, selectively diminished in Alzheimer brains, suggest synapse loss as a primary event in early-onset form (type I) and demyelination in late-onset form (type II), J. Neurochem. 62: 1039-1047.

Swaab, D.F., Raadsheer, F.C., Endert, E.F., Hofman, M.A., Kamphorst, W.C., and Ravid, R., 1994, Increases in cortisol levels in aging and Alzheimer's disease in postmortem cerebrospinal fluid, Neuroendocrinology 6: 681-687.

Takahashi, L.K., 1998, Prenatal stress: Consequences of glucocorticoids on hippocampal development and function, Int. J. Devl. Neurosci. 16: 199-207.

Taubes, G., 1996, Misfolding the way to disease, Science 271: 1493-1495.

Thomas, P.J., Qu, B.-H., and Pedersen, P.L., 1995, Defective protein folding as a basis of human disease, Trends Biol. Sci. 20: 456-459.

Vallée, M., Maccari, S., Dellu, F., Simon, H., LeMoal, M., and Mayo, W., 1999, Long-term effects of prenatal stress and postnatal handling on age-related glucocorticoid secretion and cognitive performance: a longitudinal study in the rat, Eur. J. Neurosci. 11: 2906-2916.

Valverde, A.M., Teruel, T., Navarro, P., Benito, M., and Lorenzo, M., 1998, Tumor necrosis factor-α causes insulin receptor substrate-2-mediated insulin resistence and inhibits insulin-induced adipogenesis in fetal brown adipocytes, Endodrinology 139: 1229-1238.

Verde, C., Pascale, M.C., Martive, G., Lotti, L.V., Torrisi, M.R., Helenius, A., and Bonatti, S., 1995, Effect of ATP depletion and DTT on the transport of membrane proteins from the endoplasmic reticulum and the intermediate compartment to the Golgi complex, Eur. J. Cell Biol. 67: 267-274.

Virgin, C.E. jr., Ha, T.P.T., Packan, D.R., Tombaugh, G.C. Yang, S.H., Horner, H.C., and Sapolsky, R.M., 1991, Glucocorticoids inhibit glucose transport and glutamate uptake in hippocampal astrocytes: implications for glucocorticoid neurotoxicity, J. Neurochem. 57: 1422-1428.

von der Kammer, H., Mayhaus, M., Albrecht, C., Enderich, J. Wegner, M., and Nitsch, R.M., 1998, Muscarinic acetylcholine receptors activate expression of the Egr gene family of transcription factors, J. Biol. Chem. 273: 14538-14544.

Watanabe, C.M.H., Wolffram, S., Ader, P., Rimbach, G., Packer, L., Maguire, J.J., Schultz, P.G., and Gohil, K., 2001, The in vivo neuromodulatory effects of the herbal medicine ginkgo biloba, Proc. Natl. Acad. Sci. USA 98: 6577-6580.

Wolozin, B., Kellman, W., Ruosseau, P., Celesia, G.G., and Siegel, 2000, Decreased prevalence of Alzheimer disease associated with 3-hydroxy-3-methylglutaryl Coenzmye A reductase inhibitors, Arch. Neurol. 57: 1439-1443.

Wozniak, M., Rydzewski, B., Baker, S.P., and Raizada, M.K., 1993, The cellular and physiological actions of insulin in the central nervous system, Neurochem. Int. 22: 1-10.

Wurtman, R.J., 1992, Choline metabolism as a basis for the selective vulnerability of cholinergic neurons, Trends Neurosci. 15: 117-122.

NEUROPROTECTION: A Realistic Goal for Aged Brain?

Laura Calzà* and Luciana Giardino

1. INTRODUCTION

Neuron survival during adult life is the end-product of a delicate balance between intrinsic properties of neurons and environmental factors. Intrinsic properties of a neuron derive from genetic background, but also from all stimuli that affect neurons through the life, from prenatal development till aging. In fact, being most of neurons perennial cells, this means that a neuron in an eighty-year old man is eighty year old. According to this, it accumulates age-related molecular alteration that are common in all cell types, such as slowing down of many metabolic processes, protein and membrane turnover, etc., to possible vascular pathologies compromising energy supply, to selective vulnerability of specific classes to unknown agents, that could include environmental pollutants, diet-derived factors, etc.[1] Environmental factors also include systemic signals which can invade the central nervous system, such as hormones, cytokines, other molecules, and cells that can cross the blood-brain-barrier under normal conditions and during pathological events, and local signals acting in the chemical milieu surrounding neurons. Moreover, structure and function of old brain are the result of a balance between rather selective deficits and compensatory attempts of neural networks, which are developed over a long period of time. Thus, observation of old, normal or pathological brains, is not adequate for understanding biological events that can lead to neurodegeneration and, thus, to identify protective strategies.

The aged rat brain, widely considered as an experimental model for research on ageing, also well represents a model in which central transmitter-identified pathways can be investigated in a condition of "functional reserve". Due to structural and functional progressive decline of synaptic performance in several pathways, the reactive capacity of the system can be tested in conditions far apart from what observed in young and adult animals, for whom a great functional reserve is available.

---

*Laura Calzà, DIMORFIPA, University of Bologna, Via Tolara di Sopra 50, 40064 Ozzano Emilia (Bologna), Italy, Email lcalza@vet.unibo.it

In this chapter, data from animal experiments will be joined to clinical data, in the attempt to illustrate few steps in a neurobiological framework in which neuroprotection through drug therapy should be comprise.

2. NEUROTRASMITTER HYPOTHESIS FOR BRAIN AGING

The attempt to correlate the decline of selective brain functions to the deficit of a specific transmitter has been a major effort in the ageing brain research over the past two decades, as in the case of the cholinergic hypothesis for Alzheimer's disease (AD)[2] and the dopamine theory for Parkinson's disease (PD).[3] The post-mortem neurochemical analysis of the AD brain revealed a severe decline of acetylcholine content in the basal forebrain and in the target areas, including the cerebral cortex, and in the hippocampus. These findings, together with the role of acetylcholine in memory organization and the specific cognitive deficit of AD patients, have supported the attempts to increase acetylcholine availability through administration of acetylcholine esterase inhibitors. Similarly, the dopamine deficit observed in Parkinson patients justified treatments aimed at restoring dopamine content in the basal ganglia through precursor administration or through cerebral implant of dopamine-producing cells or dopamine-releasing devices.

However specific, the transmitter deficit alone cannot account for all clinical and neurobiological aspects of these diseases, and therapies based on neurotransmitter restorations are now considered symptomatic therapies, in many cases ineffective to interfere with disease progression. Moreover, the debate whether the neurochemical deficit is the primary event or the final result of a very long process is still open. The symptoms related to the transmitter deficit, as for dopamine in Parkinson disease, are clear when the transmitter decline is severe, 80% below the concentration found in age-matching control brains, thus suggesting a large functional reserve for this system in the human brain.

The hypothesis that a single transmitter is responsible for the control of complex functions such as cognition, but also movement and posture has been now substituted by the notion of multiple transmitter interactions in the neural network supporting these functions.[4] According to this hypothesis, a selective deficit leads to a compensatory neurochemical, anatomical and functional adjustment of the entire network. Only when this attempt has failed, the symptoms of deficit become clinically evident.

3. AGING OF THE COLINERGIC SYSTEM IN THE BASAL FOREBRAIN

The progressive biochemical and functional decline of the cholinergic system is a typical feature of brain aging in different animal species including humans and rodents.[5] Moreover, several age-related neurodegenerative diseases, including dementia, are characterized by a severe damage of the cholinergic system, which leads to cognitive and memory deficits. Maintenance of mature phenotype and survival of cholinergic neurons is dependent on nerve growth factor (NGF) availability[6,7] and the cellular effect of NGF is mediated by high and low affinity receptors.[8] In particular, the expression of the NGF low affinity receptor p75 ($p75^{LNGFR}$) is associated with cholinergic neurons survival and its up-regulation parallels with trophic support.[9,10] Moreover, nNOS is normally present in cholinergic neurons within the basal forebrain[11] to generate NO, and increased availability of NO in this area has been associated with neuroprotective effects.[12] Neurotrophin and neurotrophin receptor content undergo a complex age-regulation in specific regions of the brain[13] and responsiveness of cholinergic neurons to NGF administration is also affected by aging.[14,15,16,17]

Hypotheses on mechanisms underlying age-related vulnerability of cholinergic neurons concern acetylcholine metabolism itself,[18] neurotrophic support,[19] cytoskeleton alterations,[20] target loss[21] and vascular dysfunctions.[22] In particular, recent evidences support a causal link between a dysfunction of cholinegic system in the basal forebrain and the pathogenic events that characterize AD, notably amyloid-beta peptide.[2]

However, due to results from animal studies, the role of cholinergic deficit as the only responsible for age-related cognitive decline has been disputed,[23] moreover preservation of cholinergic function in old age might only indirectly sustain cognitive function.[24] In this context, the cholinergic synapse might not be the only appropriate target for the treatment of age-related cognitive deficit. More suitably, preservation of cholinergic neurons during adult life might represent an alternative strategy to prevent or delay age-related cognitive impairment. Indeed, several compounds having no direct effect on cholinergic synapses proved effective in reducing cognitive impairment due to cholinergic damage in animals. This effect has been observed using peptides like vasoactive intestinal peptide,[25] pituitary adenylate cyclase-activating polypeptide,[26] growth factors and cytokines including glial derived neurotrophic factor, ciliary neurotrophic factor, fibroblast growth factor,[27] and leukemia inhibitory factor.[28]

These results turned back to human diseases, namely dementia[29] and in fact it is now accepted that a constellation of pathogenic mechanisms, including inflammation[30] and damaged microcirculation,[31] takes part in the clinical onset and progression of Alzheimer disease. The recent re-evaluation of the vascular component in Alzheimer's disease is worthy of mention. Vagnucci and Li[32] have proposed that the vascular endothelial cell has a central role in the progressive destruction of cortical neurons in Alzheimer's disease. In Alzheimer's disease, the brain endothelium secretes the precursor substrate for the beta-amyloid plaque and a neurotoxic peptide that selectively kills cortical neurons. Large populations of endothelial cells are activated by angiogenesis due to brain hypoxia and inflammation. Results of epidemiological studies have shown that long-term use of non-steroidal anti-inflammatory drugs, statins, histamine H2-receptor blockers, or calcium-channel blockers seems to prevent Alzheimer's disease. Authors believe that this benefit is largely due to these drugs' ability to inhibit angiogenesis. If Alzheimer's disease is an angiogenesis-dependent disorder, then development of antiangiogenic drugs targeting the abnormal brain endothelial cell might be able to prevent and treat this disease. Interestingly, we have recently proposed that NGF could be a physiological regulator of angiogenesis in post-natal nervous tissue, by regulating the synthesis of vasoactive (nNOS) and angiogenentic (Vascular Endothelial Growth Factor) molecules in neuron in need-dependent manner, only.[33]

4. THE BASAL GANGLIA IN AGING

The basal ganglia are reciprocally connected with the substantia nigra in the mesencephalon. Dopamine (DA) is the major transmitter of the mesostriatal pathway, whereas strio-pallidal and strio-nigral pathways, believed to play an inhibitory role on dopaminergic neurones, are GABAergic. It is well known that with age there is a progressive alteration of dopamine (DA) system in a number of species including rats and humans. This deficit consists in behavioural, biochemical and molecular modifications.[34] In fact, during physiological and pathological aging there is a documented relationship between the decreased extrapyramidal DA-ergic function and the deficit of psychomotor behaviours.[35] Several neurochemical changes occur, including a decrease in the number of mesencephalic DAergic neurons, of DA content, DA uptake and DA receptor densities.[36] On the contrary, the GAD mRNA expression is unaltered in old rats both in term of number of expressing neurons and mRNA content in single neurons. Moreover, a change of the threshold for GABA release induced by dopamine receptor stimulation has been described in old rats.[37]

One interesting aspect in age-related decline of DA-receptors in the basal ganglia, is that the left-right asymmetry of DA receptor distribution is lost in old rats.[36] Several behavioral studies have demonstrated brain asymmetries in humans and animals.[38] The dopaminergic system is involved in a number of lateralised behaviors including the spontaneous side preference, initial direction of movement in the open field test, side preference in the Y-maze and T-maze test and stress responsiveness. Apparently, this is not a random phenomenon, thus supporting the hypothesis of an ontogenetic role of DA brain asymmetry that could play an important functional role during the central organization of behavior.

Finally, DA acts as a powerful regulator of different aspects of cognitive brain functions also through its possible influence on the later cortical specification, particularly of pre-frontal cortical areas.[39] Nieoullon[39] suggests that the characteristic extension of the DAergic cortical innervation in the rostro-caudal direction during the last stages of evolution in mammals can also be related to the appearance of progressively more developed cognitive capacities. Such an extension of cortical DA innervation could be related to increased processing of cortical information through basal ganglia, either during the course of evolution or development. DA has thus to be considered as a key neuroregulator which contributes to behavioral adaptation and to anticipatory processes necessary for preparing voluntary action consequent upon intention. All together, it can be suggested that a correlation exists between DA innervation and expression of cognitive capacities. Altering the dopaminergic transmission could, therefore, contribute to cognitive impairment, and this could have important implication in degenerative diseases affecting the dopaminergic system, including Parkinson disease.

5. SENSORY-MOTOR INTEGRATION IN AGING: VESTIBULAR COMPENSATION

The sensory-motor integration underlying reflex and voluntary movements as well as automatic stabilization and orientation in space, is altered in elderly subjects, determining a frequent impairment of posture and locomotion.[40,41] Threshold and central representation of sensory input are modified: presbiopia, altered pain perception, presbiacusia among others sensory alterations are common features in advanced age. Moreover, the incidence of vestibular symptoms like dizziness, tinnitus, and nausea are common also in the absence of clear anatomical damage of the inner ear, vestibular nerve or central vestibular pathways. However, age-related morphological and physiological alterations of the peripheral and central structures involved in these functions, i.e. the motor unit,[42] the sensory end-organs and central vestibular pathways[43,44] have been also described.

The lesion of the vestibular end organ evokes static and dynamic symptoms, which spontaneously regress during a complex process known as "vestibular compensation." Vestibular compensation occurs due to complex integration of molecular, anatomical, functional and behavioral adjustments in the central nervous system following the damage of peripheral vestibular inputs.[45] Several processes are synergically involved: molecular and cellular mechanisms, leading to the restoration of the spontaneous resting discharge at the level of vestibular nuclei; anatomical and functional adjustments of the neural pathways responsible for vestibular reflex; behavioral strategies, triggered by inputs from sensory, motor or integration areas. This functional recovery leads to the spontaneous regression and disappearance of most dynamic and static symptoms observed in all animal species after vestibular lesion, through a complex rearrangement of sensory inputs integration and behavioral strategies that makes vestibular compensation one of the most clear example of "functional plasticity".[46] As in other diseases, the first therapeutic attempt has been to identify neurotransmitters involved in vestibular compensation, in the attempt to enhance or depress specific systems.[47] According to this, the GABAergic system has been particularly investigated, due to its major role in the control of vestibular function.[47,48,49] GABAergic nerve endings densely innervate the vestibular nuclei complex.[50] This projection is mainly responsible for the inhibitory input from cerebellar Purkinje cells to the vestibular nuclei complex. GABA is also a major transmitter in the vestibulo-ocular reflex system and modulates the vestibulospinal reflex and GABAergic pathways convey vestibular information to the inferior olive.

A working hypothesis suggests that a major role in vestibular compensation is played by substitution of vestibular inputs with other sensory modalities, through the development of different behavioral strategies. This process involves brain areas other than the vestibular nuclei complex. In fact, following a vestibular lesion various brain nuclei are activated at different times during functional recovery, as indicated by Fos histochemistry.[51] These integrated events lead to a re-learning process in which the central nervous system re-establish posture regulation and central movement plane elaboration. To this purpose inputs from other sources are used as substitute sensory information and integrated into new behavioral strategies.[52,53,54] The improvement of vestibular compensation by drugs having positive effects on learning and memory processes, like ACTH fragments,[55,56,57,58] NMDA receptor agonists,[59] and delay in vestibular compensation induced by glucocorticoid receptor antagonist,[60] strongly support this theory. More recently, a direct role of hippocampus has been suggested, due to the vestibular-hippocampal interactions through thalamus and parietal cortex.[61] Moreover, organized vestibular rehabilitation therapy programs have been successfully introduced in the acute and chronic treatment of vestibular diseases. These programs involve habituation exercises, postural control exercises and general conditioning activities designed to facilitate central nervous system compensation.[62,63,64,65]

Data reported above make clear while pharmacology based on single transmitter regulation is unsatisfactory to accelerate vestibular compensation. Therefore, attention has been devoted to compounds aimed to potentiate neural plasticity following vestibular lesion, also coupled to rehabilitative approaches and to learning and memory strategies.[66,67] The pharmacological treatment of vertigo points to three major goals: first, the symptomatic reduction of both the illusion of motion, using drugs having vestibular "suppressant" properties such as anticholinergics and antihistaminics; second, the improvement of the accompanying neurovegetative and psycho-affective signs with antidopaminergic drugs; in the third place the enhancement of the central process of the vestibular compensation.[68] Apart from the already mentioned ACTH-derived peptides, few other molecules have been evaluated, like Ginkgo Biloba extracts which accelerate vestibular compensation,[69,70] calcium channel antagonists and inhibitors of Ca^{++}-dependent enzymes.[71,72]

The behavioral recovery after vestibular end-organ lesion is severely affected by aging in humans as in other animal species. We have reported that old rats suffer from more severe symptoms after labyrinthectomy and recovery is slower than in adult animals, and incomplete.[73] Among the investigated behavioral indexes of vestibular lesion and compensation, equilibrium-orientation seemed to be most severely affected by aging. Moreover, lesion-induced alteration of GABAergic system, evaluated in terms of synthesis enzyme mRNA expression and benzodiazepine receptor density, appears to be age-dependent and parallels with behavioral recovery. The age-related difference in vestibular compensation cannot be univocally explained. A neuronal loss in the caudal part of the medial vestibular nucleus in different animal species, including human, has been recently described.[74] The loss of neurones greater than 50% in old subjects can partially explain the faulty compensation of unilateral vestibular deficits by elderly subjects. Moreover, extravestibular neural networks involved in learning and sensory substitution play a major role in vestibular compensation; therefore also age-related modifications of cognitive processes, including learning and memory, should be taken into account. In our experimental paradigm, the recovery from the vestibular lesion is investigated in term of sensory and behavioral postural strategies and not merely in term of simple vestibular reflex. Spontaneous posture and movement, postural reflexes, equilibrium-orientation capability are indeed due to labyrinthine and non-labyrinthine inputs, which include sensory modalities (proprioceptive, tactile and visual inputs) and extrapyramidal components as well, involved in the elaboration of motor and posture behaviour.

6. NICERGOLINE

Nicergoline is an ergoline derivative used in the treatment of senile mental impairment. It also produces positive effects on psychometric performance in human.[75,76,77] It was initially considered as a vasoactive drug and mainly prescribed for cerebrovascular disorders. In experimental animals, nicergoline treatment increases cerebral metabolism, regional cerebral blood flow and oxygen perfusion in post-ischemic[78,79] and post-hypoxic brain.[80,81] Chronic administration improves monoamine turnover and cholinergic function by reversing the acetylcholine deficit in aged rats.[82] However, more recent studies have indicated other possible mechanism action of nicergoline. The drug also favors signal transduction by increasing phosphoinositide turnover[83] and protein C translocation.[84] *In vitro*, nicergoline has shown a protective effect against apoptosis, using different models including NGF deprivation in PC12 cells,[85] glutathione depletion in GT1-7 cells,[86] H_2O_2-induced neurotoxicity in B50 cells[87] and superoxide burst generation by activated neurotrophils.[88] Finally, nicergoline stimulates PKC-mediated alfa-secretase amyloid precursor protein (APP) processing in cultured human neuroblastoma SH-SY5Y cells,[89] resulting in an enhanced production of non-amyloidogenic N-terminal soluble fragment of APP.[90]

These animal studies suggested other actions, which has provided a rationale for the use of nicergoline for the treatment of various forms of dementia, including AD. A systemic review of 14 clinical data included in Cochrane Database have presented consistent results on positive effects of nicergoline on cognition and behavior in various forms of dementia, including AD, and these effects are supported by an effect on clinical global impression.[91] In particular, there was a difference in favor of the active treatment in reducing the behavioral symptoms described by the Sandoz Clinical Assessment Geriatric Scale. The therapeutic effects of nicergoline seem to be evident by 2 months of treatment and maintained for 6 months. Cognitive assessment has been performed in a moderate number of patients with the MMSE and the ADAS-Cog and also for these tests there was a difference between treatment and control groups on the MMSE, favoring nicergoline treatment.

6.1. Nicergoline and Age-Related Vulnerability of Cholinergic Neurons in the Basal Forebrain

Consistently with the idea that cognitive decline associated with old age is no more considered to be related to one single transmitter, and that neuroprotection could be a key step in interfering also with neurodegenerative diseases progression,[92] we have investigated the effect of long-term treatment with nicergoline on aging and age-related vulnerability of cholinergic neurons in rats.[93] We provided evidence of protective effect of the drug from cholinergic neurons degeneration induced by nerve growth factor deprivation. Nerve growth factor deprivation was induced by colchicine administration in rats 13 and 18 months old. Colchicine induces a rapid and substantial down-regulation of choline acetyltransferase messenger RNA level in the basal forebrain in untreated adult, middle-aged and old rats. Colchicine failed to cause these effects in old rats chronically treated with nicergoline. Moreover, a concomitant increase of both NGF and brain derived neurotrophic factor (BDNF) content was measured in the basal forebrain of old, nicergoline-treated rats. Additionally, the level of messenger RNA for the brain isoform of nitric oxide synthase (nNOS) in neurons of the basal forebrain was also increased in these animals. Based on the present findings, nicergoline proved to be an effective drug for preventing neuronal vulnerability due to experimentally induced NGF deprivation. Although with no indications on possible molecular mechanisms, this study proves that nicergoline may well be included in the group of molecules that mitigates increased vulnerability exhibited by basal forebrain cholinergic neurons in aging and age-related disorders. Our findings support the hypothesis that long-term treatment with nicergoline ameliorates the trophic support of the cholinergic system in the basal forebrain and protects against neurotoxic lesion induced by colchicine in middle age and very old rats.

6.2. Nicergoline and Vestibular Compensation in Aging

As previously discussed, it is reasonable to suppose that several transmitter-identified pathways in different brain regions differently affected by aging are involved in vestibular compensation. Moreover, functions like learning and memory play an important role in vestibular compensation. Therefore, compounds that help to restore a spontaneous adjustment of this balance could represent at the moment an appropriate strategy to improve vestibular compensation.

Thus, we have considered that vestibular compensation in old animals should be a possible target for nicergoline facilitation of spontaneous recovery. In fact, nicergoline is effective in reducing age-dependent severity of behavioral symptoms of vestibular deficit: posture, locomotion, equilibrium, nystagmus rate and falls in the roto-rod test observed after vestibular lesion in old animals were improved and spontaneous recovery was facilitated by nicergoline treatment.[93,94] This behavioral effect is associated to a re-balancing of age- and lesion-induced alteration in GAD mRNA expression. In nicergoline treated rats, fully compensated from a behavioral point of view, GAD mRNA level was comparable to young, compensated animals. Nicergoline treatment tended to restore age-induced alteration of the GABAergic transmission, both in control and lesioned rats. In lesioned rats, nicergoline treatment affected primarily the cerebello-vestibular connection and the vestibular intrinsic GABAergic system. Generally, GAD mRNA expression is altered in the medial vestibular nucleus and in the cerebellum of bi-labyrinthectomized old uncompensated rats and on the lesioned side of hemilabyrinthectomized old animals; nicergoline treatment was shown to balance these alterations. As already observed, the neurochemical indexes of the GABAergic function mirror the improved behavioral profile.

The down-regulation of GAD mRNA expression induced by nicergoline in old unlesioned rats is also interesting in view of the benzodiazepine-GABAergic hypothesis of brain aging.[95,96] According to this hypothesis, the increased benzodiazepine-GABAergic tone in aged brain interferes with the axonal transport, intracellular and intercellular molecular trafficking which guarantees the normal metabolic/trophic glial/neuronal relationship. Nicergoline could then contribute to restore this normal trafficking of trophic substance, which is indispensable for survival and activity, for instance of cholinergic neurones.

7. FUTURE DIRECTIONS: A ROLE FOR NO IN BRAIN AGING QUALITY AND FUNCTIONAL REPAIR?

Nitric oxide (NO), a diffusible gas with a powerful vasodilatory effect, produced by various cell types including neurons, is involved in several processes linked to brain ageing quality. The molecule is a powerful comodulator of glutamatergic synapse, acting as retrograde messenger at NMDA receptor level.[97,98] It has been suggested that the repeated activation of NMDA receptor increases NO production by the post-synaptic neuron, which retrogradely diffuses to the presynaptic terminal, therefore increasing synaptic efficiency. This mechanism seems to be crucially involved in memory processes, namely in long-term potentiation. Moreover, NO has been indicated as a crucial modulator of blood-brain perfusion, being responsible for activity-regulated regional vasodilatation in the brain.[99] This diffusible mediator is also implicated in neurodegenerative processes related to ischemia/hypoxia where its role seems to be dual, neuroprotective and toxic, according to the producing cell type, to the oxidation-reduction state and to its final amount.[100,101] Thus, NO is under extensive investigation in brain aging quality and in age-related brain damage.[102]

Moreover, since the original definition of NO as "Janus faces" by Snyder in 1993,[100] several reports have dealt with the subject of neuroprotective *vs* neurotoxic effects of NO in selective neural populations.[102] We have recently reported that the exogenous administration of the NO precursor L-Arginine, resulting in increased NO production, protects from colchicine-induced degeneration of cholinergic neurons.[103] Likewise the same treatment also protects myocardial,[104] skeletal muscle[105] and the brain[106] from ischemic and reperfusion injury. Moreover, NGF infusion robustly and specifically increases nNOS expression in cholinergic neurons within the basal forebrain,[107] thus supporting the hypothesis that increased nNOS expression results in protection of cholinergic neurons. NO was also found to protect PC12 cells and striatal and cortical NO-producing neurons from apoptosis induced by NGF[108] and others growth factor deprivation.[109]

Results from our experiments have demonstrated that chronic treatment with nicergoline restored the reduced expression of nNOS mRNA in single neurons in the caudato-putamen in old rats and also increased the nNOS mRNA expression in other brain areas, like cerebral cortex, where nNOS level of single cells is not affected by ageing.[110] This effect could play a role in the mechanism of action of nicergoline. Specifically, the adjustment of neuronal NOS availability could influence cerebral vascular regulation during brain activation. NO is in fact a potent cerebral vasodilator, it is synthesized by active neurons, is highly diffusible and short lived[99] and it is assumed that the increased NOS availability is a good indicator for NO production.[111] This effect could also help to explain the increased cognitive performance described in aged rats treated with nicergoline.[112] An increased NO production by neurons that normally synthesize NO could also act as a protective mechanism. For example, it has been shown that NGF injection increases NO production by cholinergic neurons in the basal forebrain[113] protecting the same neurons from the retrograde degeneration induced by axotomy.

Altered NO production has been also described during vestibular compensation. NO-mediate signaling in flocculus could constitute a possible driving force in the inhibition of resting discharge in the medial vestibular nucleus at the initial stage of vestibular compensation.[114] In addition the up-regulation of NO production by the flocculus facilitates vestibular compensation.[115] Moreover, NO production is necessary to reach the compensated state following hemylabirhintectomy in the frog,[116] probably due to the role of NO as mediator of NMDA receptor activity: studies on vestibular compensation have suggested that the up-regulation of NMDA receptors in the deafferented vestibular nuclei complex could be important in vestibular compensation.[59,117]

Although it is clear that NO can lead to damaging or protective actions in the central nervous system, beneficial effects of NO are re-evaluating. In addition to its prominent roles of the regulation of cerebral blood flow and the modulation of cell-to-cell communication in the brain, recent *in vitro* and *in vivo* results indicated that NO is a potent antioxidative agent and protects against selective neurotoxic challenge.[118]

8. CONCLUSIONS

Etiology of age-related neurodegenerative diseases remains elusive and although neuroprotective strategies to prevent or delay the age-related decline in brain performance and cerebral degeneration are under active investigation, effective therapies are not yet available. However, there are multiple clinical conditions in which neuroprotection would be a crucial component in therapy and they are fairly different in term of pathogenic mechanism. By definition, neuroprotection is an effect that may result in salvage, recovery or regeneration of the nervous system, its cells, structure and function.[92] We suggest that four different categories of neuronal damage should be considered:

1. Acute diseases directly compromising central nervous system (CNS), like stroke and trauma. Here brain and spinal cord are acutely destroyed by non-neural pathologies and the aim of neuroprotection would be to stop the progression of the molecular events triggered by trauma and ischemia in the affected areas and to support healthy neurons in areas surrounding the lesion. Ischaemia is well investigated in animal models. A desirable neuroprotectant would, in theory, antagonize multiple injury mechanisms, which are considered to be triggered by excess in gluatamate release.[119] Actually, there are in ischaemia two different area, the ischaemic one and the ischaemic penumbra, a ring of hypoperfused zone surrounding the region of complete infarction. The penumbral zone is a functionally silent tissue, which is able to regain its function if promptly re-perfused. Acute stroke treatment aims to preserve the ischaemic penumbra, protect neurons against further ischaemia and enhance brain plasticity to maximize recovery.[120] In traumatic brain and spinal cord injury, preclinical studies have shown that treatment to limit secondary cell damage can significantly improve outcome after traumatic brain injury.[121] In contrast, neuroprotection trials in human traumatic brain injury have failed to convincingly demonstrate therapeutic benefit. Recent literature has begun to address this discrepancy between preclinical and clinical trials. Perhaps the most important recent observations relate to the potential role of apoptosis in secondary brain injury. Because apoptosis peaks more than 24 h after injury, concepts about the therapeutic window for traumatic brain injury treatment have changed. Apoptosis and necrosis are in delicate balance and inhibition of one cell death pathway may enhance the other. In this group of conditions, also epilepsy could be included. Although the debate on the capacity of simple seizures to induce neuronal injury is still ongoing, no doubt persists on the disastrous effects of prolonged episodes of status epilepticus.[122]

2. Chronic neurodegenerative disease, like Parkinson's and Alzheimer's disease. Here the point is slow-down the progression of still unknown molecular events that specifically affect neurons that are already sick. One negative point is the late diagnosis. In fact, when symptoms appear in diseases like Alzheimer and Parkinson, the brain is already severely compromised. Attempts for neuroprotection in these diseases have included, among others, antioxidants[123] and drug-mediated stimulation of endogenous mechanisms responsible for the detoxification,[124] Ginkgo biloba,[125] neurotrophins.[126]

3. Neural damage secondary to other disease, like in multiple sclerosis. Multiple sclerosis is a demyelinating disease in humans characterized by extensive inflammation and gliosis in the central nervous system.[127] Its main pathological feature is widespread demyelination and oligodendrocyte degeneration. MS has therefore been interpreted as an oligodendrocyte disease. More recently, the possibility of neural damage has been raised.

In particular, it is accepted that axonal loss occurs in different areas of the central nervous system including the spinal cord and corpus callosum;[128] this lesion is involved in permanent disability characterizing the later chronic progressive stage of MS.[129] Disabilities include motor disorders but also cognitive and affective disturbances. Cognitive dysfunction is a major cause of disability in patients with MS. The prevalence of cognitive dysfunction is estimated at 45 to 65%. Natural history studies suggest that once cognitive dysfunction develops in a patient with MS, it is not likely to remit.[130] MRI, MRS and MRS imaging studies have provided evidence that axonal loss and neural damage in MS can be both substantial and early and these noninvasive measures have demonstrated direct correlations between axonal changes, neural damage and disability.[131] Histopathological studies have confirmed this clinical evidence, indicating that both atrophy and decreased density contribute to substantial axonal loss in brain and spinal cord[132] and that the highest incidence of acute axonal injury is found during active demyelination as well as in patients with secondary progressive MS. Acute axonal damage is associated with an axonal conduction block, ion channel redistribution and interruption of axoplasmic flow, including also trophic factors.[133] Thus, the topic of neuronal damage in MS is emerged during the recent years as a possible substrate for both permanent motor disabilities and cognitive problems. Imaging studies have in fact indicated that while normal brain atrophy is estimated at around 0.1–0.2% per year in normal subjects, it increases to approximately 1% in patients with progressive MS.[134] This could be a "lucky condition" in term of neuroprotection: diagnosis is clear before neuronal damage occurs.

4. Protection of the brain in major surgery. The neurological complications of cardiac surgery are associated with significantly increased mortality, morbidity and resource utilization.[135] Manifestations of injuries are broad, ranging from neuro-cognitive dysfunction to frank stroke. A high percentage of patients who undergo cardiac surgery experience persistent cognitive decline and the same processes that injure the brain also appear to cause dysfunction of other vital organs. Many variables have been found to be indicative or risk for peri-operative neurological injury, but the predictive models are more useful for stroke risk than for neuro-cognitive dysfunction.[136] Neurological complications in cardiac surgery occur in elderly, when the functional reserve is over, but also in child, for example in case of congenital heart repairs. In fact, although advances in infant cardiac surgery have resulted in a dramatic decline in mortality rates, neurological morbidity remains an important concern,[137] and long-term consequences, e.g., in brain aging quality, are not yet known. Neuroprotection may best be accomplished during cardiac surgery because, in contrast to non-surgical situations, potential agents can be administered before the neurological insult occurs. This concept should be also expanded to other major surgeries, like for example prosthesic surgery in elderly patients. Although pharmacological neuroprotection may, in the future, offer some of these patients an improved outcome, it is unlikely that any single agent will prevent neurological injury.

REFERENCES

1. S.U. Dani, A. Hori and G.F. Walte, *Principles of neural aging* (Elsevier, Amsterdam, 1997).
2. D.S. Auld, T.J. Kornecook, S. Bastianetto and R. Quirion, Alzheimer's disease and the basal forebrain cholinergic system: relations to beta-amyloid peptides, cognition, and treatment strategies, *Prog. Neurobiol.* **68**(3), 209-245 (2002).

3. C.E. Clarke and M. Guttman, Dopamine agonist monotherapy in Parkinson's disease, *Lancet* **360**(9347), 1767-1769 (2002).
4. S.C. Li and S. Sikstrom, Integrative neurocomputational perspectives on cognitive aging, neuromodulation, and representation, *Neurosci. Biobehav. Rev.* **26**(7):795-808 (2002).
5. S.B. Dunnett and H.C. Fibiger, Role of forebrain cholinergic systems in learning and memory: relevance to the cognitive deficits of aging and Alzheimer's dementia, *Prog. Brain Res.* **98**, 413-420 (1993).
6. A.C. Cuello, D. Maysinger and L. Garofalo, Trophic factor effects on cholinergic innervation in the cerebral cortex of the adult rat brain, *Mol. Neurobiol.* **6**, 451-461 (1992).
7. P.A. Lapchak, Nerve growth factor pharmacology: application to the treatment of cholinergic neurodegeneration in Alzheimer's disease, *Exp. Neurol.* **124**, 16-20 (1993).
8. M. Barbacid, The Trk family of neurotrophin receptors, *J. Neurobiol.* **25**, 1386-1403 (1994).
9. Z. Kokaia, G. Andsberg, A. Martinez-Serrano and O. Lindvall, Focal cerebral ischemia in rats induces expression of P75 neurotrophin receptor in resistant striatal cholinergic neurons, *Neuroscience* **84**, 1113-1125 (1998).
10. D.A. Peterson, H.A. Dickinson-Anson, J.T. Leppert, K.F. Lee and F.H. Gage, Central neuronal loss and behavioral impairment in mice lacking neurotrophin receptor p75, *J. Comp. Neurol.* **404**, 1-20 (1999).
11. T. Sobreviela, S. Jaffar and E.J. Mufson, Tyrosine kinase A, galanin and nitric oxide synthase within basal forebrain neurons in the rat, *Neuroscience* **87**, 447-461 (1998).
12. L. Giardino, A. Giuliani and L. Calzà, Exogenous administration of L-arginine protects cholinergic neurons from colchicine neurotoxicity, *NeuroReport* **11**, 1769-1772 (2000).
13. R.J. Rylett and L.R. Williams, Role of neurotrophins in cholinergic-neurone function in the adult and aged CNS, *Trends Neurosci.* **17**, 486-490 (1994).
14. W. Fischer, A. Bjorklund, K. Chen and F.H. Gage, NGF improves spatial memory in aged rodents as a function of age, *J. Neuroscience* **11**, 1889-1906 (1991).
15. A.L. Markowska, V.E. Koliatsos, S.J. Breckler, D.L. Price and D.S. Olton, Human nerve growth factor improves spatial memory in aged but not in young rats, *J. Neuroscience* **14**, 4815-4824 (1994).
16. R.J. Rylett, S. Goddard, B.M. Schmidt and L.R. Williams, Acetylcholine synthesis and release following continuous intracerebral administration of NGF in adult and aged Fischer-344 rats, *J. Neuroscience* **13**, 3956-3963 (1993).
17. B.A. Urschel and C.E. Hulsebosch, Distribution and relative density of p75 nerve growth factor receptors in the rat brain as a function of age and treatment with antibodies to nerve growth factor, *Brain Res.* **591**, 223-238 (1992).
18. C.G. Gottfries, Neurochemical aspects on aging and diseases with cognitive impairment, *J. Neurosci. Res.* 27, 541-547 (1990).
19. R. Schliebs, S. Rossner and V. Bigl, Immunolesion by 192IgG-saporin of rat basal forebrain cholinergic system: a useful tool to produce cortical cholinergic dysfunction, *Prog. Brain Res.* **109**, 253-264 (1996).
20. K.A. Jellinger and C. Banche, Neuropathology of Alzheimer's disease: a critical update, *J. Neural Transm. Suppl.* **54**, 77-95 (1998).
21. J.D. Cooper and M.V. Sofroniew, Increased vulnerability of septal cholinergic neurons to partial loss of target neurons in aged rats, *Neuroscience* **75**, 29-35 (1996).
22. J.C. de la Torre and G.B. Stefano, Evidence that Alzheimer's disease is a microvascular disorder: the role of constitutive nitric oxide, *Brain Res. Rev.* **34**, 119-136 (2000).
23. J. Stemmelin, C. Lazarus, S. Cassel, C. Kelche and J.-C. Cassel, Immunohistochemical and neurochemical correlates of learning deficits in aged rats, *Neuroscience* **96**, 275-289 (2000).
24. M.G. Baxter, K.M. Frick, D.L. Price, S.J. Breckler, A.L. Markowska and L.K. Gorman, Presynaptic markers of cholinergic function in the rat brain: relationship with age and cognitive status, *Neuroscience* **89**, 771-780 (1999).
25. Gozes, A. Bardea, A. Reshef, R. Zamostiano, S. Zhukovsky, S. Rubinraut, M. Fridkin and D.E. Brenneman, Neuroprotective strategy for Alzheimer disease: Intranasal administration of a fatty neuropeptide, *Proc. Natl. Acad. Sci. USA* **93**, 427-432 (1996).
26. N. Takei, E. Torres, A. Yuhara, H. Jongsma, C. Otto, L. Korhonen, Y. Abiru, Y. Skoglosa, G., G. Schutz, H. Hatanaka, M.V. Sofroniew and D. Lindholm, Pituitary adenylate cyclase-activating polypeptide promotes the survival of basal forebrain cholinergic neurons *in vitro* and *in vivo*: comparison with effects of nerve growth factor, *Eur. J. Neuroscience* **12**, 2273-2280 (2000).
27. E.J. Mufson, J.S. Kroin, T.J. Sendera and T. Sobreviela, Distribution and retrograde transport of trophic factors in the central nervous system: functional implications for the treatment of neurodegenerative diseaes, *Prog. Neurobiol.* **57**, 451-484 (1999).

28. M.K. Panni, J. Atkinson and M.V. Sofroniew, Leukaemia inhibitory factor prevents loss of p75-nerve growth factor receptor immunoreactivity in medial septal neurons following fimbria-fornix lesions, *Neuroscience* **89**, 1113-1121 (1999).
29. W. Hartig, A. Bauer, K. Brauer, J. Grosche, T. Hortobagyi, B. Penke, R. Schliebs, T. Harkany, Functional recovery of cholinergic basal forebrain neurons under disease conditions: old problems, new solutions?, *Rev. Neurosci.* **13**(2):95-165 (2002).
30. H.L. Weiner and D.J. Selkoe, Inflammation and therapeutic vaccination in CNS diseases, *Nature* **420**(6917), 879-884 (2002).
31. J.C. de la Torre, Vascular basis of Alzheimer's pathogenesis, *Ann. NY Acad. Sci.* **977**, 196-215 (2002).
32. A.H. Jr Vagnucci and W.W. Li, Alzheimer's disease and angiogenesis, *Lancet* **361**(9357), 605-8 (2003).
33. L. Calzà, L. Giardino, A. Giuliani, L. Aloe and R. Levi-Montalcini, Nerve growth factor control of neuronal expression of angiogenic and vasoactive factors, *Proc. Natl. Acad. Sci. USA* **98**, 4160-4165 (2001).
34. D.L. Felten, S.Y. Felten, K. Steece-Collier, I. Date and J.A.Clemens, Age-related decline in the dopaminergic nigrostriatal system: the oxidative hypothesis and protective strategies, *Ann. Neurol.* **32**, S133-136 (1992).
35. A.M. Murray and J.L.Waddington, Age-related changes in the regulation of behavior by D-1:D-2 dopamine receptor interactions, *Neurobiol. Aging* **12**(5), 431-435 (1991).
36. L. Giardino, Right-left asymmetry of DA1 and DA2 receptor density is lost in the basal ganglia of old rats, *Brain Res.* **720**, 235-238 (1996).
37. Porras and F. Mora, Dopamine--glutamate--GABA interactions and ageing: studies in the striatum of the conscious rat, *Eur. J. Neurosci.* **7**(11), 2183-2188 (1995).
38. N. Koshikawa, Role of the nucleus accumbens and the striatum in the production of turning behaviour in intact rats, *Rev. Neurosci.* **5**(4), 331-346 (1994).
39. Nieoullon, Dopamine and the regulation of cognition and attention, *Prog. Neurobiol.* **67**(1), 53-83 (2002).
40. R.P. Di Fabio and A. Emasithi, Aging and the mechanisms underlying head and postural control during voluntary motion, *Phys. Ther.* **77**, 458-475 (1997).
41. A.B. Schultz, J.A. Ashton-Miller and N.B. Alexander, What leads to age and gender differences in balance maintenance and recovery?, *Muscle Nerve Suppl.* **5**, S60-S64 (1997).
42. L. Larsson and T. Ansved, Effects of ageing on the motor unit, *Progress Neurobiol.* **45**, 397-458 (1995).
43. R.W. Baloh, Dizziness in older people, *J. Am. Geriatric. Soc.* **40**, 713-721 (1992).
44. A. Karsarkas, Dizziness in aging: A retrospective study of 1194 cases, *Otolaryngol. Head Neck Surg.* **110**, 296-301 (1994).
45. C.L. Darlington and P.F. Smith, Molecular mechanism of recovery from vestibular damage in mammals: recent advances, *Prog. Neurobiol.* **62**, 313-325 (2000).
46. N. Dieringer, 'Vestibular compensation': neural plasticity and its relations to functional recovery after labyrinthine lesions in frogs and other vertebrates, *Prog. Neurobiol.* **46**(2-3), 97-129 (1995).
47. P.F. Smith and C.L. Darlington, Pharmacology of the vestibular system, *Baillieres Clin. Neurol.* **3**(3), 467-484 (1994).
48. L. Calza, L. Giardino, M. Zanni and G. Galetti, Muscarinic and gamma-aminobutyric acid-ergic receptor changes during vestibular compensation. A quantitative autoradiographic study of the vestibular nuclei complex in the rat, *Eur. Arch. Otorhinolaryngol.* **249**(1), 34-39 (1992).
49. C. de Waele, M. Muhlethaler and P.P. Vidal, Neurochemistry of the central vestibular pathways, *Brain Res. Rev.* **20**(1), 24-46 (1995).
50. M. Zanni, L. Giardino, L. Toschi, G. Galetti and L. Calzà, Distribution of neurotransmitters, neuropeptides, and receptors in the vestibular nuclei complex of the rat: An immunocytochemical, in situ hybridization and quantitative receptor autoradiographic study, *Brain Res. Bull.* **36**, 443-452 (1995).
51. C. Cirelli, M. Pompeiano, P. D'Ascanio, P. Arrighi and O. Pompeiano, c-fos Expression in the rat brain after unilateral labyrinthectomy and its relation to the uncompensated and compensated stages, *Neuroscience* **70**(2), 515-46 (1996).
52. P.F. Smith and I.S. Curthoys, Mechanisms of recovery following unilateral labyrinthectomy: a review, *Brain Res. Rev.* **14**(2), 155-80 (1989).
53. C. L. Darlington, H. Flohr and P.F. Smith, Molecular mechanisms of brainstem plasticity. The vestibular compensation model, *Mol. Neurobiol.* **5**(2-4), 355-368 (1991).
54. P.F. Smith and C.L. Darlington, Neurochemical mechanisms of recovery from peripheral vestibular lesions (vestibular compensation), *Brain Res Rev.* **16**(2), 117-33 (1991).
55. U. Luneburg and H. Flohr, Effects of melanocortins on vestibular compensation, *Prog. Brain Res.* **76**, 421-429 (1988).
56. D.P. Gilchrist, P.F. Smith and C.L. Darlington, ACTH(4-10) accelerates ocular motor recovery in the guinea pig following vestibular deafferentation, *Neurosci. Lett.* **118**(1), 14-16 (1990).

57. D.P. Gilchrist, C.L. Darlington and P.F. Smith, A dose-response analysis of the beneficial effects of the ACTH-(4-9) analogue, Org 2766, on behavioural recovery following unilateral labyrinthectomy in guinea-pig, *Br. J. Pharmacol.* **111**(1), 358-63 (1994).
58. D.P. Gilchrist, C.L. Darlington and P.F. Smith, Evidence that short ACTH fragments enhance vestibular compensation via direct action on the ipsilateral vestibular nucleus, *Neuroreport.* **7**(9), 1489-1492 (1996).
59. P.F. Smith and C.L. Darlington, The contribution of N-methyl-D-aspartate receptors to lesion-induced plasticity in the vestibular nucleus, *Prog. Neurobiol.* **53**(5), 517-31 (1997).
60. S.A. Cameron and M.B. Dutia, Lesion-induced plasticity in rat vestibular nucleus neurones dependent on glucocorticoid receptor activation, *J. Physiol.* **518**(Pt 1), 151-8 (1999).
61. P.F. Smith, Vestibular-hippocampal interactions, *Hippocampus* **7**(5), 465-471 (1997).
62. C.A. Foster, Vestibular rehabilitation. *Baillieres Clin. Neurol.* **3**(3), 577-592 (1994).
63. N.T. Shepard and S.A. Telian, Programmatic vestibular rehabilitation, *Otolaryngol. Head Neck Surg.* **112**(1), 173-82 (1995).
64. S.A. Telian and N.T. Shepard, Update on vestibular rehabilitation therapy, *Otolaryngol. Clin. North Am.* **29**(2), 359-71 (1996).
65. L. Yardley, S. Beech, L. Zander, T. Evans and J. Weinman, A randomized controlled trial of exercise therapy for dizziness and vertigo in primary care, *Br. J. Gen. Pract.* **48**(429), 1136-1140 (1998).
66. F.B. Horak, C. Jones-Rycewicz, F.O. Black and A. Shumway-Cook, Effects of vestibular rehabilitation on dizziness and imbalance, *Otolaryngol. Head Neck Surg.* **106**, 175-180 (1992).
67. RW. Baloh, Vertigo, *Lancet* **352**, 1841-1846 (1998).
68. O. Rascol, T.C. Hain, C. Brefel, M. Benazet, M. Clanet and J.L. Montastruc, Antivertigo medications and drug-induced vertigo. A pharmacological review, *Drugs* **50**, 777-791 (1995).
69. P.F. Smith and C.L. Darlington, Can vestibular compensation be enhanced by drug treatment? A review of recent evidence, *J. Vestib. Res.* **4**, 169-179 (1994).
70. B. Tighilet and M. Lacour, Pharmacological activity of the Ginkgo bilboa extract (Egb761) on equilibrium function recovery in the unilateral vestibular neurectomized cat, *J. Vestib. Res.* **5**, 187-200 (1995).
71. D.P. Gilchrist, C.L. Darlington and P.F. Smith, Effects of flunarizine on ocular motor and postural compensation following peripheral vestibular deafferentation in the guinea pig, *Pharmacol. Biochem. Behav.* **44**(1), 99-105 (1993).
72. A.J. Sansom, P.F. Smith and C.L. Darlington, Evidence that L-type calcium channels do not contribute to static vestibular function in the guinea pig vestibular nucleus, *Brain Res.* **630**(1-2), 349-352 (1993).
73. L. Giardino, M. Zanni, O. Pignataro and L. Calzà, Plasticity of gabaergic system during aging: focus on vestibular compensation and possible pharmacological intervention, *Brain Res.* **929**, 76-86 (2002).
74. J.C. Alvarez, C. Diaz, C. Suarez, J.A. Fernandez, C. Gonzalez del Rey, A. Navarro and J. Tolivia, Neuronal loss in human medial vestibular nucleus, *Anat. Rec.* **251**, 431-438 (1998).
75. R.D. Venn, Review of clinical studies with ergots in gerontology, *Adv. Biochem. Psychopharmacol.* **23**, 363-377 (1980).
76. P.L. Canonico, M.A. Sortino, N. Carfagna, S. Cavallaro, F. Pamparana, K. Annoni, E. Wong and C. Post, Pharmacological basis for the clinical effects of nicergoline in dementia, *Geriatria* **5**(Suppl VIII), 24-48 (1996).
77. R.G. Fariello, Treatment of impaired cognition with nootropic drugs: nicergoline versus the state of the art, *Funct. Neurol.* **12**, 221-225 (1997).
78. M. Le Poncin-Lafitte, C. Grosdemouge, D. Duterte and J.R. Rapin, Simultaneous study of haemodynamic, metabolic and behavioural sequelae in a model of cerebral ischaemia in aged rats: effects of nicergoline, *Gerontology* **30**, 109-119 (1984).
79. U. Schindler, D.K. Rush and S. Fielding, Nootropic drugs: animal models for studying effects on cognition. *Drug Dev. Res.* **4**, 567-576 (1984).
80. B. Saletu, J. Grunberger, L. Linzmayer and P. Anderer, Brain protection of nicergoline against hypoxia: EEG brain mapping and psychometry, *J. Neural Transm. Park. Dis. Dement. Sect.* **2**, 305-325 (1990).
81. K. Shintomi, Pharmacological study of nicergoline: effects on regional cerebral blood flow, *Arzneimittelforschung* **41**, 885-890 (1991).
82. N. Carfagna, A. Clemente, S. Cavanus, D. Damiani,M. Gerna, P. Salmoiraghi, B. Cattaneo and C. Post, Modulation of hippocampal Ach release by chronic nicergoline treatment in freely moving young and aged rats, *Neurosci. Lett.* **197**, 195-198 (1995).
83. N. Carfagna, S. Cavanus, D. Damiani,M. Salmoiraghi, R. Fariello and C. Post, Modulation of phosphoinositide turnover by chronic nicergoline in rat brain, *Neurosci. Lett.* **209**, 189-192 (1996).
84. , A. Caputi, M. Di Luca, L. Pastorino, F. Colciaghi, N. Carfagna, E. Wong, C. Post and F. Cattabeni, Nicergoline and its metabolite induce translocation of PKC isoforms in selective rat brain areas, *Neurosci. Res. Commun.* **23**, 159-167 (1998).

85. P.L. Canonico, M.A. Sortino, N. Carfagna, S. Cavallaro, F. Pamparana, K. Annoni, E. Wong and C. Post, Pharmacological basis for the clinical effects of nicergoline in dementia, *Geriatria* **5**(Suppl.VIII), 24-48 (1996).
86. M.A. Sortino, A. Battaglia, F. Pamparana, N. Carfagna, C. Post and P.L. Canonico, Neuroprotective effects of nicergoline in immortalized neurons, *Eur. J. Pharmacol.* **368**, 285-290 (1999).
87. E. Iwata, M. Miyazaki, M. Asanuma, A. Iida and N. Ogawa, Protective effects of nicergoline against hydrogen peroxide toxicity in rat neuronal cell line, *Neurosci. Lett.* **251**, 49-52. (1998).
88. M. Tanaka, T. Yoshida, K. Okamoto and S. Hirai, Antioxidant properties of nicergoline: inhibition of brain autooxidation and superoxide production neurophils in rats, *Neurosci. Lett.* **284**, 68-72 (1998).
89. A. Cedazo-Minguez, L. Bonecchi, B. Winblad, C. Post, E.H. Wong, R.F. Cowburn and L. Benatti, Nicergoline stimulates protein kinase C mediated alpha-secretase processing of the amyloid precursor protein in cultured human neuroblastoma SH-SY5Y cells, *Neurochem. Int.* **35**, 307-315 (1999).
90. F. Checler, Processing of the b-amyloid precursor protein and itsa regulation in Alzheimer's disease, *J. Neurochem.* **65**, 1431-1444 (1995).
91. M. Fioravanti, L. Flicker, Efficacy of nicergoline in dementia and other age associated forms of cognitive impairment, *Cochrane Database Syst Rev.* **4**, CD003159 (2001).
92. F.J.Vajda, Neuroprotection and neurodegenerative disease, *J. Clin. Neurosci.* **9**(1), 4-8 (2002).
93. L. Giardino, A. Giuliani, A. Battaglia, N. Carfagna, L. Aloe and L. Calzà, Neuroprotection and aging of the cholinergic system: a role for the ergoline derivative nicergoline (Sermion), *Neuroscience* **109**, 487-497 (2002).
94. L. Rampello and F. Drago, Nicergoline facilitates vestibular compensation in aged male rats with unilateral labyrinthectomy, *Neurosci. Lett.* **267**(2), 93-6 (1999).
95. T.J. Marczynski, J. Artwohl and B. Marczynska, Chronic administration of flumazenil increases life span and protects rats from age-related loss of cognitive functions: a benzodiazepine/GABAergic hypothesis of brain aging, Neurobiol. Aging 15(1), 69-84 (1994).
96. T.J. Marczynski, GABAergic deafferentation hypothesis of brain aging and Alzheimer's disease; pharmacologic profile of the benzodiazepine antagonist, flumazenil, *Rev. Neurosci.* **6**(3), 221-258 (1995).
97. J. Garthwaite and C.L. Boulton, Nitric oxide signaling in the central nervous system, *Annu. Rev. Physiol.***57**, 683-706 (1995).
98. S.R Vincent, Nitric oxide: a radical neurotransmitter in the central nervous system, *Prog. Neurobiol.* **42**(1), 129-160 (1994).
99. C. Iadecola, Regulation of the cerebral microcirculation during neural activity: is nitric oxide the missing link?, *Trends Neurosci.* **16**, 206-214 (1993).
100. S.H. Snyder, Janus faces of nitric oxide, *Nature* **364**, 577 (1993).
101. T. Dalkara, T. Yoshida, K. Irikura and M.A. Moskowitz, Dual role of nitric oxide in focal cerebral ischemia, *Neuropharmacology* **33**(11), 1447-1452 (1994).
102. L. Calzà, S. Ceccatelli and L. Giardino, NO and brain aging, *Perspectives in Brain Aging Res.* **1**, 10-16 (1996).
103. L. Giardino, A. Giuliani and L. Calza, Exogenous administration of L-arginine protects cholinergic neurons from colchicine neurotoxicity, *Neuroreport* **11**(8), 1769-7172 (2000).
104. J. Pernow and Q.D. Wang, The role of the L-arginine/nitric oxide pathway in myocardial ischaemic and reperfusion injury, *Acta Physiol. Scand.* **167**, 151-159 (1999).
105. I. Huk, J. Nanobashvili, C. Neumayer, A. Punz, M. Mueller, K. Afkhampou, M. Mittlboeck, U. Losert, P. Polterauer, E. Roth, S. Patton and T. Malinski, L-arginine treatment alters the kinetics of nitric oxide and superoxide release and reduces ischemia/reperfusion injury in skeletal muscle, *Circulation* **96**, 667-675 (1997).
106. E. Roth, The impact of L-arginine-nitric oxide metabolism on ischemia/reperfusion injury, *Curr. Opin. Clin. Nutr. Metab. Care* **1**, 97-99 (1998).
107. D.M. Holtzman, J. Kilbridge, D.S. Bredt, S.M. Black, Y. Li, D.O. Clary, L.F. Reichardt and W.C. Mobley, NOS induction by NGF in basal forebrain cholinergic neurons: evidence for regulation of brain NOS by a neurotrophin, *Neurobiol. Dis.* **1**, 51-60 (1994).
108. Y. M. Kim, H.T. Chung, S.S. Kim, J.A. Han, Y.M. Yoo, K.M. Kim, G.H. Lee, H.Y. Yun, A. Green, J. Li, R.L. Simmons and T.R. Billiar, Nitric oxide protects PC12 cells from serum deprivation-induced apoptosis by cGMP-dependent inhibition of caspase signaling, *J. Neurosci.* **19**, 6740-6747 (1999).
109. M.I. Behrens, J.Y. Koh and M.C. Muller, NADPH diaphorase-containing striatal or cortical neurons are resistant to apoptosis, *Neurobiol. Dis.* **3**, 72-75 (1996).
110. L. Giardino, M. Zanni, M. Pozza, A. Battaglia, L. Calzà and O. Pignataro, Nitric oxide synthase mRNA regulation by nicergoline treatment in the brain of old rats, *Eur. Soc. Clin. Neuropharmacol.* **P7**, 71 (1997).

111. A. Rengasamy and R.A. Johns, Regulation of nitric oxide synthase by nitric oxide, *Mol. Pharmacol.* **44**, 124-128 (1993).
112. R.A. McArthur, N. Carfagna, L. Banfi, S. Cavanus, M.A. Cervini, R. Fariello and C. Post, Effects of nicergoline on age-related decrements in radial maze performance and acetylcholine levels, *Brain Res. Bull.* **43**, 305-311 (1997).
113. D.M. Holtzman, J. Kilbridge, D.S. Bredt, S.M. Black, Y. Li, D.O. Clary, L.F. Reichardt and W.C. Mobley,. NOS induction by NGF in basal forebrain cholinergic neurones: evidence for regulation of brain NOS by a neurotrophin, *Neurobiol. Dis.* **1**, 51-60 (1994).
114. T. Kitahara, N. Takeda, T. Kubo and Kiyama H. Nitric oxide in the flocculus works the inhibitory neural circuits after unilateral labyrinthectomy, *Brain Res.* **815**, 405-409 (1999).
115. T. Kitahara, N. Takeda, P.C. Emson, T. Kubo and H. Kiyama, Changes in nitric oxide synthase-like immunoreactivities in unipolar brush cells in the rat cerebellar flocculus after unilateral labyrinthectomy, *Brain Res.* **765**, 1-6 (1997).
116. G. Flugel, S. Holm and H. Flohr, Chronic inhibition of nitric oxide synthase prevents functional recovery following vestibular lesions, in: *The biology of nitric oxide. 3 Physiological and clinical aspects*, edited by S. Moncada, M. Feelisch, R. Busse and E.A. Higgs (Portland Press, London, 1994), pp. 381-387.
117. P.F. Smith, C. de Waele, P.P. Vidal and C.L. Darlington, Excitatory amino acid receptors in normal and abnormal vestibular function, *Mol. Neurobiol.* **5**(2-4), 369-87 (1991).
118. C.C. Chiueh, Neuroprotective properties of nitric oxide, *Ann. N.Y. Acad. Sci.* **890**, 301-311 (1999).
119. M.D. Ginsberg, Adventures in the pathophysiology of brain ischemia: penumbra, gene expression, neuroprotection: the 2002 Thomas Willis Lecture, *Stroke* **34**(1), 214-223 (2003).
120. M.M. Brown, Brain attack: a new approach to stroke, *Clin. Med.* **2**(1), 60-5 (2002).
121. A.I. Faden, Neuroprotection and traumatic brain injury: theoretical option or realistic proposition, *Curr. Opin. Neurol.* **15**(6), 707-712 (2002).
122. Arzimanoglou, E. Hirsch, A. Nehlig, P. Castelnau, P. Gressens and A. Pereira de Vasconcelos, Epilepsy and neuroprotection: an illustrated review, *Epileptic Disord.* **4**(3), 173-82 (2002).
123. Moosmann and C. Behl, Antioxidants as treatment for neurodegenerative disorders, *Expert Opin. Investig. Drugs* **11**(10), 1407-1435 (2002).
124. Drukarch and F.L. van Muiswinkel, Neuroprotection for Parkinson's disease: a new approach for a new millennium, *Expert Opin. Investig. Drugs* 10(10), 1855-1868 (2001).
125. Y. Luo, Ginkgo biloba neuroprotection: Therapeutic implications in Alzheimer's disease, *J. Alzheimers Dis.* **3**(4), 401-407 (2001).
126. M.V. Sofroniew, C.L. Howe and W.C. Mobley, Nerve growth factor signaling, neuroprotection, and neural repair, *Annu. Rev. Neurosci.* **24**, 1217-1281 (2001)
127. J.L. Seeburger and J.E. Springer, Experimental rationale for the therapeutic use of neurotrophins in amyotrophic lateral sclerosis, *Exp. Neurol.* **124**(1), 64-72 (1993).
128. B. Ferguson, M.K. Matyszak, M.M. Esiri and V.H. Perry, Axonal damage in acute multiple sclerosis lesions, *Brain* **120**(Pt 3), 393-9 (1997).
129. P. Rieckmann and K.J. Smith, Multiple sclerosis: more than inflammation and demyelination, *Trends Neurosci.* **24**(8), 435-437 (2001).
129. B. Bagert, P. Camplair and D. Bourdette, Cognitive dysfunction in multiple sclerosis: natural history, pathophysiology and management, *CNS Drugs* **16**(7), 445-55 (2002).
130. O. Ciccarelli, E. Giugni, A. Paolillo, C. Mainero, C. Gasperini, S. Bastianello and C. Pozzilli, Magnetic resonance outcome of new enhancing lesions in patients with relapsing-remitting multiple sclerosis, *Eur. J. Neurol.* *6(4), 455-459 (1999).*
131. N. Evangelou, D. Konz, M.M. Esiri, S. Smith, J. Palace and P.M. Matthews, Regional axonal loss in the corpus callosum correlates with cerebral white matter lesion volume and distribution in multiple sclerosis, *Brain* **123**(Pt 9), 1845-1849 (2000).
132. J.T. Povlishock, Traumatically induced axonal injury: pathogenesis and pathobiological implications, *Brain Pathol. 2(1), 1-12 (1992).*
133. C. Bjartmar and B.D. Trapp, Axonal and neuronal degeneration in multiple sclerosis: mechanisms and functional consequences, *Curr. Opin. Neurol.* **14**(3), 271-278 (2001).
134. J.E. Arrowsmith, H.P. Grocott, J.G. Reves and M.F. Newman, Central nervous system complications of cardiac surgery, *Br. J. Anaesth.* **84**(3), 378-93 (2000).
135. C.W. Jr Hogue, T.M. Sundt 3rd, M. Goldberg, H. Barner and V.G., Davila-Roman, Neurological complications of cardiac surgery: the need for new paradigms in prevention and treatment, Semin. Thorac. Cardiovasc. Surg. Apr;11(2):105-15 (1999).
136. A.J. du Plessis and M.V. Johnston, The pursuit of effective neuroprotection during infant cardiac surgery, *Semin. Pediatr. Neurol.* **6**(1), 55-63 (1999).

KYNURENINES IN NEURODEGENERATIVE DISORDERS: THERAPEUTIC CONSIDERATION

Péter Klivényi, József Toldi and László Vécsei[1]

1. INTRODUCTION

The kynurenine pathway is a major route for the conversion of tryptophane to NAD and NADP (Figure 1), leading to production of a number of biologically active molecules with neuroactive properties. During the last decades, the interest in kynurenines has been emerged as two major metabolites of this pathway, quinolinic acid (QUIN) and kynurenic acid (KYNA), act on glutamate receptors. QUIN was shown to be an agonist of the N-methyl-D-aspartate (NMDA) type of glutamate receptors. KYNA was shown to be an antagonist the same NMDA receptors with a high affinity to the glycin coagonist site. The NMDA receptor activation has been implicated in many neurological disorders such as stroke, brain injury, Parkinson's disease, Huntington's disease and multiple sclerosis. The receptor antagonists reduce the excitotoxic damage both in vivo[1,2,3] and in vitro[4] and could be used against neurodegenerative disorders. However the classical antagonists have some adverse effects, that limit their clinical use such as memory and learning impairment, psychosis and cell deaths.[5] Influencing the kynurenine pathway provides an option to increase the neuroprotective capacity and decrease the concentration of neurotoxic metabolites. On the other hand, the impariment of kynurenine system has been implicated in several neurological disorders such as stroke, brain injury, Parkinson's disease, Huntington's disease and multiple sclerosis.

In this overview the characteristics of the pathway and its metabolites will be summarized.

[1] Péter Klivényi, Department of Neurology, University of Szeged POB 427, H-6701, Szeged, Hungary. József Toldi, Department of Comparative Physiology, University of Szeged POB 533, H-6701, Szeged, Hungary. László Vécsei, Neurology Research Group of the Hungarian Academy of Sciences and University of Szeged, POB 427, H-6701, Szeged, Hungary.

2. MAJOR NEUROACTIVE METABOLITES OF KYNURENINE PATHWAY

2.1. L-Kynurenine

L-kynurenine (L-KYN) is the central compound of the pathway and can be metabolized to KYNA and QUIN. L-KYN is ubiquitously present in the mammalian body. The brain accumulates L-KYN seven times more effectively then most of the peripheral organs. Only in the brain it is transported through the blood-brain-barrier by neutral amino acid carrier.[6] Brain regional analysis revealed difference between the areas of highest (cortex) and lowest (cerebellum) uptake. Systemic administration of L-KYN decreases the blood pressure[7] and increases the concentrations of QUIN. The increased QUIN concentration does not results neurotoxicity. The observed neurotoxicity is probably related to the active metabolites, which can penetrate to the CSF and serves as a precursor for QUIN synthesis. The local injection of QUIN into the brain produce convulsions.[8,9]

2.2 Kynurenic Acid

KYNA is a major metabolite of the pathway. The role of KYNA came to the frontline of interest after the ionotrophic excitatory amino acid receptor antagonist properties have been discovered in early 1980s.[10] It was the first descripition of the amino acid antagonism by KYNA. Few years later, Kessler et al. observed that KYNA could displace glycin from its strichnine-insensitive binding site.[11,12] It turned out that KYNA had complicated actions on the NMDA receptor complex: in low concentrations would act selectively on the glycin site, whereas higher concentration would act directly at the NMDA recognition site, moreover on non-NMDA receptors.[13,14,15] However, in vitro studies showed that 10^{-5} M KYNA is necessary to antagonize the effects of glutamate.[16] In different pathological conditions, including seizures and stroke the elevation of KYNA concentration is sufficient to inhibit these receptors and modulate the glutaminerg mechanism. Since the concentration of KYNA in the mammalian brain (10 to 150 nmol) is lower then those able to interact with these receptors,[17] it has been a debate whether the changes in the level of KYNA occuring in physiological conditions are sufficient to have any effect on the brain function. The experimental observations however suggest that minimal increase in the level of KYNA is sufficient to inhibit excitatory synaptic function.[18] Besides the possible blockade of postsynaptic NMDA receptors, presynaptic nicotinic acetyl-choline receptors may also contribute to the inhibitory effects of kynurenate on glutamate release especially at low concentration. All these results suggest that KYNA could play a role in regulation of glutaminerg neurotransmission. Intracerebroventricularly administered KYNA is known to dose-dependently induce characteristic behaviour in the rat: increased stereotypy and ataxia.[19,20] In vivo and in vitro data suggest that the regulation of the levels of extracellular kynurenic acid appears to occur at the level of L-KYN uptake and/or kynurenine transaminase. The conversion of L-KYN to KYNA depends on oxygenation and glucose concentration. Systemic administration of L-KYN produces increased level of KYNA in the striatum, but in contrast, systemic administration of KYNA results only a modest elevation in the striatum. No catabolic enzymes or cellular reuptake systems of KYNA exist in mammals,[21] so KYNA is extruded from the brain by a probenecid-sensitive carrier system.[17] Once formed in glial cells, KYNA is released into the extracellular space. Acidic metabolites, including the dopamine metabolite homovanilic acid or 5-hydroxyindoleacetic acid are actively transported from the brain by this probenecid-sensitive pump.[22] More recently, QUIN was also shown to be extruded from the brain by the same carrier system.[23]

2.3 Quinolinic Acid

QUIN is another natural major compound of the kynurenine pathway at concentration similar to that of KYNA (50-100 nmol). In pathological conditions dramatic increase of its concentrations has been demonstrated.[24] Despite that the QUIN is a weak agonist at the NMDA receptors, in vivo excitotoxiciy is similar to NMDA. Several factors can enhance the neurotoxicity including toxic free radical generation.[25] A number of studies have demonstrated that increased plasma (and therefore, brain) levels of tryptophan not only augment of brain levels of serotonin, but elevates the brain concentration of QUIN.

3-hydroxykynurenine

3-HK is also an intracellular neurotoxin by producing free radicals. No physiological function has been related to 3-HK so far, but this metabolites could contribute to the neuronal cell death under pathological conditions. In mammalian brain 3-HK is present in nM concentration, but in pathological condition its concentration could reach the μM range, where it acts as an excitotoxin.[26]

3. ENZYMES OF THE KYNURENINE PATHWAY

The enzymes involved in the kynurenine pathway are summarized in Figure 1. The tryptophane is cleaved by a heme-containing enzyme, indolamine 2,3 dioxigenase (IDO). This enzyme is present in the most mammalian organ including central nervous system. The gene of this enzyme contains two interferon-stimulated elements and a gamma-interferon-activated sequence,[27,28] which indicates immunological control of the gene expression. The cleavage of the tryptophane by IDO leads to the formyl-kynurenine formation which is rapidly metabolized to L-kynurenine (L-KYN) by formamidase. L-KYN serves as a substrate for several enzymes including kynurenine-3-hydroxylase, kynurenineasee, and kynurenine aminotransferase.

Kynurenine 3-hydroxylase is a flavin containing monooxygenase, which is localized in the outer membrane of the mitochondria.[29] The reaction results 3-hydroxy-kynurenine. The enzyme has high affinity for its substrate, suggesting that under physiological conditions it metabolized most of the available kynurenine to 3-hydroxy-kynurenine.

Kynureinase is a pyridoxal phosphate-dependent enzyme located in the cytosol and hydrolyzed the L-KYN and 3 HK to L-alanine and anthranilic acid (or hydro-anthranilic acid). The affinity of the enzyme is 10-fold higher for 3-HK than for L-KYN.[30]

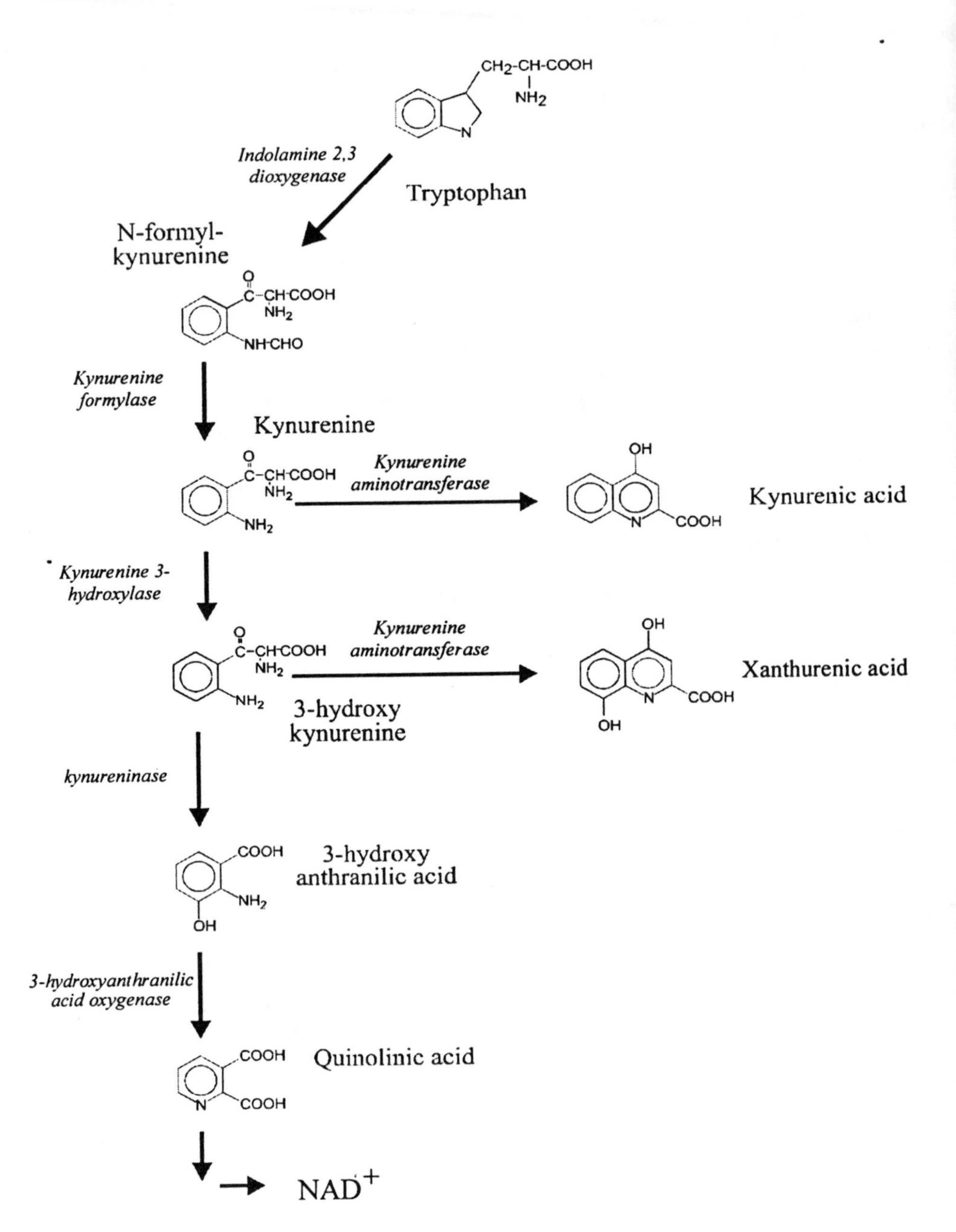

Figure 1. The kynurenine pathway.

KYNA is produced by the irreversible transamination of KYN. In the human and rat brain two distinct KATs are mainly responsible for the synthesis of KYNA (KAT I, KAT II).[31,32] These enzymes differ in pH optimum and substrate specificity. KAT I (E.C. 2.6.1.14), has an optimal pH of 9.5 to 10, prefers pyruvate as a co-substrate and it is inhibited by glutamine. The other enzyme, kynurenine aminotransferase II (KAT II) has a neutral pH optimum (pH 7.3), preferentially recognizes KYN as a substrate, shows no preference for pyruvate, and it is not sensitive to inhibition by glutamine. It is demonstrated, that in most brain regions KYNA derives primarily from KAT II activity under physiologic circumstances.[33] Under pathological conditions a massive increase in KAT I immunostaining have been observed in the neurons in the affected brain areas.[34,35] Both KATs have a K_m in the millimolar range, suggesting that KYN bioavailability is the rate limiting step for KYNA biosynthesis.

All of the enzymes required for the biosynthesis of QUIN and KYNA are detectable in the brain, but their cerebral activities are much lower than in the peripheral tissue.[15] In the CNS kynurenine pathway depends on the KYN originating from the blood. It is estimated that 60% of KYN in the brain derives from peripheral sources under physiological conditions.[36] Both KYN and 3-HK are transported easily through the blood-brain barrier by the same large neutral amino acid carrier as tryptophan.[6] In contrast, because of their polar structure and lack of transport mechanism, KYNA and QUIN poorly penetrate the blood-brain barrier.

Immunocytochemical localization have revealed that the kynurenic pathway enzymes are predominantly localized in glial cells (astrocytes and microglial cells).[36,37,38,39] Neurons sporadically express KATs, kynurenine-hydroxylase and QPRT. Astrocytes appear to be the primary source of KYNA in the brain,[40] whereas microglial cells produce metabolites of the QUIN branch of the pathway.[41]

4. ALTERATIONS OF THE KYNURENINE PATHWAY UNDER PATHOLOGICAL CONDITIONS

The role of kynurenine metabolites in many neurological disorders has been hypothesized. In general, upregulation of the cerebral KP is observed in response to focal injury resulting in elevated levels of QUIN, 3-HK and KYNA in the brain.[36] However, in infections, ischaemia, traumatic brain injury, microglial cells are activated leading to dramatic increases in the level of QUIN. Changes in KYNA synthesis are relatively modest in these states due to the low KAT activity of microglia. Thus, an imbalance arises in the KP, favoring the neurotoxins, 3-HK and QUIN over the neuroprotectant KYNA.

4.1. Diseases with Elevated QUIN Level

4.1.1. AIDS Dementia Complex

Both human and animal studies have shown a substantial increase in the level of QUIN in the CSF and in the brain of patients with AIDS-dementia complex,[42] and this elevation correlated with the severity of cognitive and motor decline.[42] In the brain parenchyma QUIN levels were elevated by 300-fold.[43] There was no correlation between blood QUIN levels and CNS QUIN levels in terminal AIDS patients, suggesting that almost all of the elevated QUIN in the brain synthesised in the CNS and not transported from the blood.[42,43] Similar significant increases in QUIN levels in the CNS have been described in patients with Lyme disease.[44]

4.1.2. Multiple Sclerosis (MS)

In patients with MS, the CSF levels of KYNA were significantly decreased during remissions, which suggests that alteration of kynurenine system may be implicated in relapsing-remitting MS.[45] The levels of 3-HK and QUIN, are elevated in the spinal cord in an animal model of MS (experimental allergic encephalomyelitis).[46,47] Since QUIN has been shown to be toxic to oligodendrocytes,[48] elevated levels of QUIN might be involved in the disease pathology characterized by demyelination. These results provide evidence of the alterations in the, kynurenine pathway in MS. Interesting observation that interferon-beta(1b), in pharmacologically relevant concentrations induces kynurenine metabolism resulting quinolinic acid in human macrophages and this may be a limiting factor in its efficacy of IFN-beta(1b), in the treatment of MS.[49]

4.1.3. Huntington's Disease (HD)

QUIN have been hypothesized to play an important role in HD, since there are major similarities between the neurochemistry and histopathology of the disease and the effects of intrastriatally administrated QUIN.[50] Therefore, animals with striatal QUIN lesions is used as an animal model of the disease. However, no changes in QUIN levels in the CSF have been detected in HD patients.[51,52,53] The KYNA and 3-HK levels were increased in the neostriatum and cortex of early stage of HD patients and in transgenic mice containing full-length mutant huntingtin.[54] These results demonstrate that an increased activity of QUIN branch of the pathway compared to the KYNA branch in early stage HD, that might contribute to the pathophysiology.[55]

4.1.4. Parkinson's Disease

The L-kyn/trp-ratios were increased both in serum and CSF of patients with PD as compared to controls. Serum tryptophan was lower in patients with Parkinson's disease.[56] The level of 3-HK is significantly increased in the putamen and substantia nigra in patients with Parkinson's disease, which could contribute to the neuronal loss in the disease.[57]

4.1.5. Ischaemia

In a gerbil model of global cerebral ischemia a delayed increase in QUIN was seen in the brain, rising 50-fold their basal value after several days.[58] No change occured in the blood and in brain areas with an uninterrupted blood supply.[59] The changes in QUIN levels were accompanied by increased activity of all kynurenine metabolic enzymes except for kynurenine aminotransferase. These observations suggest that QUIN is produced locally in the brain, and indeed QUIN-immunoreactive microglia were observed in the gerbil hippocampus after transient cerebral ischemia.[60,61]

4.1.6. Traumatic Brain/Spinal Cord Injury

In humans, the level of QUIN in the CSF is increased after traumatic brain injury and the levels correlated with the mortality.[62] In rats and guinea pigs, experimental compression injury of the spinal cord evoked changes in QUIN and L-KYN levels in the affected areas, accompanied by the invasion of mononuclear phagocytes.[63,64]

4.1.7. Epilepsy

Intracerebroventricular injection of L-KYN, 3-HK and QUIN can produce seizures,[65] whereas KYNA protects from seizures induced by QUIN.[66] These observations suggest connections between kynurenines and seizures. The notion, that KYNA plays an important role as an endogenous anticonvulsant, was supported by the finding that application of KAT inhibitors such as γ-acetilenic-GABA or aminooxyacetic acid is associated with seizures and excitotoxic neurodegeneration.[67,68] Furthermore, shifting brain kynurenine metabolism in the hippocampus toward the KYNA branch by pharmacological manipulation, has pronounced anticonvulsant effects against sound-induced seizures in DBA/2 mice and electroshock-induced seizures in rats.[69] Conversely, implicating QUIN as an endogenous convulsant is supported by the finding, that elevated levels of QUIN have been reported in the brains of epileptic E1 mice, accompanied by an increased expression and activity 3-hydroxyanthranilic acid oxygenase.[69,70,71] These observations suggest that decreased concentration of KYNA or an increased level of 3-HK or QUIN in the brain could increase the susceptibility to seizures. However, no differences in the level of QUIN were reported in the CSF or brain of epileptic patients,[72] and the concentration of KYNA in the CSF did not differ from control levels in epileptic children, except for patients with West-syndrome and infantile spasms, in whom the levels of CSF KYNA were significantly lower.[73,74] It is interesting to note that there is a transient elevation of KYNA in the brain after seizures evoked by various convulsants which could be interpreted as an attempt of the brain to protect itself against overstimulation of EAA receptors.[75]

4.2. Diseases with Elevated KYNA Levels or KYNA/QUIN Ratio

4.2.1. Schizophrenia

Recently it has been shown that the level of L-KYN and KYNA is increased in the postmortem samples of the prefrontal cortex (Brodmann area 9) of schizophrenic patients.[76] This area has often been identified as functionally and morphologically altered in this disease. No abnormality of the KP was detected in other cortical areas from the same patients. There are several reasons to assume that a hypofunction of glutamatergic neurotransmission plays a critical role in schizophrenia.[77] These data demonstrate an impairment of brain kynurenine pathway metabolism in schizophrenia, resulting in elevated kynurenate levels and suggesting a possible concomitant reduction in glutamate receptor function.

4.2.2. Alzheimer's Disease

Alzheimer's disease is the major dementing disorder of the elderly people. There are evidence that the kynurenine system is being associated with this disease. The level of KYNA was shown to be significantly increased in the putamen and caudate nucleus of Alzheimer's disease patients, and was accompanied by a significant increase of KAT I activities in both nuclei.[78] Blockade of NMDA receptors by KYNA may be responsible for impaired memory, learning and cognition in these patients. It is demonstrated that a cleavage product of amyloid precursor protein, induces production of QUIN, in neurotoxic concentrations, by macrophages and microglia. A major aspect of QUIN toxicity is lipid peroxidation and markers of lipid peroxidation are found in Alzheimer's disease. These data indicate that QUIN may be a factors in the pathogenesis of Alzheimer's disease.

4.2.3. Down's Syndrome

Compared with controls the level of KYNA was increased in both the frontal and the temporal cortex of Down's syndrome patients, that could explain the cognitive deficits of these patients.[79]

5. THERAPEUTIC PERSPECTIVES

As the glutamate receptors are involved in many neurological disorders neuroprotective abilities of the KYNA have been tested. KYNA itself poorly penetrates the blood-brain barrier, so the protective effects of KYNA is limited by its low CNS availability. Chemically related drugs with better penetration and higher potency have been developed. These drugs do not have serious side-effects associated with NMDA receptor channel blockers. To improve the BBB penetration pro-drugs of KYNA can be used, which easily enter the brain, and are hydrolyzed in the CNS to active compounds. Manipulating the pathway by inhibiting the activity of different enzymes is another therapeutic possibility. With this strategy, one can divert L-KYN metabolism to the KYNA or QUIN branch.

5.1. KYNA Analogues

It has been reported that glycine-binding site antagonists of NMDA receptor may not cause serious adverse effects associated with the glutamate binding site or channel antagonists.[80] Substitution KYNA with halogen atoms yielded 7-chloro-kynurenic acid and 5,7-dichloro-kynurenic acid.[81] Replacement of the 4-hydroxy group of KYNA with amido substituents resulted in more potent analogues such as MDL 100,748[82] and L-689,560.[83] Substitution with a phenyl group at position 3 of the KYNA nucleus yielded molecules, such as MDL 104,653[84] or L-701,324.[85]

KYNA analogues in which the six-membered nitrogen containing ring was replaced by a 5-carbon ring provided a series of indole analogues, which retained their activity on the NMDA glycine site.[86,87] One of these compounds is gavestinel (GV150526A; Glaxo, 1993). The Glycine Antagonist in Neuroprotection (GAIN)

Americas trial, is a randomized, double-blind placebo-controlled phase III trial, examined the efficacy of gavestinel as a neuropotective therapy in acute ischemic stroke.[88] Gavestinel administered after an acute ischemic stroke failed to improve functional outcome.

5.2. Prodrugs of KYNA and of KYNA Analogues

L-KYN easily penetrates the brain by the neutral amino acid transporter.[6] Systemic administration of L-KYN dose-dependently elevates the level of KYNA in the brain. L-KYN are not capable of selectively increasing levels of the neuroprotectant KYNA.

Systemic administration of L-KYN have been shown to reduce brain damage evoked by ischemia or local injection of NMDA in neonatal rats[89] and to antagonize pentylenetetrazol- and NMDA-induced seizures in mice (13), but had little activity on kainate-induced seizures in rats.[90] However, the therapeutic potential of systemic L-KYN is limited, which is supported by the finding L-KYN administration did not sufficient to provide neuroprotection against toxic injuries.[91] One explanation for this poor efficacy is that L-KYN is metabolized to 3-HK and QUIN, which could aggravate neurotoxicity.

L-4-chlorokynurenine or 4,6-dichlorokynurenines are also easily transported to the brain and are converted to 7-chlorokynurenic acid and to 5,7-dichlorokynurenic acid, respectively.[92] The conversion, which takes place primarily in astrocytes, is catalyzed by KAT I and KAT II.[93] One of the metabolite of 4-chlorokynurenine is a potent and selective inhibitor of 3-hydroxyanthranilic acid-oxygenase, and thus the application of 4-chlorokynurenine results in a reduction of QUIN synthesis and exerting NMDA receptor blockade. Because of its dual action, 4-chlorokynurenine could be particularly advantageous in clinical situations which are linked to an alteration of the kynurenine system. 4-chlorokynurenine provides protection against QUIN, kainate and malonate induced neurotoxicity[94,54] and antagonizes convulsions.[95]

Another prodrug approach uses conjugates of KYNA and its analogues to overcome the blood-brain barrier. D-glucose and D-galactose esters of 7-chlorokynurenic acid have been synthesized, assuming that the conjugates are recognized by the glucose transporter, cross the blood-brain barrier and are hydrolyzed in the brain to generate 7-chlorokynurenic acid[95]. Systemic administration of this conjugate were protective against seizures induced by NMDA in mice. Further studies are needed to characterize the therapeutic efficacy of these drugs.

5.3. Enzyme Inhibitions

The goal of this manipulation is to shift the kynurenine metabolism towards the neuroprotective KYNA by pharmacologically inhibiting enzymes of QUIN synthesis. Inhibitors of kynurenine 3-hydroxylase cause marked elevation of KYNA and attenuation of 3-HK and QUIN formation.

5.3.1. Kynurenine 3-hydroxylase Inhibitors

Recently potent and selective inhibitors have been developed. Meta-nitrobenzoylalanine (mNBA) causes a dose-dependent increase of extracellular KYNA in the brain. Modification of the molecules yielded (R,S)-3,4-dichlorobenzoylalanine (PNU

156561) and 3,4-dimethoxy-N-[4-(3-nitrophenyl)thiazol-2-yl]benzenesulfonamide (Ro-61-8048), which are currently the most potent non-competitive inhibitors of the kynurenine 3-hydroxylase.[96]

Systemic administration of kynurenine 3-hydroxylase inhibitors causes a reduction of spontaneous locomotor activity and antagonizes seizures.[97] Besides elevating KYNA levels in the brain, mNBA and Ro-61-8048 significantly reduce blood and brain concentration of QUIN in pathological states, but under physiological conditions, no change is observed in the levels of QUIN,[98] This observation indicates, that administration of kynurenine 3-hydroxylase inhibitors may be beneficial in that disorders, where elevations of QUIN occur. Kynurenine 3-hydroxylase inhibitors have also been tested as neuroprotective agents in models of brain ischaemia. Systemic administration of mNBA and Ro-61-8048 caused a significant protection in a rat focal ischemia and a gerbil global ischemia model.[99] Substantial increase in the level of KYNA is detected in the brain after administration of kynurenine 3-hydroxylase inhibitors, and the resulting KYNA concentration is still lower, then those that are able to control NMDA receptor function.[100] Considering this, kynurenine 3-hydroxylase inhibitors may reduced the neuronal death by decreasing the local synthesis of 3-HK and QUIN.

5.3.2. Kynureninase Inhibitors

Ortho-methoxybenzoylalanine (oMBA) is the prototype inhibitor of kynurenase, while S-aryl-L-cysteine-S,S-dioxides is more potent inhibitor of the enzyme. These compounds reduce QUIN synthesis and increases the amount of KYNA in the brain *in vivo*.[68]

5.3.3. 3-hydroxyanthranilic Acid-oxygenase Inhibitors

4-chloro-3-hydroxyanthranilic acid were the first compounds described to inhibit 3-hydroxyanthranilic acid oxygenase.[101] This drug are potent, selective and competitive blocker of this enzyme[102] and reduce QUIN formation. The 4-chloro-3-hydroxyanthranilic acid decreases QUIN accumulation after experimental spinal cord injury.[62]

5.3.4. Enzyme Inhibition Decreasing the Concentration of KYNA

KYNA might play an important role in cognitive processes and kynurenine aminotransferase inhibitors might be useful as cognition enhancers. Furthermore, elevated KYNA levels in schizophrenia, Alzheimer's disease or Down's syndrome might be attenuated with the use of these drugs.

5.3.5. Kynurenine Aminotransferase Inhibitors

Large portion of KYNA present in the brain derives from KAT II activity. Therefore, it is expected that the inhibition of this enzyme would decrease the level of cerebral KYNA. No preferential KAT I inhibitor has been identified so far, but α-aminoadipate, quisqualate and DL-5-bromocriptine selectivily block KAT II *in vitro*. However, all known KAT II inhibitors interfere with other cellular targets, it is impossible to study the behavioural consequences of decreased KYNA content in the brain. Development of novel and selective KAT II inhibitors are desirable to study the physiological and pathological importance of KYNA in the CNS.

REFERENCES

1. Simon RP, Young RS, Stout S, Cheng J: Inhibition of excitatory neurotransmission with kynurenate reduces brain edema in neonatal anoxia. Neurosci. Lett. **71**,361-364, (1986).
2. Gill R, Foster AC, Woodruff GN: Systemic administration of MK-801 protects against ischemia-induced hippocampal neurodegeneration in the gerbil. J. Neurosci. **7**, 3343–3349, (1987).
3. Faden AI, Demediuk P, Panter SS, Vink R: The role of excitatory amino acids and NMDA receptors in traumatic brain injury. Science **244**,798-800, (1989).
4. Choi DW, Koh JY, Peters S: Pharmacology of glutamate neurotoxicity in cortical cell culture: attenuation by NMDA antagonists. J. Neurosci. **8**, 185–196, (1988).
5. Muir KW, Lees KR: Clinical experience with excitatory amino acid antagonist drugs. Stroke **26**, 503-513, (1995).
6. Fukui S, Schwarcz R, Rapoport SI, Takada Y, Smith QR: Blood-barrier transport of kynurenines: implications for brain synthesis and metabolism. J. Neurochem. **56**, 2007–2017, (1991).
7. Lapin IP: Depressor effect of kynurenine and its metabolites in rats. Life Sci. **19**, 1479–1484, (1976).
8. Lapin IP: Kynurenines and seizures. Epilepsia **22**, 257–265, (1981).
9. Pinelli A, Ossi C, Colombo R, Tofanetti O, Spazzi L: Experimental convulsions in rats induced by intraventricular administration of kynurenine and structurally related compounds. Neuropharmacology **23**, 333-337, (1984).
10. Perkins, MN, Stone TW: An iontophoretic investigation of the actions of convulsant kynurenines and their interaction with the endogenous excitant quinolinic acid. Brain Res. **247**, 184–187, (1982).
11. Kessler M, Baudry M, Terramani T, Lynch G: Complex interactions between a glycin binding site and NMDA receptors. Soc. Neurosci. Abst. **13**,760, (1987).
12. Kessler M, Terramani T, Lynch G, Baudry M: A glycin site associated with NMDA receptors: characterisation and identification of new class of antagonists. J. Neurochem. **52**:1319-1328, (1989).
13. Birch PJ, Grossman CJ, Hayes AG: Kynurenate and Fg9041 both competetive and non-competetive antagonist actions at excitatory amino acid receptors. Eur. J. Pharmacol. **151**,313-316, (1988).
14. Birch PJ, Grossman CJ, Hayes AG: Kynurenic acid antagonises responses to NMDA via an action at the strychnine-insensitive glycin receptor. Eur. J. Pharmacol. **154**,85-88, (1988).
15. Stone TW: Neuropharmacology of quinolinic and kynurenic acids. Pharmacol. Rev. **45**, 309–379, (1993).
16. Füvesi J, et al: unpublished observations.
17. Moroni F, Russi P, Carla V, Lombardi G: Kynurenic acid is present in the rat brain and its content increases during development and aging processes. Neurosci. Lett. **94**, 145–150, (1988).
18. Moroni F, Alesiani N, Galli A, Mori F, Pecorani R, Carla V, Cherici G, Pellicciari R: Thiokynurenates-a new group of antagonists of the glycin modulatory site of the NMDA receptors. Eur. J. Pharmacol. **375**,87-100, (1991).
19. Vécsei L, Beal MF: Intracerebroventricular injection of kynurenic acid, but not kynurenine, induces ataxia and stereotyped behavior in rats. Brain Res. Bull. **25**, 623-627, (1990).
20. Vécsei L, Miller J, MacGarvey U, Beal MF: Kynurenine and probenecid inhibit pentylenetetrazol- and NMDLA-induced seizures and increase kynurenic acid concentrations in the brain. Brain Res. Bull. **28**, 233–238, (1992).
21. Turski WA, Schwarcz R: On the disposition of intrahippocampally injected kynurenic acid in the rat. Exp. Brain. Res. **71**, 563-567, (1998).
22. Dedek J, Baumes R, Tien-Duc N, Gomeni R, Korf J: Turnover of free and conjugated (sulphonyloxy) dihydroxyphenylacetic acid and homovanillic acid in rat striatum. J. Neurochem. **33**, 687-695, (1979).
23. Morrison PF, Morishige GM, Beagles KE, Heyes MP: Quinolinic acid is extruded from the brain by a probenecid-sensitive carrier system: a quantitative analysis. J. Neurochem. **72**, 2135-2144, (1999).
24. Stone TW: Kynurenines in the CNS: from endogenous obscurity to therapeutic importance. Prog. Neurobiol. **64**, 185-218, (2001).
25. Rios C, Santamaria A: Quinolinic acid is a potent lipid peroxidant in rat brain homogenates. Neurochem. Res. **16**, 1139–1143, (1991).
26. Guidetti P, Schwarcz R: 3-Hydroxykynurenine potentiates quinolinate but not NMDA toxicity in the rat striatum. Eur. J. Neurosci.**11**, 3857-3863, (1999).

27. Tone S, Takikawa O, Habara-Ohkubo A, Kadoya A, Yoshida R, Kido R: Primary structure of human indoleamine 2,3-dioxygenase deduced from the nucleotide sequence of its cDNA. Nucleic Acid. Res. **18**, 367, (1990).
28. Dai W, Gupta SL: Regulation of indoleamine 2,3-dioxygenase gene expression in human fibroblasts by interferon-gamma. Upstream control region discriminates between interferon-gamma and interferon-alpha. J. Biol. Chem. **265**, 19871-19877, (1990).
29. Hayaishi O, Okamoto H.: Localization and some properties of kynurenine-3-hydroxylase and kynurenine aminotransferase. Am. J. Clin. Nutr. **24**, 805-806, (1971).
30. I Alberati-Giani D, Buchli R, Malherbe P, Broger C, Lang G, Kohler C, Lahm HW, Cesura AM: solation and expression of a cDNA clone encoding human kynureninase. Eur. J. Biochem. **239**, 460-468, (1996).
31. Toma S, Nakamura M, Tone S, Okuno E, Kido R, Breton J, Avanzi N, Cozzi L, Speciale C, Mostardini M, Gatti S, Benatti L: Cloning and recombinant expression of rat and human kynureninase. FEBS. Lett. **408**, 5-10, (1997).
32. Okuno E, Nakamura M, Schwarcz R: Two kynurenine aminotransferases in human brain. Brain Res. **542**, 307–312, (1991).
33. Guidetti P, Okuno E, Schwarcz R: Characterization of rat brain kynurenine aminotransferases I and II. J. Neurosci. Res. **50**, 457–465, (1997).
34. Csillik A, Knyihár E, Okuno E, Krisztin-Peva B, Csillik B, Vécsei L: Effect of 3-nitropropionic acid on kynurenine aminotransferase in the rat brain. Exp. Neurol. **177**, 233-241, (2002).
35. Knyihar-Csillik E, Okuno E, Vecsei L: Effects of in vivo sodium azide administration on the immunohistochemical localization of kynurenine aminotransferase in the rat brain. Neuroscience **94**, 269-277, (1999).
36. Gal EM, Sherman AD: Synthesis and metabolism of L-kynurenine in rat brain. J. Neurochem. **30**, 607-613, (1998).
37. Guidetti P, Eastman CL, Schwarcz R: Metabolism of [5-3H]kynurenine in the rat brain in vivo: evidence for the existence of a functional kynurenine pathway. J. Neurochem. **65**, 2621–2632, (1995).
38. Heyes MP, Alchim CL, Wiley A, Major EO, Saito K, Markey SP: Human microglia convert L-tryptophan into the neurotoxin quinolinic acid. Biochem. J. 320, 595–597, (1996).
39. Schwarcz R, Ceresoli G, Guidetti P: Kynurenine metabolism in the rat brain in vivo. Effect of acute excitotoxic insults. Adv. Exp. Med. Biol. **398**, 211-219, (1996).
40. Guillemin GJ, Kerr SJ, Smythe GA, Smith DG, Kapoor V, Armati PJ, Croitoru J, Brew BJ: Kynurenine pathway metabolism in human astroyctes: a paradox for neuronal protection. J. Neurochem. **78**, 842-853, (2001).
41. Lehrmann E, Molinari A, Speciale C, Schwarcz R: Immunohistochemical visualization of newly formed quinolinate in the normal and excitotoxically lesioned rat striatum. Exp. Brain. Res. **141**, 389-397, (2001).
42. Heyes MP, Brew BJ, Martin A, Price RW, Salazar AM, Sidtis JJ, Yergey JA, Mourdian MM, Sadler AE, Keilp J, Rubinow D, Markey SP: Quinolinic acid in cerebrospinal fluid and serum in HIV-1 infection: relationhip to clinical and neurologic status. Ann. Neurol. **29**, 202–209, (1991).
43. Heyes MP, Saito K, Lackner A, Wiley CA, Achim CL, Markey SP: Sources of the neurotoxin quinolinic acid in the brain of HIV-1 infected patients and retrovirus-infected macaques. FASEB. J. **12**, 881–896, (1998).
44. Halperin JJ, Heyes MP: Neuroactive kynurenines in Lyme borreliosis. Neurology **42**, 43-50, (1992).
45. Rejdak K, Bartosik-Psujek H, Dobosz B, Kocki T, Grieb P, Giovannoni G, Turski WA, Stelmasiak Z: Decreased level of kynurenic acid in cerebrospinal fluid of relapsing-onset multiple sclerosis patients. Neurosci. Lett. **331**, 63-65, (2002).
46. Flanagan EM, Erickson JB, Viveros OH, Chang SY, Reinhard Jr JF: Neurotoxin quinolinic acid is selectively elevated in spinal cords of rats with experimental allergic encephalomyelitis. J. Neurochem. **64**, 1192–1196, (1995).
47. Chiarugi A, Cozzi A, Ballerini C, Massacesi L, Moroni F: Kynurenine 3-mono-oxygenase activity and neurotoxic kynurenine metabolites increase in the spinal cord of rats with experimental allergic encephalomyelitis. Neuroscience **102**, 687-95, (2001).
48. Cammer W: Oligodendrocyte killing by quinolinic acid in vitro. Brain Res. **896**, 157-160. (2001).
49. Guillemin GJ, Kerr SJ, Pemberton LA, Smith DG, Smythe GA, Armati PJ, Brew BJ: IFN-beta1b induces kynurenine pathway metabolism in human macrophages: potential implications for multiple sclerosis treatment. J. Interferon. Cytokine. Res. **21**, 1097-1101, (2001).
50. Quinolinic acid: an endogenous metabolite that produces axon-sparing lesions in rat brain. Science **219**, 316–318, (1983).
51. Beal MF, Kowall NW, Ellison DW, Mazurek MF, Swartz KJ, Martin JB: Replication of the neurochemical characteristics of Huntington's disease by quinolinic acid. Nature **321**, 168–171, (1986).

52. Reynolds GP, Pearson SJ, Halket J, Sandler M: Brain quinolinic acid in Huntington's disease. J. Neurochem. **50**, 1959–1960, (1988).
53. Schwarcz R, Tamminga CA, Kurlan R, Shoulson I: CSF levels of quinolinic acid in Huntington's disease and schizophrenia. Ann. Neurol. **24**, 580–582, (1988).
54. Heyes MP, Saito K, Crowley JS, Davis LE, Demitrack MA, Der M, Dilling LA, Elia J, Kruesi MJP, Lackner A, Larsen SA, Lee K, Leonard HL, Markey SP, Martin A, Milstein S, Mouradian MM, Pranzatelli MR, Quearry BJ, Salazar A, Smith M, Straus SE, Sunderland T, Swedo SE, Tourtellotte WW: Quinolinic acid and kynurenine pathway metabolism in inflammatory and non-inflammatory neurologic disease. Brain **115**, 1249–1273, (1992).
55. Guidetti P, Wu HQ, Schwarcz R: In situ produced 7-chlorokynurenate provides protection against quinolinate- and malonate-induced neurotoxicity in the rat striatum. Exp. Neurol. **163**, 123-130, (2000).
56. Widner B, Leblhuber F, Fuchs D: Increased neopterin production and tryptophan degradation in advanced Parkinson's disease. J. Neural. Transm. **109**, 181-189, (2002).
57. Ogawa T, Matson WR, Beal, MF, Myers RH, Bird ED, Milbury P, Saso S: Kynurenine pathway abnormalities in Parkinson's disease. Neurology **42**, 1702–1706, (1992).
58. Heyes MP, Nowak Jr TS: Delayed increases in regional brain quinolinic acid follow transient ischemia in the gerbil. J. Cereb. Blood. Flow. Metab. **10**, 660–667, (1990).
59. Saito K, Nowak TS Jr, Markey SP, Heyes MP: Mechanism of delayed increases in kynurenine pathway metabolism in damaged brain regions following transient cerebral ischemia. J. Neurochem. **60**, 180-192, (1993).
60. Saito K, Nowak Jr TS, Markey SP, Heyes MP: Delayed increases in kynurenine pathway metabolism in damaged brain regions following transient cerebral ischemia. J. Neurochem. **60**, 180–192, (1992).
61. Barattè S, Molinari A, Veneroni O, Speciale C, Benatti L, Salvati P: Temporal and spatial changes of quinolinic acid immunoreactivity in the gerbil hippocampus following transient cerebral ischemia. Mol. Brain. Res. **59**, 50–57, (1998).
62. Sinz EH, Kochanek PM, Heyes MP, Wisniewski SR, Bell MJ, Clark RSB, Dekosky ST, Blight AR, Marion DW: Quinolinic acid is increased in CSF and associated with mortality after traumatic brain injury in humans. J. Cereb. Blood. Flow. Metab. **18**, 610–615, (1998).
63. Blight AR, Cohen TI, Saito K, Heyes MP: Quinolinic acid accumulation and functional deficits following experimental spinal cord injury. Brain **118**, 735–752, (1995).
64. Blight AR, Leroy Jr EC, Heyes MP: Quinolinic acid accumulation in injured spinal cord: time course, distribution, and species differences between rat and guinea pig. J. Neurotrauma. **14**, 89–98, (1997).
65. Lapin IP: Stimulant and convulsive effects of kynurenines injected into brain ventricles in mice. J. Neural. Transm. **42**, 37-43, (1978).
66. Foster AC, Vezzani A, French ED, Schwarcz R: Kynurenic acid blocks neurotoxicity and seizures induced in rats by the related brain metabolite quinolinic acid. Neurosci. Lett. **48**, 273-278, (1984).
67. McMaster OG, Du F, French ED, Schwarcz R: Focal injection of aminooxyacetic acid produces seizures and lesions in rat hippocampus: evidence for mediation by NMDA receptors. Exp. Neurol. **113**, 378-385, (1991).
68. McMaster OG, Baran H, Wu HQ, Du F, French ED, Schwarcz R: Gamma-acetylenic GABA produces axon-sparing neurodegeneration after focal injection into the rat hippocampus. Exp. Neurol. **124**, 184-191, (1993).
69. Carpenedo R, Chiarugi A, Russi P, Lombardi G, Carla V, Pellicciari R, Moroni F, Mattoli L: Inhibitors of kynurenine hydroxylase and kynureninase increase cerebral formation of kynurenate and have sedative and anticonvulsant activities. Neuroscience **61**, 237–244, (1994).
70. Nakagawa Y, Asai H, Miura T, Kitoh J, Mori H, Nakano K: Increased expression of the 3-hydroxyanthranilate-3,4-dioxygenase gene in brain of epilepsy-prone E1 mice. Mol. Brain. Res. **58**, 132–137, (1998).
71. Nakano K, Asai H, Kitoh J: Abnormally high activity of 3-hydroxyanthranilate 3,4-dioxygenase in brain of epilepsy-prone E1 mice. Brain. Res. **572**, 1–4, (1992).
72. Nakano K, Takahashi S, Mizobuchi M, Kuroda T, Kitoh KJ: High levels of quinolinic acid in brain of epilepsy-prone E1 mice. Brain. Res. **619**, 195–198, (1993).
73. Heyes MP, Wyler AR, Devinsky O, Yergey JA, Markey SP, Nadi NS: Quinolinic acid concentrations in brain and cerebrospinal fluid of patients with intractable complex partial seizures. Epilepsia **31**, 172–177, (1990).
74. Yamamoto H, Shindo I, Egawa B, Horiguchi K: Kynurenic acid is decreased in cerebrospinal fluid of patients with infantile spasms. Pediatr. Neurol. **10**, 9–12, (1994).
75. Yamamoto H, Murakami H, Horiguchi K, Egawa B: Studies on cerebrospinal fluid kynurenic acid concentrations in epileptic children. Brain. Dev. **17**, 327–329, (1995).

76. Schwarcz R, Rassoulpour A, Wu HQ, Medoff D, Tamminga CA, Roberts RC: Increased cortical kynurenate content in schizophrenia. Biol. Psychiatry. **50**, 521-530, (2001).
77. Tamminga CA: Schizophrenia and glutamatergic transmission. Crit. Rev. Neurobiol .**12**, 21–36, (1998).
78. Baran H, Jellinger K, Derecke L: Kynurenine metabolism in Alzheimer's disease. J. Neural. Transm. **106**, 165–181, (1999).
79. Baran H, Cairns N, Lubec B, Lubec G: Increased kynurenic acid levels and decreased brain kynurenine aminotransferase I in patients with Down syndrome. Life. Sci. **58**, 1891–1899, (1996).
80. Hargreaves RJ, Rigby M, Smith D, Hill RG: Lack of effect of L-687,414 ((+)-cis-4-methyl-HA-966), an NMDA receptor antagonist acting at the glycine site, on cerebral glucose metabolism and cortical neuronal morphology. Br. J. Pharmacol. **110**, 36–42. (1993).
81. Baron BM, Harrison BL, Miller FP, McDonald IA, Salituro FG, Schmidt CJ, Sorensen SM, White HS, Palfreyman MG: Activity of 5,7-dichlorokynurenic acid, a potent antagonist at the NMDA receptor-associated glycine binding site. Mol. Pharmacol. **38**, 554–561, (1990).
82. Baron BM, Harrison BL, McDonald IA, Meldrum BS, Palfreyman MG, Salituro FG, Siegel BW, Slone AL, Turner JP, White HS: Potent indole- and quinoline-containing NMDA antagonists at the strychnine-insensitive glycine binding site. J. Pharmacol. Exp. Ther. **262**, 947–956, (1992).
83. Leeson PD, Carling RW, Moore KW, Moseley AM, Smith JD, Stevenson G, Chan T, Baker R, Foster AC, Grimwood S, Kemp JA, Marshall GR, Hoogsteen K: 4-Amido-2-carboxytetrahydroquinolines. Structure-activity relationships for antagonism at the glycine site of the NMDA receptor. J. Med. Chem .**35**, 1954–1968, (1992).
84. Kulagowski JJ, Baker R, Curtis NR, Leeson PD, Mawer IM, Moseley AM, Ridgill MP, Rowley M, Stansfield I, Foster AC: 3'-(Arylmethyl)- and 3'-(aryloxy)-3-phenyl-4-hydroxyquinolin-2(1H)-ones: orally active antagonists of the glycine site on the NMDA receptor. J. Med.Chem. **37**, 1402-1405, (1994).
85. Bristow LJ, Flatman KL, Hutson PH, Kulagowski JJ, Leeson PD, Young L, Tricklebank MD: The atypical neuroleptic profile of the glycine/N-methyl-D-aspartate receptor antagonist, L-701324, in rodents. J. Pharmacol. Exp. Therap. **277**, 578–585, 1996.
86. Salituro FG, Tomlinson RC, Baron BB, Demeter DA, Weintraub HJR, McDonald IA: Design, synthesis and molecular modeling of 3-acylamino-2-carboxyindole NMDA receptor glycine-site antagonists. Bioorg. Med Chem. Lett. 1, **455**–460, (1991).
87. Rao TS, Gray NM, Dappen MS, Cler JA, Mick SJ, Emmett MR, Iyengar S, Monahan JB, Cordi AA, Wood PL: Indole-2-carboxylates, novel antagonists of the NMDA-associated glycine recognition site-in vivo characterization. Neuropharmacology **32**, 139–147, (1993).
88. Sacco RL, DeRosa JT, Haley EC Jr, Levin B, Ordronneau P, Phillips SJ, Rundek T, Snipes RG, Thompson JL: The Glycine Antagonist in Neuroprotection Americas Investigators. Glycine antagonist in neuroprotection for patients with acute stroke: GAIN Americas: a randomized controlled trial. JAMA **285**, 1719-1728, (2001).
89. Nozaki K, Beal MF: Neuroprotective effects of L-kynurenine on hypoxia-ischemia and NMDA lesions in neonatal rats. J. Cereb. Blood. Flow. Metab. **12**, 400–407, (1992).
90. Vécsei L, Beal MF: Influence of kynurenine treatment on open-field activity, elevated plus-maze, avoidance behaviors and seizures in rats. Pharmacol. Biochem. Behav. **37**, 71-76, (1992).
91. Wu H-Q, Guidetti P, Goodman JH, Varasi M, Ceresoli-Borroni G, Speciale C, Scharfman HE, Schwarcz R: Kynurenergic manipulations influence excitatory amino acid receptor function and excitotoxic vulnerability in the rat hippocampus in vivo. Neuroscience **97**, 243-251, (2000).
92. Hokari M, Wu H-Q, Schwarcz R, Smith QR: Facilitated brain uptake of 4-chlorokynurenine and conversion to 7-chlorokynurenic acid. Neuroreport **8**, 15–18, (1996).
93. Wu H-Q, Salituro FG, Schwarcz R: Enzyme-catalyzed production of the neuroprotective NMDA receptor antagonist 7-chlorokynurenic acid in the rat brain in vivo. Eur. J. Pharmacol. **319**, 13-20, (1997).
94. Wu H-Q, Lee S-C, Scharfman HE, Schwarcz R: L-4-chlorokynurenine attenuates kainate-induced seizures and lesions in the rat. Exp. Neurol. **177**, 222-232, (2002).
95. Battaglia G, La Russa M, Bruno V, Arenare L, Ippolito R, Copani A, Bonina F, Nicoletti F: Systemically administered D-glucose conjugates of 7-chlorokynurenic acid are centrally available and exert anticonvulsant activity in rodents. Brain. Res. **860**, 149-156, (2000).
96. Röver S, Cesura AM, Hugenin P, Kettler R, Szente A: Synthesis and biochemical evaluation of N-(4-phenylthiazol-2-yl) benzenesulfonamides as high-affinity inhibitors of kynurenine 3-hydroxylase. J. Med. Chem. **40**, 4378–4385, (1997).
97. Russi P, Alesiani M, Lombardi G, Davolio P, Pellicciari R, Moroni F: Nicotinylalanine increases the formation of kynurenic acid in the brain and antagonizes convulsions. J. Neurochem. **59**, 2076–2080, (1992).
98. Chiarugi A, Moroni F: Quinolinic acid formation in immune-activated mice: studies with (m-nitrobenzoyl)alanine (mNBA) and 3,4-dimethoxy-[N-4-(3-nitrophenyl)thiazol-2-yl]benzenesulfonamide

(Ro 61-8048), two potent and selective inhibitors of kynurenine hydroxylase. Neuropharmacology **38**, 1225–1233, (1999).
99. Cozzi A, Carpenedo R, Moroni F: Kynurenine hydroxylase inhibitors reduce ischaemic brain damage. Studies with (m-nitrobenzoyl)-alanine (mNBA) and 3,4-dimethoxy-[N-4-(nitrophenyl)thiazol-2-yl]benzenesulfonamide (Ro61-8048) in models of focal or global brain ischaemia. J. Cereb. Blood. Flow. Metab. **19**, 771–777, (1999).
100. Urenjak J, Obrenovitch TP: Kynurenine 3-hydroxylase inhibition in rats: effects on extracellular kynurenic acid concentration and N-methyl-D-aspartate-induced depolarisation in the striatum. J. Neurochem. **75**, 2427-2433, (2000).
101. Todd WP, Carpenter BK, Schwarcz R: Preparation of 4-halo-3-hydroxyanthranilates and demonstration of their inhibition of 3-hydroxanthranilate oxygenase in rat and human brain tissue. Prep. Biochem. **19**, 155-165, (1989).
102. Walsh JL, Todd WP, Carpenter BK, Schwarcz R: 4-Halo-3-hydroxyanthanilic acids: potent competitive inhibitors of 3-hydroxyanthranilic acid oxygenase in vitro. Biochem. Pharmacol. 42, 985–990, (1991).

INDEX

D

O

P

Q

R